Advances in
RADIATION BIOLOGY

Volume 12

Advances in
RADIATION BIOLOGY

Relative Radiation Sensitivities of Human Organ Systems

Edited by

JOHN T. LETT

DEPARTMENT OF RADIOLOGY
 AND RADIATION BIOLOGY
COLORADO STATE UNIVERSITY
FORT COLLINS, COLORADO

KURT I. ALTMAN

DEPARTMENT OF RADIATION BIOLOGY
 AND BIOPHYSICS
THE UNIVERSITY OF ROCHESTER
 SCHOOL OF MEDICINE
ROCHESTER, NEW YORK

Associate Editors

Ursula K. Ehmann

DEPARTMENT OF MEDICINE
VETERANS ADMINISTRATION
 MEDICAL CENTER
PALO ALTO, CALIFORNIA

Ann B. Cox

RADIATION SCIENCES DIVISION
UNITED STATES AIR FORCE
 SCHOOL OF AEROSPACE MEDICINE
BROOKS AIR FORCE BASE, TEXAS

Volume 12

1987

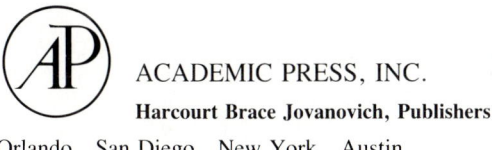

ACADEMIC PRESS, INC.

Harcourt Brace Jovanovich, Publishers

Orlando   San Diego   New York   Austin
Boston   London   Sydney   Tokyo   Toronto

577.4 (05) : 539.12.047

ADV

ACADEMIC PRESS, INC.
Orlando, Florida 32887

United Kingdom Edition published by
ACADEMIC PRESS INC. (LONDON) LTD.
24–28 Oval Road, London NW1 7DX

LIBRARY OF CONGRESS CATALOG CARD NUMBER: 64-8030

ISBN 0–12–035412–8 (alk. paper)

PRINTED IN THE UNITED STATES OF AMERICA

87 88 89 90   9 8 7 6 5 4 3 2 1

CONTENTS

Relative Radiosensitivities of the Thymus, Spleen, and Lymphohemopoietic Systems

Yosh Maruyama and Jose M. Feola

Relative Radiosensitivities of the Small and Large Intestine

Aldo Becciolini

Relative Radiosensitivities of the Oral Cavity, Larynx, Pharynx, and Esophagus

Joella F. Utley

Relative Radiation Sensitivity of the Integumentary System: Dose Response of the Epidermal, Microvascular, and Dermal Populations

John O. Archambeau

Relative Radiosensitivity of the Human Lung

Elizabeth L. Travis

Relative Radiosensitivity of Fetal Tissues

R. L. Brent, D. A. Beckman, and R. P. Jensh

Tolerance of the Central and Peripheral Nervous System to Therapeutic Irradiation

Steven A. Leibel and Glenn E. Sheline

Preface

This volume addresses the response of selected human organ and tissue systems to the exposure to ionizing radiation. It is logical that such a response should be viewed primarily from the vantage point of the radiation therapist whose goal it is to eradicate neoplastic tissues while at the same time avoiding injury to normal tissues.

Reviewers of the radiation responses of human organ systems can draw upon information gathered over a period of almost a century. When this information includes, as it has in recent years, quantitative population dynamics, e.g., of the human skin and lung, it becomes useful not only to the clinician as a guideline for therapy and the interpretation of therapeutic results, but also to the radiation biologist interested in the application of basic radiobiological data to such practical areas as radiation therapy. In situations where the quantitative dynamics of cell populations with respect to cell proliferation, cell killing, and cell death are not available, radiation effects are interpreted in terms of tissue injury that often is manifested as hypoplasia. The use of dose fractionation to optimize therapeutic results is described in detail for the human organ systems covered in this volume and interpreted, where possible, in terms of cell population dynamics, particularly with respect to cell kinetics and cell proliferation.

From a clinical point of view the radiation sensitivity of an organ can be estimated on the basis of severity and rapidity with which the effect of radiation manifests itself. Added to this is the clinical concern about the occurrence of late effects from radiation exposure, as exemplified by the human skin.

Although radiation biology has not yet been able to supply all the answers needed to support practical radiation therapy, it has been established that organ radiosensitivity depends on the turnover rate of cells in the organ, i.e., the life span of cells prior to their elimination from the organ. Thus, organs with rapid rates of cell turnover, e.g., the gastrointestinal epithelium, epidermis, and tumors, are generally more radiosensitive than organs with slow rates of cell turnover, e.g., kidney, liver, spinal cord, and dermis. Where possible, attempts have been made in the discussion of various organ systems to take cognizance of cell turnover rate as well as the other parameters, such as oxygen enhancement ratio and relative biological effectiveness, which influence the relative radiosensitivities of human organ systems.

I wish to express my special thanks and appreciation to Dr. Yosh Maruyama for his invaluable advice, comments, and discussion throughout the assembly of this volume and wish to acknowledge with thanks the helpful suggestions received from Drs. George W. Casarett and Philip Rubin. I also thank Dr. Ann B. Cox for editing the translation of one of the chapters and am indebted to my wife for her never-failing help in the stylistic editing of the volume.

KURT I. ALTMAN

Relative Radiosensitivities of the Thymus, Spleen, and Lymphohemopoietic Systems

YOSH MARUYAMA AND JOSE M. FEOLA

DEPARTMENT OF RADIATION MEDICINE
UNIVERSITY OF KENTUCKY MEDICAL CENTER
LEXINGTON, KENTUCKY 40536

I. Introduction

Recently, dramatic advances have taken place in our knowledge of the organization and functions of the thymus and lymphoid system. Some of our understanding has come from the application of radiation therapy to diseases derived from lymphoid organs. Human investigations using radiotherapy have shed light on the radiosensitivities of the system as well as on its important functional activities. Likewise, studies of cellular traffic in the lymphoid organs, i.e., the thymus, spleen, and lymph nodes, showed that extensive cellular exchange occurs between the bone marrow compartment with its hemopoietic and progenitor cells and the various central and peripheral lymphoid organs. Because this is so, the term lymphohemopoietic (LH) system is appropriate. This very selective review will recapitulate some of the ongoing research in radiotherapy and radiobiology in these closely related fields. The clinical therapeutic advances have taken place in four important settings closely related to clinical therapy of tumors derived from the LH system. These are

1. Acute lymphoblastic leukemia of childhood (Pinkel, 1979), using *sanctuary* therapy

2. Hodgkin's disease, a disease of the lymphoid system, treated using so-called "mantle" and total nodal irradiation (TNI) (Kaplan, 1980)

3. Immunosuppression for autoimmune disorders and for the conditioning of patients for homografts of various transplanted organs (Strober *et al.*, 1979a), using total lymphoid irradiation (TLI)

4. The treatment of leukemia and other blood disorders using total body

1

irradiation (TBI) in conjunction with bone marrow cell transplantation (Thomas
et al., 1975)

The nuclear holocausts from the atomic bombing of Hiroshima and Nagasaki
led to research into the effects of TBI on animals and extensive research into the
tissue and functional hemopoietic and immunological impairments in the irradiated organism. It also led to the discovery of reconstitution of heavily and
lethally irradiated animals (Jacobson *et al.,* 1949, 1954). This eventually led to
the use of deliberate high-dose TBI to doses of about 1000–1400 rad, now being
regularly used to reduce the load of leukemic cells, to destroy the bone marrow
of leukemic patients, to produce a completely tolerant state, and to condition the
patient for bone marrow transplantation (BMT) to rescue and restore the hematological integrity of the patient. TBI and BMT have now been applied to a
sufficient number of patients and long-term follow-up reported for a sufficient
period of time to draw some early conclusions regarding efficacy (Thomas *et al.,*
1975, 1978).

An important outgrowth of the wide-field irradiation studies was that a tolerant state was produced by heavy dose radiation of the central lymphoid organs.
This tolerance was identified for TBI (see Van Bekkum, 1967) and then for
thymic irradiation in mice by Maruyama and Barkley (1967). The application of
wide-field irradiation studies in Hodgkin's disease (Kaplan, 1980) led to the
recognition and realization that the successfully treated patients were severely
immunodeficient and were crippled in their ability to reject homologous tissue
grafts (Slavin *et al.,* 1977; Strober *et al.,* 1979a). The patients were thus "conditioned" for transplantation therapy for donated homografted tissues and organs.
The recognition of this tolerant state has led to the application of TLI to organ
and tissue grafting in aplastic anemia and chronic renal, liver, heart, and lung
failure (Strober *et al.,* 1979a,b).

The high sensitivity of lymphocytic neoplasms and limited ability of cells to
repair radiation damage has led to the observations that the total dose rather than
fraction number or dose rate determines therapeutic effects. Multiple daily fractions (MDF), e.g., three times per day fractionation or superfractionated schedules, have led to greatly improved results over conventional fractionation. Such
treatment schedules have improved therapy of, e.g., Burkitt's lymphoma (Norin
et al., 1971; Norin and Onyango, 1977) and patient conditioning for BMT.
Fractionated low-dose TBI (Heublein, 1932) [10 fractions of 10–15 cGy to 100–
150 cGy (Johnson, 1975; Carabell *et al.,* 1979)] can produce complete remissions of chronic LH leukemias. Low dose rate radiation to ~1000 cGy with ^{60}Co
along with chemotherapy and BMT has also led to long-term remissions of
certain lymphocytic neoplasms (Thomas *et al.,* 1975, 1978). The extreme radiosensitivity of these lymphomas has led to the complete disappearance of tumor
cells and produced long relapse-free periods of survival. This is consistent with

greater tumor cell radiosensitivity and lack of ability to repair sublethal damage (SLD) by the lymphocyte.

The principal target of the above therapies was directed at the LH system and at the thymus, spleen, lymph nodes, and other sites of the peripheral lymphoid tissues that constitute the system. Radiation used in the setting of the above groups of diseases has proved effective; lymphocyte radiosensitivity represents one of the principal reasons that the therapy was successful.

This review will be selective and will concentrate on an overview of the field and on topics and studies considered pertinent to these problems. The reader is directed to numerous other recent reviews for more specialized discussions concerning related current and clinical treatment developments, especially as related to closely allied topics. Perhaps the most important conceptual advance has been that the cytotoxic and immunosuppressive effects of radiation can be successfully applied to major groups of human disorders of the LH organs [e.g., Anderson and Warner (1976), Leone (1962), Stewart and Perez (1976), Taliaferro *et al.* (1964), Talmage (1955), Dubois *et al.* (1981), Ballou (1976), and Doria *et al.* (1982)].

II. The Lymphoid Organs

A. General

The lymphoid organs are scattered throughout the body as lymph nodes, the spleen, thymus, and other lymphoid aggregates such as the tonsils, Waldeyer's ring (in the oropharynx), appendix, Peyer's patches, and related lymphoid nodules in the intestines and abdomen. These are considered the peripheral lymphoid organs (PLO). The thymus is regarded as the central organ of cellular immunity, and prothymocytes of bone marrow origin traffic through the organ to become thymocytes or T lymphocytes before emigrating elsewhere to undergo postthymic differentiation (Gowans and McGregor, 1965; Stutman, 1978). The thymus is an anterior mediastinal organ with two lobes. It has a cortex and medullary regions with lymphocytes of the two zones of different characteristics. Cortical lymphocytes are sensitive to radiation and cortisone; medullary thymocytes, unlike cortical cells, are more radiation and cortisone resistant and resemble T lymphocytes in physical properties and function. The latter cells enter the peripheral circulation and lymphoid organs to undergo postthymic differentiation and carry out lymphocyte-mediated immune reactions throughout the body. Intermingled with lymphocytes in the lymphoid organs are multiple other cellular elements related to the lymphopoietic, hematopoietic, phagocytic, immunological, and cell-regulatory systems. The PLO include major lymphocyte aggregates such as spleen, tonsil, appendix, lymph nodes, and diffuse infiltrations of lymphocytes, e.g., in Waldeyer's ring, and Peyer's patches.

There is an active exchange of cells between the various components of the entire system and the bone marrow. Interconnecting the lymph nodes are lymphatic vessels in which the lymph nodes are scattered in discrete aggregates in special sites throughout the body (Fig. 1). The lymphoid organs are supplied by systemic blood circulation and venous drainage. A separate system of lymphatic vessels drains the lymph nodes and drains into progressively larger lymph vessels, leading eventually to the thoracic duct. The thoracic duct represents a major avenue of lymphocyte traffic for cells of the system and drains back into the great veins of the thorax (Gowans and McGregor, 1965). The biological functions of the system are extremely complex and diverse. Its regulation is likewise complex and is dependent upon central nervous system (CNS), thymus, adrenal, gonadal, renal, and other endocrine organs (Dunlap and Warren, 1942; Dougherty et al., 1964), especially adrenal glucocorticosteroid (Homo-Delarche, 1984; Claman and Mosier, 1972) receptors.

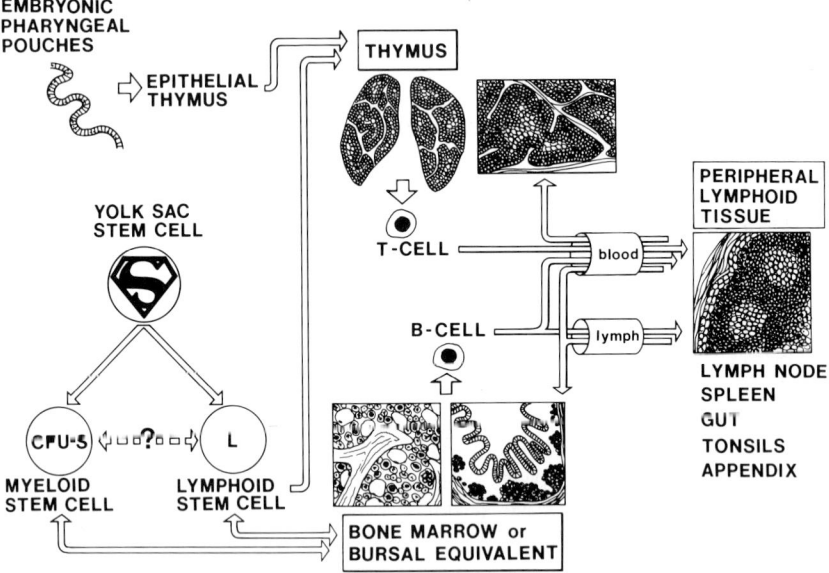

FIG. 1. Embryogenesis and distribution of lymphohemopoeitic (LH) system. A yolk sac stem cell gives rise to the myeloid system and the lymphoid system. The stem cells are closely related at an early stage in development. The embryonic foregut pouches give rise to the epithelial thymus. Further development of the thymus derives from mesenchyme (Auerbach, 1967) and from migratory cells from other LH organs. T cells seed the peripheral lymphoid organs shown to the right of the figure. The bone marrow and the bursa (in birds) or bursa equivalent organ give rise to B cells which likewise seed "downstream" to the peripheral lymphoid organization (PLO). Extensive traffic of cells takes place through the bloodstream and lymphatic-vascular systems. The embryonic liver/spleen may play an important intermediate role as a site for stem cells to sojourn between the yolk sac and subsequently the thymus, bone marrow, bursa equivalent, and PLO.

The entire system is extremely sensitive to ionizing radiation, and there is an enormous literature on this subject which extends back to the earliest days since the discovery of ionizing radiation (see, e.g., Heineke, 1904, 1905; Lacassagne and Gricouroff, 1958; Anderson and Warner, 1976). The system is sensitive to radiation of any energy, electrons, radium, and radioactive isotopes, internal and external, and β-, γ-emitting radioisotopes, neutrons, and radiation (Osgood *et al.*, 1942; Warren *et al.*, 1950; Trowell, 1952; Schrek, 1946, 1947; Lawrence and Tennant, 1937; Scott and Lawrence, 1941). While the principal function of the lymphoid system appears to be immunological (Table I), it is clear that lymphoid organs may carry out many functions, e.g., lymphopoiesis, hemopoiesis, phagocytosis of antigen and foreign substances, blood cell clearance (e.g., erythrocytes), instructing and regulating lymphocytes (B and T cells and T cell subsets) (see Cantor and Boyce, 1977), surveillance against infection, homografts, and neoplasia, and directing the functioning and differentiation (e.g., T cell repertoire, NK) of lymphocytes. B cells form antibodies and are responsible

TABLE I

CHARACTERISTICS OF B AND T LYMPHOCYTES

Origin Characteristic	Hemopoietic stem cell B Lymphocytes	Hemopoietic stem cell T Lymphocytes
Role in immune response	Humoral response	Cell-mediated reactions; lymphocyte regulation (helper, suppressor, cytoxic)
Site of differentiation and initial proliferation	Bursa of fabricius (birds); bone marrow or bursal equivalent organ (mammals)	Thymus
Frequency in peripheral blood (% of lymphocytes)	20–25%	65–80%
Lymphoid organ site	Lymph node cortex, splenic follicle, peyer's patches	Paracortical region of lymph node, periarteriolar sheath of spleen (white pulp)
Life span	Generally short (days, months)	Generally long (months, years)
Surface receptors	Immunoglobulins	T cell receptors
PHA stimulation	None	Marked with immunoblastic transformation
Pokeweed mitogen stimulation	Transformation into immunoblasts or plasmacytoid cells (uses T helpers)	Transforms
E Rosette formation (SRBC)[a]	−	+
Cytolplasmic/membrane immunoglobulin binding	+	−

[a] Sheep red blood cells.

for humoral immunity. Removal of the thymus can interfere with T cell (Miller 1961) and B cell function (Archer and Pierce, 1961).

On occasion, its activities may become perverted by uncontrolled neoplastic and malignant growth (Aisenberg, 1981) and can produce excessive antibodies and cells, which can lead to autodestruction of the organism. All these various aspects represent major fields of basic and clinical research. Virtually all have been assessed for their sensitivity to one or other of the many forms of ionizing radiation, and radiation represents one of the principal means of dissecting and studying their function(s) and therapy. This review will, of necessity, be selective and focus only on problems that have been of current interest to clinical radiation medicine. Neoplasms derived from these tissues may cause significant aberrations of immunological function (e.g., anergy, hyperglobulinemia) along B or T lymphocyte lines as proposed initially by Cooper *et al.* (1966b). Nearly all the immunological functions of the thymolymphatic tissues are affected dramatically by radiation.

B. Origins and Interrelationships of Lymphohemopoietic Organs

Lymphocytes are the most important population in the lymphoid organ, have a complex origin and traffic, and are very heterogeneous (Yoffey, 1970). The bone marrow, spleen, and thymus are very actively proliferating members of the LH system. Lymphopoiesis occurs in special sites such as the avian bursa of Fabricius, mammalian bone marrow, and thymus organ. The bone marrow possesses its own stem cell compartment, whereas the thymus is dependent upon a continual influx of precursor cells which arrive by migration through the vascular system. Beginning some time in the prenatal period and continuing subsequently throughout life, the bone marrow seeds organs such as the thymus with prothymocytes (see, e.g., Harris and Ford, 1967) and other lymphoid organs (Barg *et al.*, 1978) with a variety of precursor cells at varying stages of differentiation. Cells that undergo further maturation in the thymus in turn differentiate to migrate to other secondary lymphoid organs downstream (see Fig. 1), which constitutes the various tissues comprising the lymphoid system (Ford *et al.*, 1956, 1963, 1966; Ford, 1975; Micklem *et al.*, 1966). This process continues into early postnatal life. The postnatal thymus receives its stem cells primarily from the bone marrow, which are processed through the thymus or are produced there and then in turn leave to seed systemic lymphoid organs (Ford and Micklem, 1963). In the irradiated animal, injected bone marrow cells go to the bone marrow, spleen, and the thymus (Ford *et al.*, 1966). Some of the most exciting research on lymphocytes and lymphoid organ biology has taken place in the past two decades. There is an intimate and complex relationship between the bone marrow, the thymus, and the peripheral lymphoid systems (see Gowans and McGregor, 1965). Thymic hormones and secretions have been recognized to be important in recent years and have been found to be decreased or absent in a number of diverse conditions (Dardenne and Bach, 1981).

Early clones appearing after spleen seeding and growth have erythrocyte or granulocytic cells, but by day 12, about half the colonies have mixtures of cell types of three different cell lines (Curry and Trentin, 1967). Lymphocytes do not appear to be part of the spleen colonies, but may arise from a common ancestral origin (Wu et al., 1968; Edwards et al., 1970). Using radiation-marked chromosomes produced by sublethal irradiation of donors, it was shown that the same chromosomal abnormality was present in the spleen colony cells and in the thymus and lymph node cells (Abramson et al., 1977). Nodes containing such marker-bearing T lymphocytes can participate in immunological responses. Since the same unique marker was present in both, it was argued that the precursor cells of the thymus and lymph nodes started from a common ancestral cell or both were descendants of a common progenitor cell which has not yet been identified.

The site of production of the lymphoid precursor cells for the B cells has been sought. That site is still not clear (Wu, 1970; Abramson et al., 1977). The thymus and the bone marrow (and perhaps other sites as well) provide microenvironments which instruct the prothymocyte or pro-T lymphocyte to differentiate in specific T cell or B cell repertoires and lineages. The spleen and bone marrow provide microenvironments to instruct these cells to differentiate along hemopoietic lines (Curry and Trentin, 1967). The thymus provides the environment seeded by prothymocytes from the bone marrow to differentiate to a T thymocyte and the bone marrow or bursal equivalent to a B lymphocyte. There they acquire more specific T cell functional repertoires (e.g., helper, suppressor, killer). The latter cells then enter the circulation to migrate downstream to spleen, lymph nodes, and PLO.

In the early 1960s several independent laboratories began to work on the thymus as an organ important to immune function (Good et al., 1962; Miller, 1962; Arnason et al., 1962). The initial important experiments utilized animals in which the thymus was surgically excised and their immune function studied after thymectomy. Furth had shown many years earlier (1946) that the removal of the thymus gland prevented the development of lymphocytic leukemia. Tumors of lymphocytes (lymphomas) could be prevented by simply removing the organ shortly after the leukemogenic treatment. Metcalf (1960) showed that lymphopenia and lymphoid organ atrophy followed thymectomy. Miller (1961) showed that the removal of the thymus from newborn mice led to poor animal growth and development and inability to reject skin from donor mice of different strains, i.e., they had been rendered "tolerant" to foreign tissue grafts. The thymus in animals was thus shown to be important to cellular immunity, especially for that class of lymphocytes (thymus derived) which attacks homografts of foreign tissue, mediates delayed hypersensitivity reactions, and combats viral infections.

These thymus-derived cells, or those that traffic through thymus and mediate cell-mediated immunity, were called T cells and have distinctive surface antigens (Reif and Allen, 1964; Raff, 1969). They could give rise to thymus-derived leukemic cells (Reif and Allen, 1966; Schlesinger and Yron, 1969). Lymphocytes were also found to be important to antibody formation, i.e., humoral immunity as well. The cells mediating humoral immunity were shown to be derived from the bone marrow in animals and therefore were called B cells, i.e., bone marrow or bursal derived. B cells also have distinct markers (Pernis et al., 1970; Raff, 1970). B cell leukemias were also identified (Raff, 1970) with immunoglobin surface markers (Aisenberg and Bloch, 1972) in conditions such as chronic lymphocytic leukemia (CLL). These two functions of lymphocytes are present in the peripheral lymphoid organs, e.g., lymph nodes and spleen, and are highly regulated by other lymphocytes. The field of surface markers has advanced rapidly and has been frequently reviewed (see, e.g., Knapp, 1981).

The immunological role of the lymphocytes was first clearly identified in birds rather than mammals, using surgical removal of the bursa. Glick (1964) showed that the removal of a collection of lymphoid tissues in the cloacal region at hatching (called the bursa of Fabricius) of young birds led to an inability of the animals to form antibody later on at maturity. The removal of that organ at a critical young age led to a severe impairment of their later immunological functions and ability to form antibody (Glick, et al., 1956). Analogies between the bursa and major lymphoid organs in mammals led to similar extirpative studies of central lymphoid organs in mammals as well. The bursal studies established the impairment of antibody-forming capacity of bursectomized animals (Warner et al., 1962; Cooper et al. 1966a). The bursa gives rise to B cells in birds and the bone marrow or bursal equivalent organs, the B cell precursor cells in mammals. Archer et al. (1964) postulated that the appendix or gut-associated lymphoid tissues were the precursors of the B cells in rabbits.

It is necessary that two different cell types of different origins interact in order for a functional antibody-producing cell to be activated (Table I). For antibody synthesis, the B cell is the effector and the T cell is the helper cell. Serum antibody is released by cells of bone marrow origin. Cells are either capable of antibody production or are paralyzed by the presence or absence of thymus-derived cells. Thymus-derived cells were capable of only small amounts of antibody production which remains cell bound.

C. Repopulation Patterns in Lymphohemopoietic Organs after Irradiation

Recirculating small lymphocytes are characterized by their long life spans and ability to recirculate from blood through lymphoid tissues, to lymph, and back again to blood. They participate in a wide variety of immunological responses (Gowans *et al.*, 1962). In the rat the thoracic duct lymphocytes circulate through the various lymphoid organs, and this recirculating pool can be severely depleted by protracted thoracic duct drainage (Gowans and McGregor, 1965). Under antigenic stimulation, the small lymphocyte can dedifferentiate to large pyronophilic blast cells capable of producing instructed progeny lymphoid cells (see Gowans and McGregor, 1965). This process is accompanied by DNA synthesis which can be measured by tritiated thymidine uptake by the stimulated lymphocytes (Hirschorn *et al.*, 1963). Clonal expansion of the antigenically stimulated lymphocyte leads to acquired immunity (Burnet, 1959).

Recently, techniques for labeling cells within the adult mouse thymus and tracing labeled emigrant cells to peripheral lymphoid organs have been developed. Migrating cells arise from the medulla, are already much like mature T cells, are committed to one or other of the specialized T sublineages, and are functional. There is a unique seeding–reconstitution pattern which appears to be from two independent lineages related to medulla or cortex and medulla for T cells (Ezine *et al.*, 1984). The thymus of heavily irradiated animals was found to be recolonized by very few and sometimes only one stem cell. The cortex and medulla thus may be separately colonized, with the cortex and medulla containing the mitotically active population. It has been believed that cortical cells migrate to the medulla. Ezine and co-workers noted that both cortex and medulla can be repopulated by a single clone and that medullary seeding could be by an independent mechanism. Their data suggest that the medulla may contain a separate lineage of self-renewing cells.

Gallatin *et al.* (1983) recently described a cell surface molecule specific for mature lymphocyte homing to lymph nodes, but not to Peyer's patches, which depends on vascular endothelial cells in lymphoid organs. The recirculation of lymphocytes from blood to lymph and lymphoid organs was shown by the work of Gowans and Knight (1964) as well as by the lymphocytes' immunological activities (Gowans and McGregor, 1965; Ford, 1975). If mice are exposed to lethal whole body irradiation and are injected with syngeneic bone marrow cells carrying a chromosome marker (Ford *et al.*, 1956, 1963), the movements of cells of different tissue origins can be accurately traced, i.e., the bone marrow was repopulated by bone marrow cells. Shielded femoral marrow migrates to the spleen (Hanks, 1964); bone marrow cells can repopulate the spleen and lymph nodes and have a delayed entry into the thymus beginning at about 10 days (Ford *et al.*, 1956, 1966; Takada *et al.*, 1971).

The spleen is a major source of cells for all types of repopulation of the host (Jacobson *et al.*, 1949). The proliferation of bone marrow or spleen proceeded differently in the irradiated host (Takada and Takada, 1972) when used as donor cells. Spleen regenerative growth peaks on day 18 after irradiation and greatly overshoots the control before returning to normal size. The thymus aids spleen regeneration (Auerbach, 1963). Bone marrow cell injection or marrow shielding can greatly accelerate the regenerative process so that a peak occurs earlier. Bone marrow is the major source of repopulating cells. Silini *et al.* (1976) showed that splenectomy affected marrow CFU-S ability for repopulation only during the early postirradiation stages. The marrow pool size expanded independently and was the main source of precursor cells of the host. Spleen cell repopulation was a rapid, but transient contributor of cells to host recovery and was noted in the early period of host response to irradiation injury.

One marker of great value has been the enzyme terminal deoxynucleotidyl-transferase (TdT) discovered by Bollum (1974) and Chang (1971) to be present mainly in the thymus and at low levels in bone marrow. Since TdT was unique to the thymus, it has served as a marker of thymus origin or destined cells. TdT was also used to trace migration and movement of lymphoid cells. Coleman *et al.* (1974) and McCaffrey *et al.* (1975) found that precursor prothymocytes were present in the bone marrow and dropped precipitously after irradiation, consistent with interphase thymocyte death (Maruyama, 1978). However, those levels reappeared after irradiation, also consistent with repopulation and regeneration (Dalculsi *et al.*, 1982).

While thymic lymphocyte traffic to peripheral lymphoid organs is well established (see, e.g., Scollay *et al.*, 1980), there also appear to be mechanisms for postthymic cell population expansion and regulation. The renewal pattern of T lymphocytes and relative turnover rates have been studied using hydroxyurea (HU). Where lymphocytes are not in cycle, they are not affected by HU. With use of such methods, 50% of T lymphocytes were found to have short life spans and to drop precipitously after HU, but to be rapidly renewed in the peripheral circulation and lymphoid organs (Rocha *et al.*, 1983). Although a large fraction of the lymphocytes was not in cycle, they belonged to a rapidly renewed population capable of rapid expansion from a postthymic pool of precursor cells. The same effects were observed in thymectomized and nude mice. The effects of local radiation therapy to different regions of the body have been studied best in man and show that lymphopenia occurs and affects the distribution and subtypes in lymphocyte subsets. This is manifested by changes in T and B counts and ratios (Blomgren *et al.*, 1983; Stratton *et al.*, 1975; McLaren *et al.*, 1981; Concannon *et al.*, 1978; Raben *et al.*, 1976).

The effects and distributions differ with the anatomic site irradiated. These observations have led to the postulate that severe chronic deficiencies in immune reactivity and resistance to the spread of tumor can follow radiotherapy, especially of the chest and mediastinum (Stjernsward *et al.*, 1972).

In summary, there is considerable exchange in cellular traffic between the bone marrow, thymus, spleen, lymph nodes, and other peripheral lymphoid tissues. Injury to one organ is offset and corrected by both endogenous and migratory cells from the bone marrow and other sites. In no organ system is remote repopulation so important to recovery as the LH system.

D. Cell Kinetics of Lymphocytes/Lymphoid Tissues and Lymphomas

The postulate of Hamilton (1959) that there were two populations of lymphocytes, one short-lived and whose life span was 3–7 days and the other long-lived whose life span was weeks, months, or years, was examined by [³H]thymidine (TdR) studies. Little *et al.* (1962) and Schooley *et al.* (1964) showed that large and small lymphocytes could be quantitatively labeled by continuous infusions of [³H]TdR. TdR was the most important single agent for these studies (Cronkite *et al.*, 1959). Tritium was a stable label on TdR when incorporated into DNA; labeled TdR was stable and did not exchange with unlabeled TdR; labeled TdR was not stored in any significant amount or time in the tissues or cells in a small molecular pool. When labeled TdR was reutilized from dead or dying labeled cells (Maruyama, 1964), autoradiograms revealed grains over nuclei as a very light label rather than as a dense cellular labeling characteristic of semiconservative labeling. At one time, the reutilization of DNA was felt to be extensive, but DNA can be synthesized so rapidly that macromolecular reutilization of DNA fragments is probably unnecessary. In studies by Yoffey *et al.* (1961), extensive DNA-S labeling was evident in all lymphoid tissues, as had been found earlier by Andreasen and Ottersen (1945) using ³²P. Everett *et al.* (1964) analyzed labeling curves for lymphocytes *in vivo* and *in vitro* and found different labeling patterns with cell size. The labeling curves for small lymphocytes increased rapidly for 4–5 days and then followed a second slower labeling curve corresponding to the slower proliferating, longer-lived lymphocyte population. Of lymphocytes circulating from lymph organs in the thoracic duct (TDL), 90% were of the long-lived variety and these recirculated from blood to lymph. In thymus, spleen, and bone marrow, 80–100% of lymphocytes are of the short-lived, rapidly labeling variety, while in the TDL, blood, gut, and lymph nodes, they are mostly (10–30%) of the long-lived, slower labeling type. The rate of short-lived lymphocyte formation is very high in thymus and very low in lymph nodes, and long-lived lymphocyte numbers are very low in the thymus and much greater in the lymph nodes. In spleen, the values are intermediate. Most of the circulating small lymphocytes, whether short- or long-lived, arise from the thymus in young animals. The short-lived lymphocytes arise from the bone marrow or thymus. At one time, it was also erroneously believed that the small lymphocyte was an end stage, terminally differentiated cell. It is now known that the small lymphocytes can be stimulated to enlarge, become blastoid in appearance, and undergo cell division (Gowans *et al.*, 1962; Porter and Cooper, 1962), especially when stimulated by plant lectins or similar materials (Nowell, 1960). T cells are the predominant cell types in the thymus, TDL, and blood. In lymph node, spleen, gut, and bone marrow, B cells are more numerous (see, e.g., Fabrikant, 1975).

Sainte-Marie and Leblond (1958, 1964; Sainte-Marie and Sin, 1970) proved that lymphocyte differentiation proceeded from large to medium to small lymphocytes. The thymus is well known to have very high mitotic activity, with cell production far exceeding what would be required to maintain a stable organ size or cell population. Their research showed that the reticulum cells and subcapsular large lymphocytes of the thymus were rapidly labeled, transformed into medium-sized lymphocytes, and then into small lymphocytes. The cortical thymocytes contain numerous mitotic figures, and each reticular cell divides to give rise to one large lymphocyte and another reticular cell. The cell cycle (Howard and Pelc, 1953) time for large lymphocytes is 6–8 hr, and for medium lymphocytes 8.2 hr (Metcalf and Wiadrowski, 1966) (Fig. 2).

With antigen presentation and stimulation, actively proliferating germinal centers with high mitotic activity appear in peripheral lymphoid tissues. These centers actively form cells which lead to antibody-secreting lymphocytes. Yoffey

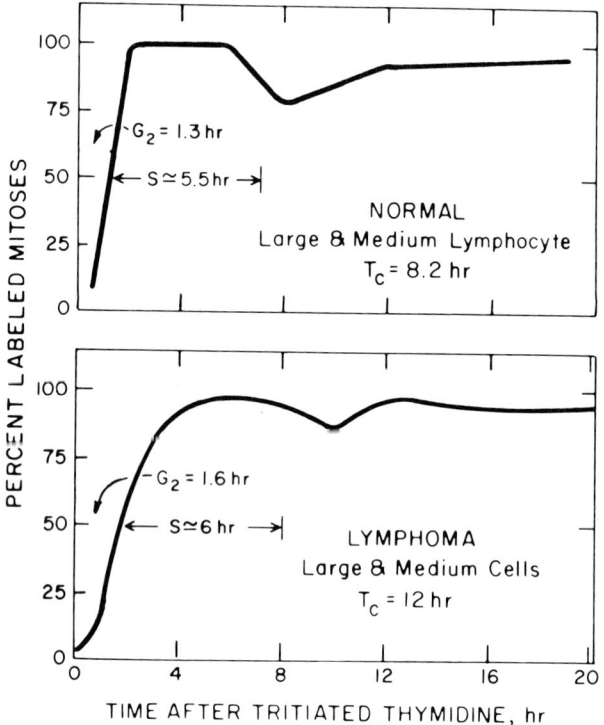

FIG. 2. Cell cycle kinetics of normal large and medium lymphocytes as studied in the thymus by percentage labeled mitosis curves. Upper curve for normal thymus; lower figure for thymic lymphoma. S, DNA-S phase; G_2 = premitosis gap period; T_c, cell cycle time. Redrawn from Metcalf and Wiadrowski (1966).

et al. (1961) found high labeling of germinal centers. Fliedner *et al.* (1964) and Hanna (1964) studied resting and stimulated spleen germinal centers. The normal unstimulated spleen is in a steady state and the whole process is set off by antigen trapping by phagocytic reticular cells. Characteristic morphological cellular changes include lymphocyte transformation and rapid increased proliferation of the stimulated lymphocyte, i.e., the large pyronophilic cell. After 1–2 days, labeled plasma cells appear, and an increase in plasma cell population soon follows. Plasma cells are specialized antibody-synthesizing cells derived from B lymphocytes.

Lymphomagenesis/leukemogenesis has been studied extensively and is a major problem of cancer research. With malignancy, it was found that neoplastic cells cycled *more slowly* (Fig. 2) than normal cells (Metcalf and Wiadrowski, 1966; Fabrikant, 1970). T_c for large lymphocytes increased to 7.6 hr from 6.8 hr (for normal counterparts) and 20.2 hr for medium lymphoma cells (from 8.2 hr for normal medium lymphocytes) by percentage labeled mitosis (PLM) labeling curves and by dilution of grain counts. Using dilution of labeled DNA in labeled lymphoma cells, the cell doubling time and $T_{1/2}$ for label dilution were found to measure cell cycle time for rapidly growing ascites lymphoma cells (Maruyama, 1963). For small lymphocytes, 50% of cells were labeled in 1.6 days, suggesting a total population replacement rate of 3.2 days. However, many were incapable of division and therefore appeared to arise from division from medium lymphocytes. For the C57BL radiation lymphoma/leukemia, T_c increases to 17 hr from control cell cycle rates of 9 hr (Fabrikant, 1975). Larger numbers of labeled cells appeared with an increase in the proportion of large lymphoma cells, but there is also an increased proliferation rate of large, medium, and small lymphoma cells. A higher growth fraction leads to expansion of the lymphoma cell population.

For growth of transplant lymphomas, many studies of cell kinetics of ascites lymphoma are available and have been reviewed.

When lymphoma/leukemia tumors grow to kill the host animal, this follows a period of exponential tumor growth to some critical or lethal cell population number (Maruyama and Brown, 1964). The latter model was independently proposed for the LSA lymphoma (Maruyama and Brown, 1964) and for the L1210 leukemia (Wilcox, 1967; Skipper *et al.*, 1964). The exponential leukemia/lymphoma growth model has proved to be extremely valuable to the study of leukemia/lymphoma growth, behavior, and cause of host death in leukemia and lymphomas. Characteristically, organ infiltration and size increase of principal LH organs (e.g., liver and spleen) at death in animals with lymphoma are similar at the time of host death (Maruyama, 1968a,b). Phasic periodic growth with bursts of DNA-S activity and organ size increases are observed for liver and spleen (Maruyama and Lee, 1980). The ultimate size in the terminal animals ends in a greatly enlarged organ at the time of death. Mean survival time of host can thus be used to accurately study the growth rate of lymphoma/leukemia tumors, since regardless of the size of initial inoculum, cell growth appears to cease at some common lethal cell number (Maruyama and Brown, 1967).

III. Lymphocyte Sensitivity to Radiation

A. General

One observation of the early radiobiologist was the radiation response of the lymphoid organs. Degeneration of the lymphoid tissues was one of the dramatic changes observed produced by radiation and was regarded as one of the characteristic responses of lymphocytes. X-Irradiation response occurs rapidly and in a period of hours rather than following the pattern of delayed death characteristic of epithelial cells. Lacassagne and Gricouroff (1958) showed that lymphocyte response could be studied easily *in vitro*. These methods were utilized by Osgood *et al.* (1942) and Trowell (1952) to study lymphocyte response *in vitro*. That small and large lymphocytes respond differently was also recognized very early. Small lymphocytes were much more sensitive than the large lymphocytes to ionizing radiation (see Bloom, 1949).

A careful study of thymocytes using dark-field illumination showed that response to radiation led to the formation of intranuclear vacuoles, karyorrhexis, and discharge of nuclear debris into the cytoplasm. This formed in a short period of several hours after irradiation and led to a dead pyknotic nucleus, followed by cytolysis in ~6 hr (Schrek, 1947). The changes caused by radiation were apparently the same as other causes of cell death, and radiation activated a preexistent cell death mechanism. The response of lymphocytes in the blood was also rapid. It is the earliest cell to decrease in the peripheral blood after total body irradiation or man of animals (see Cronkite and Bond, 1960). Lymphopenia occurred within a few hours after small doses and persisted for 7–10 days longer.

Lymph nodes, thymus, and spleen are very sensitive to irradiation. The small lymphocytes in lymphoid tissue are observed to be one of the most radiosensitive cells in the body. Medium lymphocytes are almost equally sensitive. Lymphocytes in the blood, spleen, tonsils, and intestines as well as nodes and thymus respond similarly to X rays or atomic radiation (Oughterson and Warren, 1956). Degenerative changes are evident in 2–3 hr. Cellular and nuclear debris and pyknotic changes are evident in the cells in 30 min. Cell lysis and loss occur and reach a maximum in 6 hr, with extensive cellular debris choking the lymphoid organs. By 24–72 hr regeneration and repair are actively in progress, and after 3 weeks regeneration is complete. Fifty rads produces minimal changes, and at 600–800 rads most lymphoid organs are destroyed until regeneration occurs in about 3 weeks. With regeneration, extensive mitotic activity is observed followed by replacement with medium and large lymphocytes. Similar changes are produced by injection of radioactive phosphate ^{32}P (Scott and Lawrence, 1941;

Warren *et al.*, 1950). With still higher radiation doses, tissue destruction can be permanent and sometimes irreversible.

Trowell (1952) carefully studied dose–response and radiosensitivity for lymphocytes using ionizing radiation. He found the earliest visible change was clumping of the nuclear chromatin. Nuclear vacuolization follows and then pyknotic degeneration and nuclear fragmentation. The cytoplasm disintegrates, fragments (karyorrhexis), and disappears when nuclear degeneration is far advanced. Death occurs within 1–2 hr of irradiation (''rapid''), with cell death, cytological and nuclear degeneration, and nonspecific autolysis of the dead cell. Dose–response curves for irradiation and pyknotic count at 5–6 hr show a shoulderless one-hit curve for lymph nodes *in vivo,* lymphocytes *in vitro,* and blood lymphocytes. Intestinal mucosa lymphocytes *in vivo* showed a shouldered, less steep dose–response curve (Fig. 3). Medium and small lymphocytes had very similar sensitivities. Low oxygen tension greatly reduced the radiosensitivity of the cultured cells.

The lymphocyte represents a morphologically unique cell of the lymphoid system. It is one of the major cells of the thymus, spleen, and lymph nodes and represents one of the most important and heterogeneous functional cells known. Its response to radiation is of special interest and has been measured by means of morphological changes, interphase death, ability to divide, stimulation by PHA or antigen, and DNA synthesis (Nowell, 1960; MacKinney *et al.*, 1962; Hirschhorn *et al.*, 1963). These latter measures give an indication of different patterns of radiation injury. Because of a unique interphase death, it is possible to determine a survival curve by morphological and pyknotic nuclear changes. Cortical lymphocytes are much more sensitive than medullary lymphocytes. In addition to one-hit inactivation, the initial response curves differ in slope, with a low D_0 for the more sensitive cortical cells. The medullary lymphocytes exhibit a multicomponent survival curve, with 7% in a more radioresistant state than the cortex lymphocytes. Both Trowell (1952, 1961) and Kallman and Kohn (1955) concluded that the thymus contains two independently reacting lymphocyte populations with different sensitivities. Trowell's cell counts were made 48 hr after thoracic irradiation; Kallman and Kohn allowed maximal regression between 2 and 5 days and used organ weight to measure cell death. Many others have also found more radioresistant lymphocyte populations (see, e.g., Rixon, 1967; Sato and Sakka, 1969; Sharp and Thomas, 1975; Kadish and Basch, 1975; Kataoka and Sato, 1975; Madhvanath *et al.*, 1976; Knox *et al.*, 1982).

In recent years, many approaches have been used to characterize and define different subpopulations of lymphocytes. By every assay, the lymphocyte has been established to be radiosensitive. Morphological cell death has been used to study lymphocyte death after neutrons (Jackson *et al.*, 1968; Hedges and Hornsey, 1978) and has been thoroughly documented as a form of cell death

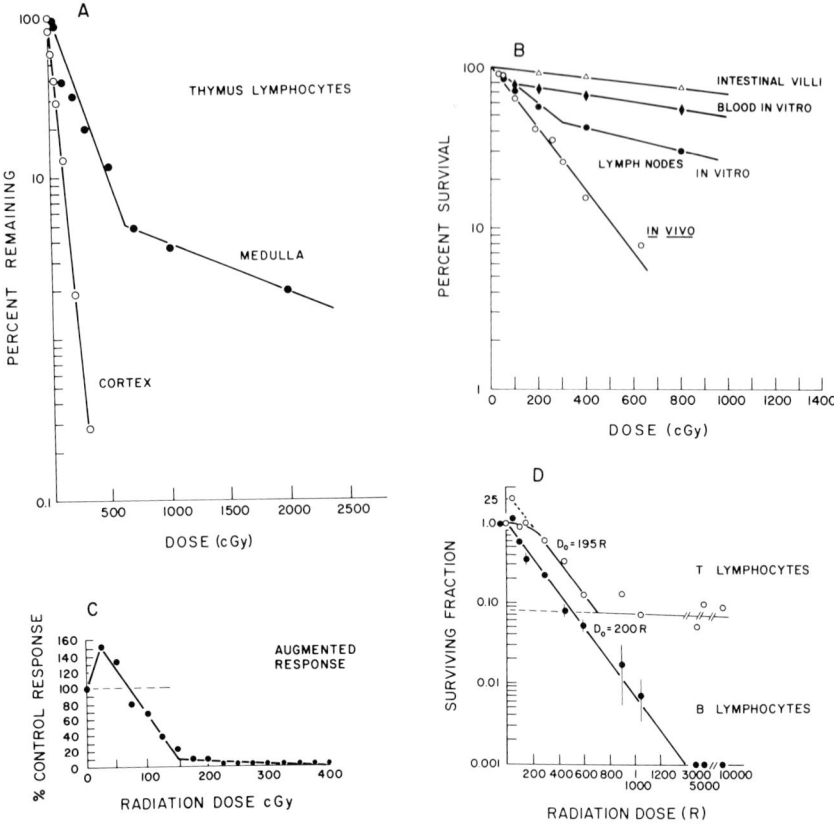

Fig. 3. Dose–response survival curves for lymphocytes. (A) Dose–response for cortex and medulla lymphocytes by pyknotic changes of nuclei. Cortex curve is one hit and exponential. Medulla lymphocyte curve is biexponential, with an initial steep and second less steep curve (drawn after Trowell, 1952). (B) Dose–response curves by pyknotic change for lymphocytes *in vivo* and in the intestines (after Trowell, 1961). (C) Augmented response curve for lymphocyte functional response with dose. At low dose some functional endpoints can exceed that of control (see Anderson *et al.*, 1980). (D) T and B lymphocyte dose curves by surface markers and persistence. B cell curve is exponential; T lymphocyte curve is multihit and biexponential (after Kataoka and Sato, 1975).

occurring within a period of several hours and taking place without entering mitosis. A peak of nuclear debris in lymphoid tissue occurs within 3 hr. Lymphocyte numbers in the circulating blood drop precipitously and rapidly and fall to 25% of normal values within 4 hr after 100 R. Those cells surviving radiation may show persistent morphological changes for years afterward (see, e.g., Buckton *et al.*, 1967). Anderson and Warner (1976) have stressed the heterogeneity of lymphocytes and the importance of the endpoint under study. Morphological changes and sensitivity are different for B and T lymphocytes (Olson *et al.*, 1979) and affect optical density distribution (Anderson *et al.*, 1975) or electron microscopy (Smith *et al.*, 1967). One of the important features of lymphocyte response is that of interphase cell death.

The effects of radiation on lymphocytes of thymus, lymph nodes, spleen, and peripheral lymphocyte counts were recognized very early in mammalian radiobiology. For example, the pattern of radiosensitivity and rapid shrinkage of thymus and lymphoid organs was noted early. The earliest workers and many others studied the histological effects of radiation on lymphoid organs. The thymus shows rapid involution after X irradiation. Thymic irradiation affects multiple organs and glands and produces rapid falls in circulating lymphocytes. More recently, total body irradiation or total lymphoid irradiation has been shown to severely affect immune function and allow homografts of a variety of tissues in man.

The lymphocytes of the thymus can be destroyed by single or fractionated doses of radiation. Sensitivity is different in the cortical or medullary regions and greater in the cortex (Trowell, 1952, 1961). There is rapid dissolution of the cortical thymocyte zone, but the medullary regions in contact with the reticuloepithelial regions and lymphocytes in the gut lymph nodes are less responsive. Destruction can occur after even small doses of radiation, and the permanence of destruction is dose dependent. In order to permanently ablate the gland, large doses of radiation must be given. To abolish its role as a central lymphoid organ, a single dose of at least 3000 cGy must be given. If total ablation of the organ is not achieved, the organ will eventually regenerate, presumably from the more resistant reticuloepithelial remnants, and eventually regenerate a morphologically intact thymus gland (Maruyama and Barclay, 1967). Localized thymus irradiation must be combined with total body irradiation in order to abolish the circulating T cells and the flow of prothymocytes from seeding the thymus from the bone marrow.

Functional destruction of the thymus and the thymus-dependent portions of the thymolymphoid organs has now been established by recent studies of total lymphoid irradiation (Slavin *et al.*, 1977; Strober *et al.*, 1979a,b). In these studies, shaped irradiation fields were constructed to selectively irradiate the thymus and mediastinum as well as the lymph node-bearing regions of the body. Using fractionated techniques, high-dose irradiation was given. The animals from a variety of species, including man, have now been shown to have severe deficits in immune function so that, e.g., organ homografts (Strober *et al.*, 1979a,b) or short-term remissions of autoimmune diseases are produced (Trentham *et al.*, 1981; Strober *et al.*, 1979b). Repopulation of lymphoid organs follows irradiation, depending upon dose. Depletion of lymphocytes is dose dependent and at appropriately high doses is permanent. Regeneration begins in 3–5 days and initially involves the regeneration of cortical lymphocyte followed by germinal center formation. Cortical cells then migrate centrally into the medullary zones. Regeneration is more rapid after local than after total body irradiation (Benninghoff *et al.*, 1969). Restoration is more rapid by bone marrow shielding or replacement (in the case of total body irradiation). The presence of an intact thymus is necessary for restoration of immune reactivity.

Direct studies of radiosensitivities of lymphocytes have concentrated on DNA synthesis or residual functional activities. These include tests for blastogenic transformation for antigen- or lectin-stimulated cells. Cells of B or humoral functional activity are felt to be more sensitive than T or cellular immune functional activity (see Anderson and Warner, 1976). Other biochemical consequences of lymphocyte irradiation have been studied in detail, e.g., plasma membrane, biosynthesis of nuclei acid, energy metabolism (see Altman *et al.*, 1970), DNA polymerases and TdT of thymus (Maruyama, 1978), chromosomes (Buckton et al., 1967), and cytotoxicity (Gerber 1984). Stimulation of DNA synthesis by antigen greatly increases the radioresistance of lymphocytes and thus may lead to an erroneous view of great radioresistance of lymphocytes. Hypoxia also greatly increases the radioresistance of lymphocytes (Schrek, 1946; Vos, 1967). Vos showed a D_0 of 85 R for oxic cells and 230 R for anoxic cells, leading to an oxygen enhancement ratio (OER) of 2.7, a value also found by Blackett (1965). There was SLD repair of lymph node cells mediating graft-versus-host (GVH) disease.

The effect of ionizing radiation on lymphocyte immunological function depends on which step in the various immune reactions is being studied, T or B cells, different lymphocyte subtypes, and the level of differentiation of the cell. These also depend upon radiation dose, type of radiation, dose rate, and time of irradiation relative to antigen injection, and can be manifested as an altered lag phase for response and height of antibody levels. In primary responses which require cooperative reactions between multiple cell types, there will be responses different from carrier cell functions. The logarithmic, late logarithmic, plateau, and declining phases of antibody production are only minimally to moderately radiosensitive. The secondary responses are generally more radioresistant. All functional activities of T lymphocytes, i.e., helper, suppressor, and cytotoxic, are radiosensitive in the precursor cell stage and radioresistant in the effector cell phase and also between T and B lymphocytes (see review by Anderson and Warner, 1976; Pazdernik and Nishimura, 1978; Kataoka and Sado, 1975). Those based upon primary responses are generally sensitive compared to secondary responses, consistent with a tissue and cell type of great radiosensitivity, limited repair capacity, and rapid cellular death and dissolution.

Owing to diversity and heterogeneity of lymphocyte types, detailed understanding of subtype radiation responses is still in an early stage. Recent advances have enhanced our appreciation of the capacity of the mature lymphocyte for cell division and that it is a reversibly differentiated cell. After plant lectin phytohemagglutinin (PHA) became available, Nowell (1960) showed that it was an initiator of mitosis in cultures of lymphocytes. MacKinney *et al.* (1962) showed by morphological analysis that lymphocytes entered DNA synthesis, reverted to become swollen, large less differentiated mononuclear cells, and avidly took up tritiated thymidine for 24–72 hr. Boyum (1968) developed a technique for isolation of lymphocytes using a 1 *g* gravity field and X-ray contrast media. Miller and Phillips (1969), Peterson and Evans (1967), and Shortman (1972) developed sedimentation procedures for the analysis of differentiating living cells, and these were used for analyses of lymphocytes (Miller *et al.*, 1975). Isolated lymphocytes were found to be stimulated to divide by antigens such as tuberculin (Pearmain *et al.*, 1963), viruses (Elves *et al.*, 1963), and foreign lymphocytes (Bain *et al.*, 1964; Hirschhorn *et al.*, 1963; Hume and Weldemann, 1980). Very recently, colony formation for lymphocytes in semisolid medium has shown biexponential survival curves to radiation, suggesting at least two cell components of differing radiosensitivity (Knox *et al.*, 1982). Dose–response curves based upon colony formation show no shoulder curves at low doses with two exponential slopes. The biological significance of these complex curves is still unknown and being studied for T, B, and other types of lymphocytes. Other means of selecting cells are being studied, e.g., Sephadex G-10 columns (Ly and Mishell, 1974) and fluorescence-activated cell sorting (Herzenberg and Herzenberg, 1978). Selective culture methods for B and T lymphocytes have also been developed (see Metcalf, 1977). These portend a continuing prodigious literature on lymphocyte response to radiation and other agents for the future.

Radiation response of different types of lymphocytes and lymphoid organs appears to be complex and heterogeneous. There are radiosensitive large, medium, and small lymphocyte populations, T and B cells (D_0 ~50–100 cGy), thymic macrophages of reduced sensitivity (D_0 ~2000 cGy), reticulum cells, hemopoietic cells, and so on (see Sharp and Crouse, 1981).

Future investigations of radiation response(s) will require the use of more selective tests, *in vivo* assays, well-fractionated and characterized cell populations, appropriate growth fractors, and controlled cell-to-cell contact of selected cell populations.

B. Lymphocyte Radiation Survival Curves

For lymphocytes, multihit inactivation models of radiation cell survival (Puck and Marcus, 1956) may not provide an accurate or complete description of radiation response. Particularly for normal tissue and cell response, heterogeneity of cell subpopulations leads to complex dose–response curves.

For cell death, rapid cell dissolution by an interphase mechanism occurs with pyknotic changes of the cells. There is also rapid shrinkage of lymphoid organs accompanied by rapid dissolution of lymphocytes (Trowell, 1952, 1961). These changes occur within a period of hours and do not require either mitotic failure or an entry into mitosis mechanism (Altman *et al.*, 1970). Cell changes occur with extraordinarily low radiation doses of 2–5 R (Stefani and Schrek, 1964). With low dose rate (LDR) irradiation, no dose rate effect is observed (Konings, 1981). In fact, an inverse dose rate effect can be observed. Closely related pluripotent hematopoietic CFU-S cells show little or no dose rate effects whether of bone marrow (Frindel *et al.*, 1972; Puro *et al.*, 1974; Fu *et al.*, 1975; Krebs and Jones, 1972; Ainsworth *et al.*, 1976) or spleen (Maruyama *et al.*, 1983a,b) origin. No dose rate effect by LDR irradiation occurs for spleen weight loss or thymus weight loss (Feola *et al.*, 1983). No potentially lethal damage (PLD) recovery occurs for the closely related hematopoietic CFU-S (Thomas *et al.*, 1982).

Trowell's (1952, 1961) pyknotic cell death data are plotted in Fig. 3A and B for cortical and medullary lymphocytes of the thymus and for lymphocytes in a variety of environments, e.g., lymph node *in vivo* or mesenteric lymph nodes. One-hit exponential cell survival curves were observed in each site, but biexponential survival curves with sensitive and radioresistant subpopulations were also noted. These patterns of one-hit survival curves showing biexponential slopes with radiosensitive and radioresistant components are now well established and have been shown for photon and high linear energy transfer, LET radiations by many workers for lymphocyte response (Rixon, 1967; Sato and Sakka, 1969; Jackson *et al.*, 1968, 1969; Geraci *et al.*, 1974; Kataoka and Sato, 1975; Anderson and Warner, 1976; Madhvanath *et al.*, 1976; Hedges and Hornsey, 1978). These sensitive and resistant components may represent heterogeneity in radiosensitivity, different subpopulations, or different physiological states of the cell (e.g., DNA synthesis). More recently, in addition to the morphological pyknotic cell survival curve, assays are performed based upon cell ability to participate in immune reactions, survival of cell type (e.g., B or T lymphocyte), lymphocyte blastogenesis, entry into DNA synthesis, lymphocyte stimulation by different mitogens (Anderson and Warner, 1976), and those based upon cell proliferative ability and colony formation (>40 cells) or function. Culture methods for T and B cells have also been developed.

Colony growth of T lymphocytes was achieved and described by Shen *et al.* (1977) and has been used for survival curve determination. Clonal assays of lymphocytes have now been developed for B and T lymphocytes. Techniques allow the growth of B cells in semisolid agar in the presence of conditioned media, mitogenic agents, and 2-mercaptoethanol (MEA) (Metcalf, 1977; Fibach *et al.*, 1976). B cell growth requires the presence of PHA. T lymphocytes will also grow in soft agar with the appropriate mitogen (PHA) used for stimulation of cell growth for human and mouse cells (Rozenszajn *et al.*, 1975; Sredni *et al.*, 1976; Shen *et al.*, 1977; Wilson *et al.*, 1980; Wilson and Dalton, 1976). The methods used single-phase semisolid culture systems containing concanavalin A (Con A) or PHA which produced colonies greater than 40 cells, with cells

capable of E rosette formation, absence of surface immunoglobulins, and proliferation inhibition by anti-T but not anti-B antisera. Knox *et al.* (1981) developed methods for culture from whole blood. The clonal growth assay used for
human lymphocyte radiosensitivity (Knox *et al.*, 1982) lacks a shoulder at doses
less than 75–100 R and followed a biexponential radiosensitive and radioresistant curve (Fig. 4B). This supports the postulated multiple subpopulation concept
developed from pyknotic assays for lymphocytes and weight change assays of
lymphoid organs. Lymphocyte stimulation survival curves are different and
probably reflect age–response changes (Fig. 4A).

Figures 3 and 4 show the variety of dose–response curves that have been
observed with each type of assay. Within each assay, marked variability has been
noted, with a large spectrum of responses observed. This has also been found by
the clinical application of radiotherapy to lymphoid tumors where an extremely
wide and diverse array of schedules, doses, fractionation schemes, field sizes,
and shapes has been used. High-energy photons from linear accelerators have
proved useful by providing deeply penetrating flat field irradiation or electrons
that treat only the body surfaces. High LET fast neutrons may also have a relative
biological effectiveness (RBE) of only 1.0 (Hedges and Hornsey, 1978; Geraci *et
al.*, 1974) and therefore are probably no better than photons for most lymphoid
tumor therapy.

Still further complexities of lymphocyte radiosensitivity are observed for *in
vitro* lymphocyte response based upon functional response (Anderson and Warner, 1976). In assays using primed and unprimed spleen cells enriched for T and B
cells assayed for ability to respond to sheep blood cells (SRBC), the dose–
response curves showed two components: (1) augmented response components
(Fig. 3C) associated with very low doses (0–25 cGy) of radiation, and (2) rapid
loss component persistent to doses of ~10,000 R for both T and B cells. Pazdernik and Nishimura (1978) and Anderson *et al.* (1977) have also reported differences in B and T cell radiosensitivities in regard to organ depletion rates. An
augmentation response has also been observed (see Anderson and Lefkovits,
1980) and may be related to radiosensitive, interacting subpopulation response,
while the radioresistant component is related to a resistant subpopulation and/or
function. Such responses may be observed *in vitro* and *in vivo* (Taliaferro and
Taliaferro, 1970). These effects may be due to differences in radiation effects on
interacting T and B cell subpopulations.

Figure 3D shows the dose–response characteristics of persisting T and B
lymphocyte remaining in the spleens of mice 3 days after graded doses of X-rays
(Kataoka and Sato, 1975). Survival was measured by surface markers on surviving cells. The B lymphocytes had a survival curve with $n = 1.0$ and $D_0 = 200$ R;
for T lymphocytes, $n = 2.5$ and $D_0 = 195$ R. For T lymphocytes 8% were highly
radioresistant. Rapid interphase cell death allowed survival curves to be determined based upon cell survivors at 3 days. These data closely resemble the

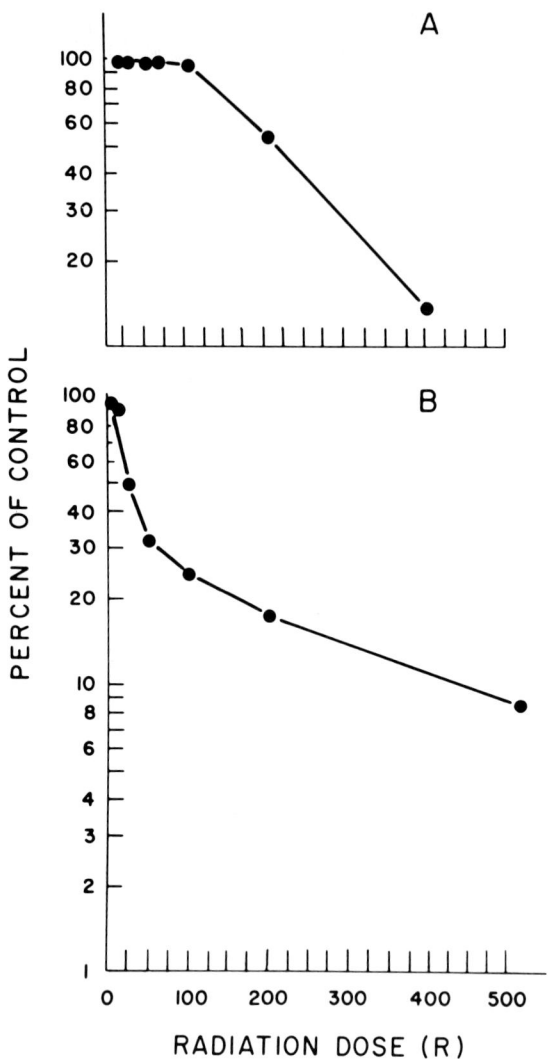

FIG. 4. Dose–response survival curves for lymphocytes. (A) Survival curve by lymphocyte stimulation assays shows a shoulder followed by an exponential declining curve. (B) Survival curve by colony-forming ability shows biexponential curves, with an initial one-hit dose–response curve with more radioresistant tail. (Redrawn after Knox *et al.*, 1982.)

colony-forming ability survival curves of Knox *et al.* (1982), but a separate parameter was measured. Puck (1966) concluded that the nucleated cell depletion curves for thymus and spleen reflected inhibition of cell multiplication and blocked cell turnover rates.

Numerous normal and many malignant lymphocytes exhibit one-hit inactivation and no SLD or dose rate-dependent repair. A human cell line which secretes immunoglobulins has been reported to have multihit inactivation and to exhibit SLD repair on split-dose study (Drewinko and Humphrey, 1971). A mouse lymphoma, the P388, exhibits SLD recovery (Belli *et al.*, 1967). However, some lymphoma cell lines show low extrapolation numbers and some may also have very low D_0 values (50–100 cGy). Chronic LDR irradiation may also lead to selection of cell lines capable of SLD repair or induce adaptive capability (Juraskova and Drasil, 1984). Similar observations were reported by Courtenay (1965).

C. Interphase Death of Lymphoid Cells and Tissues

The nature of the initial radiation damage is identical for cells and affects the critical target molecules, directly or indirectly, through cellular water. However, the manner in which lymphocytes die is quite different in mammalian and human radiobiology. A strikingly diverse number of acute biochemical and degenerative changes accompany this process of cell death (Trowell, 1952; Lacassagne and Gricouroff, 1958; Altman *et al.*, 1970; Schrek, 1961).

In general, three types of cell death occur for mammalian cells in response to radiation. The most widely studied is mitotic or delayed death and occurs for most cells of epithelial, mesenchymal, or adenomatous cell origin some time after irradiation. This may manifest itself as mitotic failure with even small doses of radiation, and the failure to divide, or reproductive death, has become the classic definition of cell death since the work of Puck and Marcus (1956). Thus, a cell that fails to form a colony (and in the classic definition, at least ≥50 cells) was regarded as "dead." The other major form of cell death is known as intermitotic or interphase death and cell and nuclear lysis and dissolution occurs very soon, within 3 to 6 hr, after irradiation. This type of cell death is characteristic of lymphoid cells and organs containing large lymphocyte populations such as lymph nodes, spleen, and thymus, and cell death occurs independently of mitosis. Nuclear pyknosis and karyorrhexis occur shortly after irradiation injury, and in organs comprised of large proportions of these cells, maximum and dramatic shrinkage and atrophy occur within one to several days after injury and thereby have led to the use of the term radiosensitive. For a neoplasm derived from lymphoid organs, a mixed death pattern may be seen, and one can observe both interphase death of cells as well as delayed cell death and division failure. Some of the characteristics were shown in Table II. The process of interphase

TABLE II

CHARACTERISTICS OF REPRODUCTIVE DEATH AND INTERPHASE DEATH[a]

Characteristic	Reproductive death		Interphase death
	Rapidly dividing cells		Nondividing or slowly dividing cells
Type cells	Mammalian cells, e.g., epithelial cells (e.g., HeLa), sarcoma, adenocarcinoma, fibroblasts		Small lymphocytes, thymocytes, young oocytes, neuroblasts, spleen, lymph node, thymus cells
Mean lethal dose	$D_0 = 50–20$ cGy (around 100 cGy)		Several centigrays to several hundred (or more) centigrays for certain cells
Survival curve	Sigmoidal		Nonsigmoidal
	Feature	Time	Histological damage
Time of death	Delayed over hours, days, weeks, or months (can be after multiple divisions)	Immediate changes, first hour (latent period)	Extensive histological damage; nuclear pyknosis
	Delayed at mitosis Numerous mitotic cell and chromosome abnormalities	Early changes: 1–3 hr (interphase death)	Aggregation of chromatin near nuclear membrane; changes in nuclear membrane and in mitochondria; "homogenization" (loss of structure) of nuclei, release of histones into cytoplasm; pyknosis of nuclei
	Giant cells Detachment from surface	Intermediate changes, 2–24 hr	Phagocytosis of lymphocytes and cellular debris by macrophages and reticulum cells, removal of dead lymphocytes from organ and shrinkage of organ
		Late changes: 1–10 days (repeat)	Low doses: Regeneration; large monocytoid lymphoblasts reappear first, mature lymphocytes 1–2 days later High doses: Fibrosis by connective tissue Very high doses: No organ regeneration
Death and cell division	Associated with cell division		Not associated with cell division
Mitotic changes	Marked mitotic accumulation (reversible/irreversible)		Not characteristic

[a]Modified from Altman et al. (1970) and Kelly (1961).

death is dependent upon metabolic activity, particularly for that interfering with respiratory activity (Scaife and Broher, 1976; Altman *et al.,* 1970).

As pointed out elsewhere, there have been many reviews that have lumped diverse species such as mouse with hamster and human, as well as different tissues of origin, e.g., leukemias, with adenocarcinomas, sarcomas, and epithelial carcinomas. These studies represent oversimplifications which are not likely to be relevant to specific human tumor problems in the future. While these differences also relate to lymphoid tissues and tumors, several markers of lymphoid tissue radiosensitivity appear to be common for all mammalian species, regardless of diversity, and to be a characteristic of lymphoid cells. Differences will very likely become more pronounced as newer research techniques focus on and characterize in greater detail the different specific properties and responses of lymphocytes. Several common features noted at this time are (1) interphase cell death, (2) one-hit cell killing, (3) absence of or minimal SLD recovery, (4) greater radiosensitivity of resting cells, and (5) marked increase in radioresistance of stimulated cells. While this review will cover many species and attempt to relate the observations to human therapy, much more research is needed before a better understanding emerges.

The major theories of RBE and radiation effects are also touched on by observations on lymphocytes. The reproductive failure model of Puck and Marcus (1956) is based upon (usually) multihit events in the cell target. A unique one-hit cell death mechanism is observed for lymphocytes wherein interphase or pyknotic cell death occurs without any relationship to mitosis and at very small photon doses (Stefani and Schrek, 1964). This allows direct cell counts a few hours after injury, with a determination of cell viability or death to be made. Cell death occurs following a one-hit mechanism, with rapid (interphase) cell death, early dissolution of dead cells, and prompt shrinkage of lymphoid organs. Thus, it is less dependent on cell division and reproductive cell death. Pyknotic cell death may occur by activation of lysosomal enzymes (Bacq and Alexander, 1961) and nuclear disruption, but does not appear to require a multihit mechanism. Likewise, small size of dose or low dose rate do not appear to alter the observed effects or enhance recovery.

RBE for neutron versus photon radiation for pyknotic cell death can be as small as 1.0 (Geraci *et al.,* 1974; Hedges and Hornsey, 1978). For ^{252}Cf neutrons, RBE by thymic weight loss was 3.5, with no dose rate effect. One-hit killing was noted for photon radiation (Feola *et al.,* 1983). Madhvanath *et al.* (1976) reported RBEs for human blood lymphocyte interphase death to radiations of a wide range of LET from ^{60}Co radiation to ^{40}Ar (i.e., from 0.3 to 2000 keV/μm). Survival curves 5 days after irradiation were exponential in shape for all the radiation studies. Low LET radiation yielded biexponential curves with a sensitive and less sensitive population. ^{7}Li ions with an LET of 43 keV/μm had the highest RBE, 3.4, and there was a decrease above this value to 0.3 for ^{40}Ar

TABLE III

D_0, RBE, and Inactivation Cross-Sectional Values
of Heavy-Ion Beams

Radiation	LET (keV/μm)	D_0 (5 days; cGy)	RBE
^{60}Co	0.27	75	1
^{1}H (15 MeV)	3.6	51	1.5
^{4}He (80 MeV)	12	32	2.3
^{7}Li	43	22.5	3.4
^{11}B	120	30	2.5
^{12}C	190	48	1.56
^{16}O	360	81	0.93
^{10}Ne	655	121	0.63
^{40}Ar	2000	270	0.28

(Table III; Fig. 5). The OER was 2.3 for ^{60}Co, fell to 1.1 at 43 keV/μm, and was 1.0 for ^{12}C. The nature of the radiation response and RBE with LET is difficult to reconcile with observations of lymphocytes where exponential and not sigmoid cell killing is observed. These observations suggest that perhaps very novel mechanisms are involved in cell death for lymphoid cells. For photon radiation, cell death occurs with small radiation doses of 2–5 R and even at LDR (Konings, 1981). The effects observed with neutrons suggest that in addition to the high LET events, photon radiation, δ rays, and secondary particles, different biological mechanisms (e.g., membrane damage) must be present to account for the RBE and dose rate effects. Clearly, low-energy neutron radiation studied at low dose LDR represents an extreme situation. The efficiency of photons for

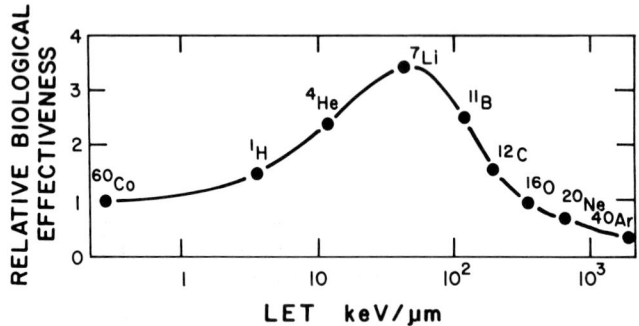

FIG. 5. Relative biological effectiveness with LET (in keV/μm) for lymphocytes by interphase death survival curves. RBE at 3.4 was highest for ^{7}Li at an LET of 43 keV/μm and decreased beyond that value. (From Madhavanath et al., 1976.)

destroying lymphoid cells and tissues is supported by established clinical efficacy of photon radiations for the therapy of lymphoid neoplasms.

Lymphocyte cell death is also greatly altered by antigenic stimulation (see Anderson and Warner, 1976) or blastogenic transformation (Schrek and Stefani, 1964), and the stimulated cell becomes much more resistant to radiation. The lymphocyte irradiated before (Rickinson and Ilbery, 1971) or during transformation sustains a mitotic progression blockade and cannot further divide and differentiate to become small lymphocytes. Thus, cell cycle progression and DNA-S activity blockade can occur and increase radioresistance after irradiation similar to other mammalian cells. Blocked proliferative activity and increased cell and target multiplicity no doubt can contribute to the complex mechanism involved in changes in radiosensitivity. However, the extent and magnitude of the changes are still not well understood and described. The cell cycle (Howard and Pelc, 1953) transition lymphocyte radiation response will need to be studied in greater detail in the future.

D. Repair of Radiation Injury

In vivo and *in vitro* studies have documented the extreme radiosensitivity of the small lymphocyte (Anderson and Warner, 1976). The variety of assays include morphology, dye exclusion, blastogenic transformation, and colony formation. One-hit exponential survival curves have been reported as well as biexponential survival curves with a radiosensitive and radioresistant subpopulation. With the multiplicity of assays and endpoints used, considerable uncertainty exists in regard to repair of radiation injury in lymphoid cells and tissue. Drewinko and Humphrey (1971) studied human antibody-forming lymphoid cells by colony-forming ability and observed typical shouldered sigmoid survival curves and repair of SLD using split-dose experiments. Evans and Norman (1968), Spiegler and Norman (1969), and Kwan and Norman (1977) concluded that lymphocytes could carry out unscheduled DNA synthesis and incorporation of thymidine into DNA consistent with repair of DNA. Lett *et al.* (1967) showed that rejoining of x ray-induced breaks occurred in DNA of leukemic cells. Scott *et al.* (1974), Fox (1979), and Fox and Fox (1979) reported DNA repair replication in different lymphoma cell lines and different radiation and alkylating agent sensitivities. Johanson *et al.* (1982) found DNA repair and rejoining was the same in sensitive or resistant cell lines. Lymphoma cell surface charges can recover after irradiation injury, except for Burkitt's lymphoma (Sato *et al.* 1972).

However, cells exhibiting interphase pyknotic death may generally not show strong dose rate or fractionation effects, and this is evident on irradiation studies of lymphocytes and lymphoid organs (e.g., thymus, lymph nodes, and spleen). Irreparable damage to thymus has been reported by Jackson *et al.* (1969, 1972) using cell number and organ DNA content. Feola *et al.* (1983) studied LDR[137] irradiation by thymic weight loss, and acute doses of ^{60}Co produced effects no different compared to LDR^{137}Cs irradiation. LDR neutron irradiation with ^{252}Cf showed greater effect as expected for high LET radiation. Many assays based upon cell membrane permeability, organ DNA content, and dye exclusion assays all respond in a manner consistent with an interphase one-hit death pattern and show little fractionation, dose rate effect, or SLD recovery (see Altman et al., 1970). These indicate more complex responses in lymphocytes and lymphoid tissues than epithelial tissues.

Blastogenic transformation assays are based upon cell cycle progression into DNA synthesis. Lymphocyte assays based upon stimulation may show survival curves that are sigmoid in shape and show recovery with dose fractionation or LDR (Fig. 4). However, changes in radiosensitivity between two dose fractions may be contributed to by cells stimulated to proliferate by antigens (e.g., GVH, activated spleen cells) (see Anderson and Warner, 1976).

The heterogeneity of lymphocyte populations and the varieties of endpoints make the situation confusing. The multihit survival curve shown by Kataoka and Sato (1975) showed that some populations of lymphocytes are capable of SLD repair with fractionated radiation or LDR. Generally, however, it appears that many lymphoid cell lines show little and sometimes no split-dose or dose-rate effect. These include several human lymphocytic lymphoma cell lines, Burkitt's lymphoma, and diffuse histiocytic lymphomas (see Johansson et al., 1982; Yau et al., 1979). Caldwell et al. (1965) found little or no SLD (or even negative) recovery for mouse lymphoma cells.

In mouse lymphoma cell lines irradiated in vivo with two doses during a rapidly proliferating stage, survival differences (i.e., SLD repair) may be observed (Belli et al., 1967, 1970). In unpublished experiments, we have observed no SLD recovery for the lymphosarcoma (LSA) lymphoma. This cell line (Silini and Maruyama, 1965) was studied when it was shown to have a low extrapolation number in early transplant generations. In in vivo thymic dose-split irradiation studies (Hanks and Kaplan, 1965), no recovery was observed in lymphoma cells in vivo. Beer (1973) and Fox (1979) also noted similar responses along with Inada et al. (1976) and Johansson et al. (1982), and little SLD recovery occurred with split-dose studies.

Cole (1967) and Kennedy et al. (1965) showed that immunocytes lack the capacity to recover from radiation injury. Thus, not only normal lymphocytes but also their neoplastic counterparts may show a lack of ability to repair SLD. In extreme situations, there may be a negative effect where the first dose of radiation actually sensitizes the survivors to the second dose (Yau et al. 1979).

While SLD recovery may not be a characteristic of lymphocytes or lymphoid tissue, human conditions in which photon radiation repair is deficient [i.e., ataxia telangiectasia (AT)] are very susceptible to lymphoma/leukemia induction. Where DNA repair is defective (Paterson et al., 1979, 1984) in human cells

(Taylor *et al.*, 1975), susceptibility to lymphomagenesis (Gotoff *et al.*, 1967; Filipovich *et al.*, 1984) is noted in addition to inordinate sensitivity to radiation therapy (Cunliffe *et al.*, 1975). Generally, however, DNA repair in mouse lymphoma lines occurs regardless of the inherent characteristics of radiation response of reproductive radiosensitivity (Johansson *et al.*, 1982).

Recent studies of chronic LDR exposure of cultured lymphoma cell lines have shown that cells gradually acquire a multihit survival curve from the original one-hit survival curve as well as SLD recovery capability. Juraskova and Drasil (1984) postulated that this was due to an adaptive response of the cell or the selection of lines capable of repair. We believe that the heterogeneity of lymphocytes, their cell kinetic status, and their status as normal or neoplastic will lead to a variety of radiation dose responses for acute, LDR, and high LET radiations and also to split-dose or dose rate effects.

E. Experimental Malignant Lymphocyte Response

Puck's (1956) development of clonal assay and a radiation survival curve for human cells led to studies of radiation response of mouse lymphoid tumors and dose–response radiation survival curves for many different experimental leukemia/lymphomas. These studies used the terminal dilution assay for 50% tumor takes (TD_{50}) using cell number to produce a tumor (Hewitt and Wilson, 1959). Furth and Kahn (1937) showed that a single transplanted cell could transfer the tumor to an isogenic host. The TD_{50} assay used the number of tumor takes in a serial dilution assay as an endpoint. Since the development of the TD_{50} assay, many additional methods have been developed. These measure both reproductive integrity and proliferative capacity of surviving cells. There has been greater attention to separating these endpoints for lymphoma cells than was originally suggested by Puck. Based upon colony formation or the cell multiplication methods, Dewey *et al.* (1963) found that survival curves were similar but not necessarily identical using the growth assay. This was confirmed by Maruyama *et al.* (1967) for lymphoma cells *in vivo* using survival time assay. McNally (1972) and Berry (1967) also showed that there could be a greater difference in OER by the use of clonal or growth assays.

The term reproductive integrity is often erroneously used and regards reproductive and proliferative capacity as identical. Colony-forming ability is the ability of a cell to proliferate to a 50-cell colony in a designated period of time. Because of growth delay, growth slowing, and changes in cell proliferation rate (small colony formation), growth assay needs to be done with surviving cell populations which grow at the same rate as control cells (Nias and Fox, 1968). When such is not the case, appropriate corrections are in order. Lymphocytes do not usually grow as a discrete colony as they are naturally highly mobile. However, using semisolid media, colonial growth has been achieved. Still, additional assay methods have been developed for tumors and include (1) growth curve extrapolation (Alexander and Mikulski, 1961); (2) spleen colony formation (Bush and Bruce, 1964); (3) survival time (Maruyama and Brown, 1964); (4) effective fraction (Maruyama and Ceman, 1970); and (5) semisolid culture media (Fox and Gilbert, 1966).

Puck's method for *in vitro* culture assay has been extended considerably from the surface attachment-colonial growth assays to the use of semisolid media (agar, methylcellulose). assays (see, e.g., Nias and Fox, 1968; Fox and Gilbert, 1966). In recent years, human lymphoma cells have also been cultured and survival curves determined in an effort to better characterize human leukemia/lymphoma cell radiosensitivity (Weichselbaum *et al.*, 1981). Altman *et al.* (1970) reviewed and tabulated leukemia/lymphoma cell lines studied by early researchers. In Table IV, additional cell lines in which cell radiosensitivity has been studied are tabulated and include both mouse and human lines. The apparent origins of the cells, type of photon radiation used, extrapolation number n, D_0 under oxic conditions, conditions of irradiation (*in vivo* or *in vitro*), and SLD repair, if reported, are noted. Early research focused on mouse leukemia/-lymphoma cell lines; more recent studies have addressed the radiation sensitivity of human lymphoma cell lines. The list is selective for lymphocytic cell lines and excludes myeloid cell lines. Histiocytic cells are noted as they could possibly be lymphoblastoid. If studies for SLD repair were carried out (fractionated radiation or LDR) and reported, it is noted.

It is clear that many of these studies show low n values. It was found very early by Caldwell *et al.* (1965) that lymphoid cell lines lack SLD repair. In recent studies, it has been evident that many lymphocyte-derived cell lines also lack SLD repair capability. Some may, in fact, exhibit negative SLD repair so that the first dose of radiation sensitizes the cells to the second dose of radiation (Yau *et al.*, 1979; Caldwell *et al.*, 1965). This characteristic has proved useful in TBI therapy studies of leukemia/lymphoma patients so that multifraction TBI can aid in destroying some leukemic cells. A few cell lines show recovery (Drewinko and Humphrey, 1973; Belli *et al.*, 1967), but may be exceptions to the majority of cells studied. It may be that the longer cells are passaged, the SLD capacity tends to increase and can lead to the selection of cell lines which are more capable of SLD repair. Under continuous irradiation, cell lines gradually develop survival curves with shoulders and larger extrapolation numbers (Juraksovic and Drasil, 1984). The L5178 has lines (resistant) that repair SLD as well as those that do not (sensitive). Thus, for lymphocytes, both normal and malignant, SLD should not be regarded as a typical or necessary characteristic.

The growth curve extrapolation assay (Alexander and Mikulski, 1961) has been useful for study of radiation survival curves for malignant lymphocytes. When corrected for growth delay and growth occurring without alteration in growth rate, it can match colony-forming assay results (Nias and Fox, 1968).

When growth rates are altered or many small colonies are produced, largely as a function of a large dose of radiation, then results can be different.

The spleen colony assay (Bush and Bruce, 1974) has had limited use for study of survival curves. This is because lymphoma cells have a tendency to migrate as well as to form confluent colonies. With the extremely rapid growth rate of the tumor cells and the many small colonies found, some uncertainty regarding the actual number of tumor colonies exists for some tumors. Some tumors do not form colonies at all. The assay has had much more application to the study of chemotherapy than to radiation biology (Bush and Bruce, 1964). The colony-forming efficiency is also low for some tumors so that the behavior of all cells is not sampled and is restricted to those possessing colony-forming ability (Wodinsky *et al.*, 1967).

The survival time assay has been used for study of tumor growth (Maruyama and Brown, 1964) and for tumor radiation response (Maruyama *et al.*, 1967). Leukemia/lymphoma cells are widely disseminated in the animal during *in vivo* growth. However, if one postulates that death occurs at some critical cell number, then one can analyze and directly check the growth rate of the tumor. For the growth endpoint as noted above, colony survival and regrowth may not give identical results. This is due to more slowly growing cells, as noted by Berry (1967), McNally (1972) and Maruyama *et al.* (1970). The survival time assay has been excellent for measuring the growth or regrowth kinetics of hematologic tumors (Bergsagel *et al.*, 1968; Wodinsky *et al.*, 1967; Tanaka and Lajhta, 1970; Wendling *et al.*, 1973; Malaise *et al.*, 1974; Juraskova, 1975, 1978; Covelli, 1981; Karlsberg *et al.*, 1981; Wilcox, 1967). With very rapidly growing tumors, it is difficult to separate survival time data using closely spaced cell doses. One can obtain dose–response survival curve data with tumors not able to form colonies. It is therefore a valuable assay for the study of many hematologic tumors. In solid tumors, one can allow tumors to grow to a specific size and determine time to reach that size. However, a determination for only a small fraction of the growing cells may lead to misleading results. For radiation response it has been found that survival curves identical to the TD_{50} assay can be obtained under certain conditions. Hypoxic tumors and OER results, however, can differ (Berry, 1967; McNally, 1972; Maruyama *et al.*, 1967).

The effective fraction method was designed to avoid the insensitivity of the TD_{50} assay to the study of a single surviving cell (Maruyama and Johnson, 1969; Maruyama and Ceman, 1970). A large number of mice are injected with the irradiated cell inoculum at the endpoint of the dilution assay. Based upon the Poisson distribution of takes and no takes, one can determine the number of viable cells at the dilution endpoint. At higher radiation doses, division and growth delays are marked (Maruyama and Ceman, 1970) consistent with small colony formation. The assays have shown that secondary host reactions can affect the single surviving cell residing in the immune competent intact host (Maruyama, 1971).

TABLE IV

LYMPHOID TUMOR CELL LINES[a,b]

Cell lines	Origin	Radiation	D_0 (cGy)	N	Method assay	SLD repair	Authors	
Human								
T1 lymphoma	Plasma cell ?	X rays	70	18	Colony	+	Drewinko and Humphrey	(1973)
U-937	N	137Cs	150	3.3	Regrowth	+	Johansson	(1983)
SKW-4	L	137Cs	80	2.0	Regrowth	−	Johansson*	
SU-DHL2	L	137Cs	120	2.4	Regrowth	−	Johansson*	
SU-DHL4	N	137Cs	120–140	1.2–1.6	Regrowth	−	Johansson*	
SU-DHL3	L	137Cs	190	1.0	Regrowth	−	Johansson*	
U-698	L	137Cs	190	1.2	Regrowth	−	Johansson*	
U-715	L	137Cs	160	1.0	Regrowth	−	Johansson*	
P3HR-1	Burkitt L	137Cs	130	1.2	Regrowth	−	Johansson*	
P3HR-1	Burkitt L		125	1.1	Regrowth	−	Inada et al.	(1976)
P3HR-1	Burkitt L		70	1.0	Regrowth		Sato et al.	(1972)
Histiocytic J-774	N	137Cs	116	0.67	In vitro	ns	Greenberger et al.	(1979)
Line 45	L	X rays	147	1.1	Colony	ns	Weichselbaum et al.	(1981)
Line 176	MM	X rays	76	4.0	Colony	ns		
Mouse								
CBA	L	60Co	162	2.0	In vivo TD50	ns	Hewitt and Wilson	(1959)
P388F	L	60Co	88.5	2.9	Agar colony TD50	+	Fox and Lajtha	(1967)
P388	L	60Co	160	1.6	In vivo TD50	ns	Berry and Andrews	(1961)
P388	L	60Co	151	4.5	In vivo TD50	ns	Belli and Andrews	(1963)

Tumor	Cell	Radiation	Dose	Ratio	Assay	Response	Investigator	Year
P388	L	^{60}Co	458	1.6	Day 2–7 tumor Day 5 tumor in vivo	+	Belli et al.	(1970)
LSA	L	X rays	170	1.2	TD_{50}	ns	Silini and Maruyama	(1965)
LSA	L	X rays	170	1.2	In vitro/in vivo Survival time	–	Maruyama et al.	(1967)
LSA	L	X rays	103	5.0	In vitro eff. fraction	ns	Maruyama and Ceman	(1970)
LSA	L	X rays	120	4.0	TD_{50} In vitro	ns	Feola	(1977)
L2	L	X rays	120	1.0	In vitro TD_{50}	ns	Feola	(1969)
C3HED	L	^{60}Co	110	1.2	In vivo TD_{50}	ns	Powers and Tolmach	(1963)
AKR	L	^{60}Co	114	0.7	Spleen colony	ns	Bush and Bruce	(1964)
L1210	L	^{60}Co	83–88	3.4	Spleen colony	ns	Hendry	(1972)
L1210	L	X ray	136	4.4	Agar colony	ns	Kann et al.	(1980)
L5178Y-R	L	X ray	85	1.5	In vitro growth	ns	Szumiel et al.	(1981)
L5178Y-R	L	X ray	61	1.05	In vitro growth	ns	Szumiel et al.	(1981)
L5178	L	X ray	100	3.8	In vitro growth	ns	Sato	(1979)
L5178	L	X ray	85	4.7	In vitro growth	–	Yau et al.	(1979)
L5178	L	X ray	100	12	Regrowth	ns + resistant line	Alexander and Mikulski	(1961)
L5178	L	X ray	45	5	Regrowth	–	Caldwell et al.	(1965)

[a]ns, Not studied: L, lymphocyte; H, histiocyte; M, monocyte; MM, multiple myeloma; *, unpublished studies.

[b]Altman et al. (1970) have a complete listing through 1970.

33

The sterilizing dose (or SD_{50}) assay (Maruyama, 1967) was adapted from the TCD_{50} assay described by Suit *et al.* (1965). Fixed numbers of cells are irradiated by graded doses of irradiation for determination of tumorigenic efficiency. This assay is useful for studying the immunogenicity of tumors by assaying the same cells in normal or sublethally irradiated immunosuppressed mice. Such assays established that syngeneic animals can resist irradiated transplant tumor and that the resistance is greater against more heavily irradiated tumor. Radiation survival curves of immunogenic lymphomas in normal hosts produce curvilinear rather than linear exponential survival curves. Porteus *et al.* (1979) have shown that there is an immunogenic threshold cell number capable of being recognized by the host. This contribution independently adds to the radiation killing of the tumor (Maruyama *et al.*, 1969).

The use of semisolid media was first reported by Fox and Gilbert (1966) for the radiation study of lymphoma cells. These methods and media have now been highly developed and used for a variety of mouse and human cell lines as well as for hemopoietic stem cells. The latter require using appropriate growth factors. For human lymphoid tumors and cell lines, a variety of assays have been used. Both the growth extrapolation method and the colony-forming assays appear to be best for this purpose. Probably the semisolid media will find extensive use for the study of leukemia/lymphomas in man in the future. Radiolabeled antibodies may also find application to therapy in leukemic/lymphomas in the future owing to the disseminated nature of the disease.

IV. Radiation Response of Hemopoietic Stem Cells

A. General Stem Cell Response

Lymphoid organs are heavily dependent on cell migration for cells, precursor cells, recovery, and repopulation. The main source of these cells is from the bone marrow. The stem cell for lymphoid tissue is closely related to the hemopoietic precursor cell, but it is likely that it is a separate, but closely related cell to the CFU-S (Wu *et al.*, 1968). The pluripotent stem cells of the bone marrow are the main cell population responsible for hemopoietic cells and ultimately differentiate along different unipotential precursor lines to give rise to different cellular elements.

1. Cell Survival

Data for bone marrow or hemopoietic CFU-S show similar radiosensitivities when treated with ionizing radiations of various types and energies. The following conclusions are drawn:

1. The radiation response of the hemopoietic CFU-S cell is characterized by single-hit multitarget sigmoid survival curves with very small shoulders and low n.

2. CFU-S cells of bone marrow origin have D_0 values ranging from 60 to 120 cGy, with most common values around 100 cGy.

3. Spleen and fetal liver contain CFU-S cells with radiation characteristics very similar to that of the bone marrow (Siminovitch *et al.*, 1965; Maruyama and Eichten, 1969).

4. CFU-S cells of fetal liver origin show higher D_0 values, of the order of 150 rad.

5. There are some minimal differences between radiosensitivity estimates obtained by exogenous or endogenous CFU-S assays.

6. For all cells assayed, extrapolation numbers are between 1 and about 2.5 i.e., very low, and possess minimal ability to repair SLD as shown by split-dose irradiation or by LDR irradiation.

2. *Repair of Radiation Damage*

Till and McCulloch (1964) were able to show that a small amount of SLD recovery followed a conditioning dose of radiation using the split-dose technique. Recovery takes place in 2–4 hr followed by a fall in survival, followed by a second rise at 6–12 hr and full recovery. SLD repair occurs as well as cell and tissue repopulation based upon the regrowth, regeneration, and migration of surviving stem cells into the tissue. By fractionation is meant the division of doses into multiple separate exposures; protraction is spreading out the period of time over which the radiation is given. Low dose rate (LDR) radiation depends upon continuous radiation exposure over a protracted period at rates of, e.g., 1–100 cGy/hr. Since some of the earliest studies in clinical radiotherapy, it was known that biological effects are reduced by either delivering the dose in multiple smaller increments (i.e., fractionation) or by increasing the time interval over which the radiation dose was given. If LDR radiation is applied to marrow or spleen cells, little, if any, recovery of SLD damage has been observed to occur (Puro and Clark, 1972; Ainsworth *et al.*, 1976; Krebs and Jones, 1972; Maruyama *et al.*, 1983a; Glasgow *et al.*, 1983; Lajtha *et al.*, 1971). Kimler *et al.* (1984) found n to be 1.1 and D_0 to be 66 cGy for cultured human bone marrow cells. Radiation beams with moderate to high LET or neutrons eliminate the shoulder region completely (Ainsworth *et al.*, 1983; Maruyama *et al.*, 1983a). Because of the limited capacity for SLD repair, LDR and split-dose radiation have been extensively applied to conditioning radiation doses used for total body irradiation in bone marrow transplant therapy. Tolerance of other organs, e.g., gut, lung, is greatly increased, but hemopoietic stem cells are ablated and immune suppression is produced by the single or multiple doses of radiation used. CFU-S have D_0 values of 50–100 cGy and, with low extrapolation numbers, represent one of the most readily destroyed cell types, with a great sensitivity of bone marrow and secondarily derived tissues.

Another type of repair is potentially lethal damage (PLD) repair and is enhanced when cells enter a less physiologically and proliferatively active phase. PLD repair has been less well studied for hemopoietic cells than for tumor cells and cultured mammalian cells. The absence of PLD or even negative repair has been reported for hemopoietic cells (Thomas and Gould, 1982).

3. CFU-S Suicide Assay

Bone marrow cell cycle times and cell production are moderated by humoral factors and regulatory mechanisms that mobilize the response of the host to the degree and severity of blood loss or tissue injury and aim to restore the integrity of the host. By studies of injury–response, it has been possible to detect the movement of pluripotent stem cells into cell cycle. An important assay to detect these responses has been the use of the "suicide" technique (Becker *et al.*, 1963, 1965). This method of study utilizes CFU-S from a variety of sources studied either *in vitro* or *in vivo* using either high specific activity tritiated thymidine (HSATT) or HU (Becker *et al.*, 1965; Vassort *et al.*, 1973). Both of these agents have been shown to be specifically toxic for DNA-S phase cells, and the uptake of these chemicals by cells that enter DNA-S phase leads to cell death and loss of clonogenic capacity. The cycling activity of cells *in vivo* is readily measured using this method, and we have used it extensively to trace CFU-S cell kinetics after perturbation by HU and HSATT given in large doses or with agents such as *Corynebacterium parvum*. These agents induce progression of cells from resting to cycling activity and can produce measurable changes in radiosensitivity and HSATT sensitivity (Maruyama *et al.*, 1979b; Maruyama and Grider, 1981). HSATT suicide assays can be used to determine dose–response curves for stimulated tissues and to determine the sensitivity of S-phase cells.

4. Repopulation of the Stem Cell Compartment

McCulloch and Till (1964) followed the changes in survival ratio with time by an endogenous spleen colony assay and derived a growth curve for endogenous colony-forming units (CFU). They showed that regrowth of the stem cells after a dose of 400 rad starts after a lag period of 1–2 days following TBI, and then increases exponentially, with a doubling time of 32 hr. This was followed by an overshoot in the stem cell number and a stationary phase. By about 20 days, the growth curve of transplanted CFU-S cells reaches stationary phase. Playfair and Cole (1965) carried out extended analysis of stem cell repopulation and found CFU-S in the bone marrow were completely restored to preirradiation levels by about 8 weeks. Repopulation proceeded more rapidly in the spleen and for many weeks the number of CFU exceeded the normal values. The infusion of bone marrow or spleen cells provides precursor stem cells to repopulate the LH systems.

Expansion of the stem cell pool takes place at different rates in different tissues. In the spleen, growth proceeds faster and reaches values above normal. This condition may persist for several weeks, but eventually the number of stem cells returns to normal levels.

B. The Hemopoietic Syndrome

After TBI in man one of the clinical syndromes observed results from bone marrow failure (i.e., the hemopoietic syndrome). Radiation can destroy the marrow and produce either complete or partial aplasia of the LH tissues (see Bond et al., 1965). These effects are dose dependent and the effects observed depend on the time at which the observations are made. The sensitivity of the various cells and organs is a function of dose and volume of tissue irradiated. The lymphocyte shows the greatest effects, most rapid decrease in number, and is the slowest to return to normal. Reduced lymphocytes leads to impaired immune responses (see Van Bekkum, 1967) and immunosuppression.

Lymphocyte sensitivity to radiation leads to the appearance of pyknotic lymphocytes in the peripheral blood. A peak is seen 3 hr after irradiation and rapidly declines thereafter. At the same time, studies of lymphoid tissues such as the thymus, lymph nodes, and spleen have shown extensive clogging of these organs with dying lymphocytes. Pyknotic lymphocytes apparently undergo degeneration in the blood and are rapidly cleared by alveolar macrophages of the lung and other tissues with reticulum cells with phagocytic function (see Bloom, 1979). Lymphocyte destruction is the basis of immunosuppression.

C. Bone Marrow Transplantation

1. General

From the extensive observations that large TBI doses produced lethality from bone marrow failure, anemia, leukopenia, and immunosuppression came the search for therapy of the acute bone marrow syndrome. The most dramatic event in the history of atomic radiation was the bombings of Hiroshima and Nagasaki in World War II. From those tragic applications of nuclear radiation and from the massive death and destruction which resulted arose detailed descriptions of the effects of radiation on man (see, e.g., Oughterson and Warren, 1956). The subsequent two decades led to much research into the biological effects of TBI and its reversal by therapy. A number of nuclear reactor accidents also led to the need for medicine to be able to treat accidental human exposures from nuclear reactor accidents (Bond et al., 1965). It was in the era immediately after the atomic bombings of Japan that Jacobson et al. (1949, 1950) discovered the protective effect of the spleen on mortality following whole body X-irradiation. By spleen shielding or by preparing homogenates of spleen which were injected into the lethally irradiated animal, it was found that one could protect against radiation death. By bone marrow cells injected intravenously, one could produce a modification of the radiation syndrome in animals (Lorenz *et al.*, 1951), and it was proposed that a radiation "chimera" was produced by the process (see Van Bekkum and DeVries, 1967). A radiation chimera was an animal that carried a foreign hemopoietic system as a result of TBI followed by transplantation of hemopoietic cells from another animal. Ford *et al.* (1956) also showed, using a cytological T6 chromosome marker, that the donor bone marrow cells had re-populated the recipient animal's lymphohemopoietic organs, resided there, and continued to produce progeny which sustained life and hemopoietic integrity of the recipient animal. It was also noted that the TBI animal was now immuno-logically crippled and was "tolerant" to homografts of bone marrow from a variety of animals. It could accept skin grafts from a variety of nonisogenic sources. The concept of radiation-induced "immunological paralysis" was thus established and a tolerant state shown. TBI conditioned the recipients for homo-grafts of a variety of tissues, specifically, e.g., bone marrow and hemopoietic tissues.

2. GVH Syndrome

Two interesting syndromes were noted after the transplantation of hemo-poietic tissue from bone marrow, spleen, or lymph nodes into a nonisogenic animal. Shortly after successful grafting of bone marrow, it was noted that an apparently well animal, sickened, became hunched up and cachectic, and then slowly wasted and died. Subsequent studies proved that the graft had reacted against the host animal and produced a curious immunological reaction whereby the grafted cells appeared to reject the tissues of the host animal. This was later called the graft-versus-host (GVH) reaction. A separate but similar state occurred with full recovery of the host. This was the regeneration of the hemopoietic and immunological system of the host with a secondary rejection of the engrafted tissues, or the host-versus-graft (HVG) reaction. Either reaction could ultimately purge the host of one or the other tissues. In the successful graft, a chimeric state was noted to be established.

In the early days of experimental bone marrow transplantation, the sug-gestion was made that a neoplasm of man which might be suitable for therapeutic bone marrow transplantation was the leukemias. For example, Hollcroft *et al.* (1953) attempted the therapy of animal leukemias by total body X-irradiation of the host to destroy the leukemic cells. This was followed by hemopoietic tissue grafts to restore the hemopoietic competence of the host. Among the early workers in this field were Barnes *et al.* (1956, 1957), DeVries and Vos (1958), Trentin (1957), and Simonsen *et al.* (1960). Thus, large total body doses of ionizing radiation led to a depopulation of the leukemic cells, and subsequent bone marrow grafting attempted to restore the host. The use of isologous marrow was noted to be less successful than that of homologous marrow (Barnes *et al.*, 1956, 1957), and a minor GVH syndrome favored the host by exerting a weak antileukemic cell action which increased both survival time and number of cures. One of the novel suggestions of this era for therapy was to use a radioprotective drug, e.g., those chemicals which were first described by Patt *et al.* (1949, 1952) to protect the normal marrow, to allow the delivery of a larger TBI dose to destroy leukemic cells and then to infuse isologous marrow to restore the competence of the host. Unfortunately, those studies were not successful in curing the leukemic animal (Schwartz, 1958). Nevertheless, the approach suggested by these early workers proved later to be usable for human cancer therapy. It is now being extensively applied to human leukemia/lymphoma therapy, especially for patients already placed into remission by chemotherapy (Thomas *et al.*, 1975, 1978, 1982) using TBI and BMT.

The treatment of bone marrow with monoclonal Thy 1.2 anti-T cell antibodies has been shown to purge it of T cells that lead to severe GVH syndrome. This treatment of donor bone marrow before transplantation reduces the severity and frequency of GVH and facilitates successful BMT (Vallera *et al.*, 1981). Such methods are now being tested extensively in a number of bone marrow transplant centers and have been found to be clinically effective.

3. Marrow Shielding

Cells migrate from shielding bone marrow as well as spleen and can extensively recolonize the thymus. Thymus and lymph node regeneration is extensive from shielding bone marrow sites. Spleen shielding or exteriorization and shielding protect against the hemopoietic death syndrome following TBI (Jacobson *et al.*, 1949). Animals shielded or restored after irradiation by isologous bone marrow cell therapy rapidly recover from the effects of irradiation. Ten days after irradiation and bone marrow cell injection, the spleen and lymph nodes are nearly completely restored. For thymus there is a secondary decrease in organ size followed by a gradual and sustained recovery after which the organ will return to normal size and morphology. This unusual feature of two-phase regeneration is quite characteristic of thymus regeneration after irradiation.

4. Marrow Shielding—Protection against Lymphoma/Leukemia

Bone marrow cell injection or bone marrow shielding will also protect against later radiation-induced leukemia development. Whole body sublethal irradiation repeated serially at weekly intervals will produce leukemia/lymphoma in virtually every animal thus treated. Kaplan *et al.* (1953) observed that lymphomas did not develop if irradiation was followed by the intravenous infusion of normal bone marrow cells or by thigh shielding during irradiation. Irradiation directed to the thymus alone did not produce tumors. The bone marrow apparently contains suppressor NK and other cells which act against emergent leukemia/lymphoma cells and are not only able to restore both the integrity of the host, but also to restrain the emergence of leukemic cells (Burton *et al.*, 1978).

V. Radiation Response of Lymphoid Organs

A. Thymus

A curious and characteristic pattern of postirradiation thymic weight changes (Fig. 6A and B) has been noted after irradiation (Takada *et al.*, 1969). Following either total body or thymus irradiation, from small doses to doses in the lethal range, the thymus showed very rapid shrinkage (Storer *et al.*, 1957; Kallman and Kohn, 1955), which reached a maximum in 1–4 days. After the initial decrease, the thymus regenerated and reached a peak recovery size in ~10–14 days. Subsequently, there was a second decline in thymus size, reaching a minimum in about 3 weeks. The second minimum was reached in 20–24 days, after which there was a second and more sustained and permanent regeneration. Brecher *et al.* (1948) described lymphoid tissue findings and noted historical descriptions to 1905. Kaplan and Brown (1952) noted the pattern in C57BL mice, a highly radiation-susceptible leukemia strain after 300 R, and contrasted it to lymph nodes and spleen changes which were much slower and showed less oscillation. Smith and Kieffer (1957) used colchicine strathmokinesis to measure dividing cells in the thymic cortex and noted a marked reduction of mitoses 24 hr after 400 R, but beginning at 2.5–3 days there was an abrupt rise in mitotic activity in the subcapsular cortical region followed by a fall; but mitotic activity was low in the medulla. DNA synthesis was very active as measured by [^{3}H]TdR uptake (Sharp and Thomas, 1975). Cross *et al.* (1964) studied immunological function after whole body irradiation. Immune reactivity of the mouse was noted whether the irradiated thymus or an unirradiated engrafted thymus were present, but followed different kinetics. Dukor *et al.* (1965) and Haran-Ghera (1965) described regeneration in thymus grafts, noted that regeneration proceeded from the capsule inward, and observed that in heavily irradiated grafts, cell appearance was delayed until after 11 days and was derived from extrathymic sites. Takada *et al.* (1969) carried out a careful study of the biphasic regeneration pattern and noted that the early decrease in size was due to the extensive death of resident cells. At the fourth day there was a very thin cortex with intensive mitotic activity of subcapsular large lymphocytes and reticular cells. Large to medium-sized cells appeared (Sato and Sakka, 1969) and later became medium to small lymphocytes. This phase was due to *in situ* precursor proliferation and maturation. Cells appeared to move to the medullary region (Sainte-Marie and Leblond, 1964). By day 12 a large thymus cortex reappeared and then underwent a second decrease in size and thinning of the cortex at about 3 weeks.

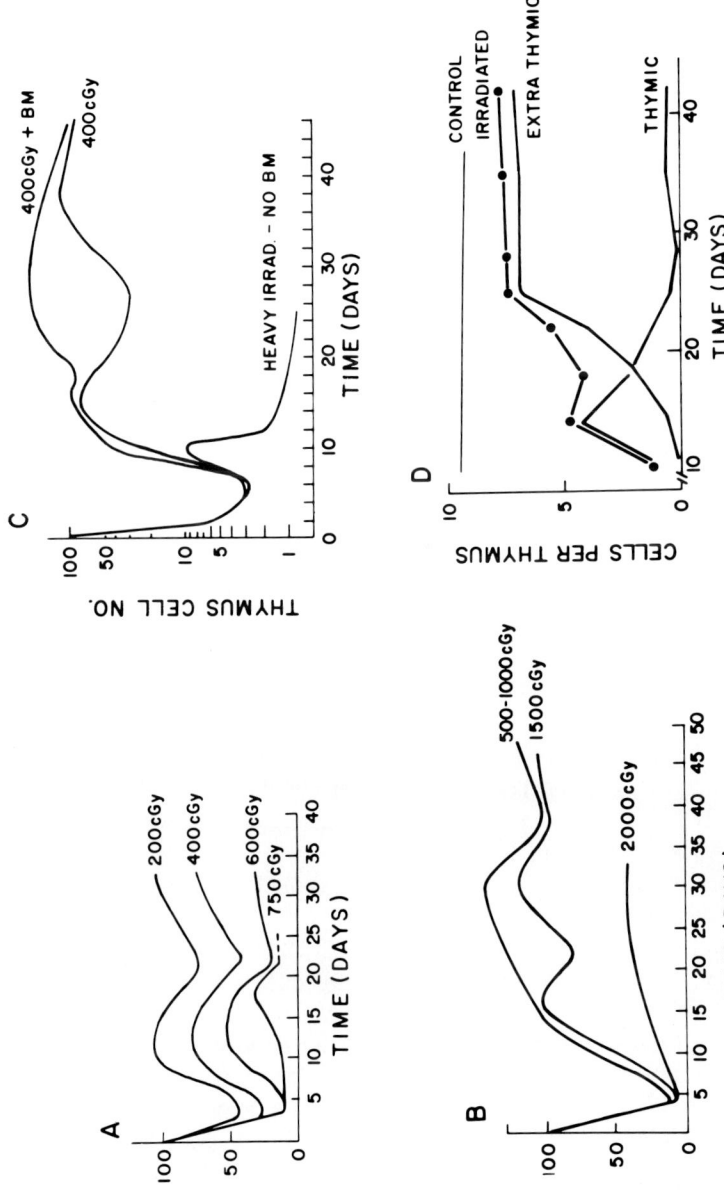

Fig. 6. (A) Typical thymus weight after TBI in percentage relative to controls, with time in days for different doses of radiation. The rapid early fall, rise at 10–15 days, second fall, and second rise are typical changes (biphasic). (B) Typical thymus weight after local thymus irradiation, with different irradiation doses and time in days afterward. Permanent gland atrophy follows high-dose irradiation. (C) Typical thymus cell number after TBI as a function of days afterward. Heavy irradiation leads to marked prolonged fall. A dose of 400 rad produces typical biphasic curve. With bone marrow transplantation regeneration is accelerated and the second dip is eliminated. (D) Source of cells for thymus after irradiation. Closed circles trace overall cell number curve for thymus. Lower curve traces cells of thymic origin; upper curve traces appearance of extrathymic cells from bone marrow. (Redrawn from Dalcubi *et al.*, 1982.)

By bone marrow shielding or bone marrow cell transplant studies, it was shown that the second fall was reduced (Fig. 6C) by providing extrathymic repopulating cells (Blomgren, 1969; Takada et al., 1969). T6T6 chromosome marker studies showed that soon after bone marrow grafting, the bone marrow had mostly donor dividing cells, but few were found in the thymus. At ~10 days, 80% of the dividing mitotic cells in the thymus were of donor origin and increased further for a month. The second fall in thymic weight was attributed to an inadequate flow of cells to maintain the thymus and required exogenous cell repopulation and the proliferation of migrating cells from the engrafted or shielded marrow sources. Sato and Sakka (1969) found biexponential survival curves for the small and large lymphocytes, with D_0 for small lymphocytes of 145, and 57 R for medium, and 57 R for large lymphocytes for the sensitive component. The resistant components had values for D_0 of 425 and 520 R, respectively.

TdT has also been used to trace the origins of the repopulating cells of the thymus (prothymocytes) from bone marrow cell injections. After TBI with large doses of ionizing radiation, DNA polymerase α, β, and TdT disappear from the thymus gland (Maruyama, 1978). Dalculsi et al. (1982) traced TdT reappearance in the thymus after transplants of bone marrow repopulating cells. In lethally irradiated mice, thymus repopulation followed a two-stage process (Fig. 6D). During the first period most of the thymic cells were derived from surviving irradiated thymocytes. After a transient increase, the endogenous cells were replaced by new ones which came from the transplanted bone marrow and were of the donor type (Micklem et al., 1966). This was shown by use of Thy 1.1 and 1.2 markers and the pattern of TdT activity. On day 10, all thymic cells were host derived, displayed a high H_2 concentration, and were associated with a 7-fold increase of TdT peak 1 activity, much higher than a normal population, consistent with thymic derivation. Between days 18 and 28, one TdT peak fell and a second rose, consistent with maturation of one cell type, an increase and rise of a migrant TdT marked cell population, with a low H_2 concentration and a high separate peak of TdT activity.

A bone marrow stem cell gives rise to erythroid, myeloid, and lymphoid cells in lymphoid organs of irradiated animals. In in vivo thymus regeneration studies, prothymocytes of bone marrow can be studied by buoyant density, sedimentation velocity, and electrophoretic mobility studies close to CFU-S (Boersma et al., 1981), but of slightly different radiation sensitivity, proliferative status, and greater sensitivity to proliferation inhibitory action of cold thymidine (Boersma, 1983). These cells lack typical T cell markers. It appears that only a few cells, of many bone marrow cells injected, can home on the thymus, and these may seed either the cortex or medullary regions (Ezine et al., 1984). In those two different regions, they give rise to two different types of T cells, with those in the cortex possessing major proliferation capacity and those in the medullary region destined to migrate and seed peripheral lymphoid organs.

A separate function of the thymus appears to elaborate trophic and growth regulatory hormones for systemic release (Dardenne and Bach, 1981). By administration of exogenous thymosin (Goldstein et al., 1970), it is possible to greatly accelerate lymphoid tissue regeneration in mice.

Following graded doses of heavy mediastinal irradiation, the thymus can be totally ablated and reduced into a fibrotic atrophic organ. This requires large, localized, single doses of 3000–3500 cGy (Maruyama and Barclay, 1967) which produce a high mortality rate. The atrophy can be long lasting or permanent (Fig. 7). When thymic irradiation was combined with TBI, the animals became tolerant to skin grafts across the sex antigen barrier. Subsequent studies by Slavin et al. (1977) and Strober et al. (1979a,b) have shown that high-dose fractionated TLI, including the thymus, can produce tolerance to homografts of other organs for the short or long term, depending upon dose and closeness of antigenic match.

Following fractionated irradiation of the thymus or whole body of the mouse, similar cyclical changes in thymic weight are observed. However, thymus repopulation proceeds more slowly than after single exposure. The second regeneration was delayed and failed to restore the organ as completely to normal. When serially irradiated bone marrow cells were injected into irradiated mice, CFU-S colony size distribution was smaller, indicating impaired growth (Blomgren, 1971). High-dose fractionated lymphoid irradiation appears to give the most prolonged suppression of the lymphoid immunological system and leads to a long-term tolerant/partially tolerant state.

The biphasic pattern of thymus recovery following whole body irradiation is well established. The population of cells leading to the initial peak has been considered relatively radioresistant and felt to be an intrathymic stem cell (Kadish and Basch, 1975) or a transient extrathymic population (Barg et al., 1978). The radioresistant subpopulation may be thymic nonlymphoid cells with many features of macrophages (Watkins and Sharp, 1979). Watkins et al. (1980) carried out TBI dose–response studies between 200 and 700 cGy and noted that the rapidity of the initial fall was similar regardless of dose, the initial secondary rise was dose dependent, the second nadir was dose independent, and the final rise was dose dependent. Using thymus-only irradiations, much larger local doses were possible (Maruyama and Barclay, 1967; Watkins et al., 1980). Between 500 and 2000 cGy, phasic fall-rise was possible, with a twofold compensating growth overshoot possible. When combined with 200–700 cGy of TBI, there was a much more sustained atrophy of thymus. There was severe impairment of immune reactivity to all antigens, depending on dose of irradiation and time of sampling. Irradiation of the infant newborn thymus region led to

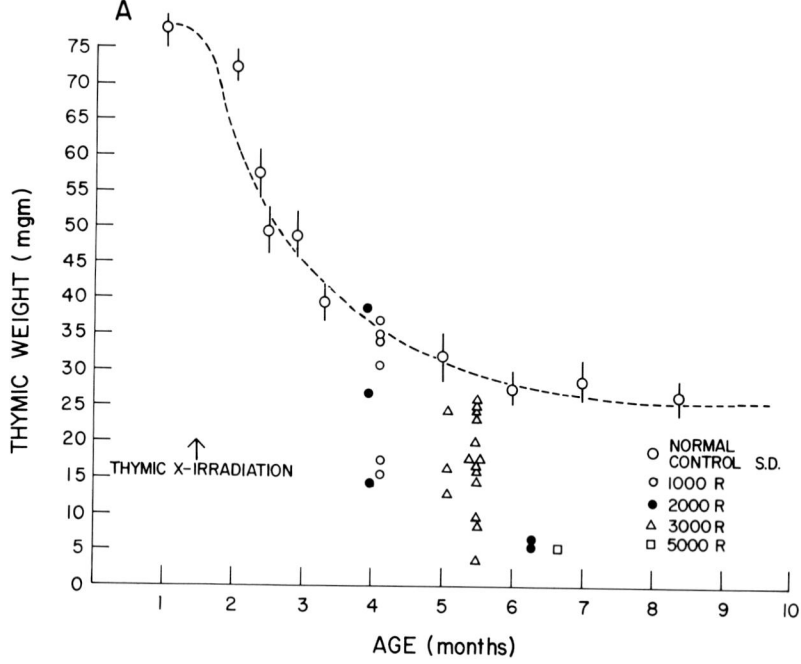

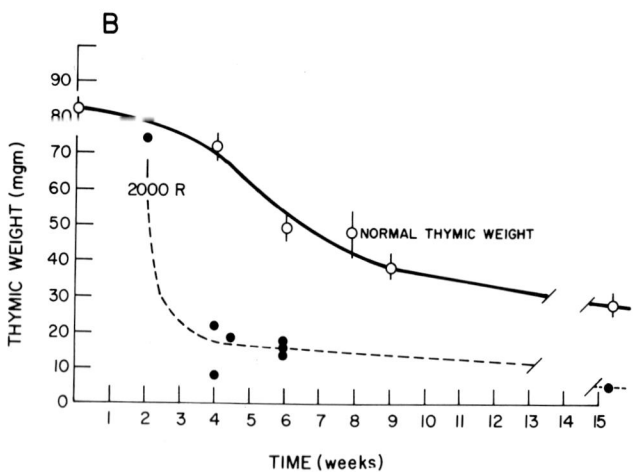

FIG. 7. Thymic weight, with mouse age in months after irradiation. Arrow indicates time of irradiation. Open circles trace control thymus weights. Various radiation doses from 1000 to 5000 rad lead to atrophy, which can be permanent at the highest doses. (B) Decline in thymus weight with time after 2000 rad irradiation. (Unpublished data, Maruyama and Barclay, 1967.)

more severe effects, with protracted weight depression, slower recovery, and depletion of the peripheral lymphoid tissues. The data show that the traffic, maturation, and export of postthymic precursor cells to the spleen, lymph nodes, and peripheral lymphoid tissues is severely impaired by radiation damage of the thymus. Similar effects are observed with thymectomy and TBI (Miller *et al.*, 1962). Sundaran *et al.* (1967) showed that a wasting syndrome developed in mice irradiated with 900 cGy with extremity shielding, but this was prevented by thymic shielding. Maruyama and Barclay (1967), Slavin *et al.* (1977), and Strober *et al.* (1979a,b) showed that a tolerant state resulted from thymic and TLI or TBI.

Thymus irradiation can lead to delayed effects in man. Thymic irradiation was a common medical practice in children at one time. It was done for X-ray enlargement of the thymus gland in young infants in association with difficult respiration, detected by chest X ray.

B. Spleen

1. General and Dose Response

The spleen is a major secondary peripheral lymphoid organ that is situated on a large abdominal vascular pedicle and represents one of the major discrete aggregations of lymphocytes in the body. It also has lymphopoietic, hemopoietic, phagocytic, and cell regulatory functions, most of which are still poorly understood. Following the removal of normal spleen by splenectomy, white cell counts tend to be elevated, and there is a persistent leukocytosis indicating a cell number regulating function. Following lethal TBI, spleen cell repopulation of the hemopoietic tissues was discovered by Jacobson *et al.* (1949, 1951) using initially spleen shielding and later spleen and bone marrow cell transplants. The spleen receives cells from the thymus and bone marrow, and those derived from the thymus and bone marrow have been carefully studied (Takada *et al.*, 1971). Thymus and bone marrow-derived lymphocytes carry out cell-mediated immune reactions and humoral antibody production.

T cells traffic from thymus to the spleen and lymph nodes and are found in the white pulp and periarteriolar regions of the spleen (Parrot and deSouza, 1966). B cells also occupy distinct and separate regions of the spleen. Bone marrow and spleen cells travel to the spleen with a high clonogenic efficiency, forming spleen colonies (CFU-S) containing erythromonomyeloid and megakaryotic platelet precursor cells (Till and McCulloch, 1961). Lymphocytes from thymus, thoracic duct, or lymph nodes do not protect against lethal irradiation and hemopoietic cells are required for repopulation. Spleen response to radiation injury has been found to be prompt, with rapid shrinkage of the organ after irradiation injury (Kallman and Kohn, 1955). But it also regenerates rapidly and produces cells quickly, releasing them to the circulation to respond to injury or immune challenge. The spleen reacts with a quick-reacting cell population which is rapidly mobilized and released to repopulate bone marrow and extrasplenic sites (Silini *et al.*, 1976). The bone marrow provides a more sustained release of cells to support the host.

The rapid shrinkage after irradiation is characteristic of all lymphoid organs and tissues. The shrinkage is rapid and exponential with time (Puck, 1966) and dose (Fig. 8). With weight loss assay (see Storer *et al.*, 1957), fraction weight loss is plotted with dose on a similar scale and gives two component survival curves, with photon irradiation showing radiosensitive and less radiosensitive components. Similar responses are observed following neutron, high LET, or ^{252}Cf irradiation. RBE values vary with the type of high LET radiation used (Ainsworth *et al.*, 1977; Maruyama *et al.*, 1984a; Montour *et al.*, 1974; Huiskamp *et al.*, 1983). Maximum shrinkage occurs after 1–3 days and regeneration takes place after maximum shrinkage following a single monophasic curve rather than the complex curves observed following thymic irradiation.

Regeneration is aided by the intravenous injection of bone marrow or spleen cells. These cells travel to the spleen and other hematopoietic organs. In the spleen, discrete spleen nodules or CFU-S are formed (Till and McCulloch, 1961). The nodules are the result of proliferating and differentiating stem cells which grow rapidly and reconstitute the elements of the hemopoietic systems following a colonial growth pattern. At large cell doses, nodules quickly become confluent and the spleen assumes a whitish bumpy appearance. There is a later gradual transition back to the reddish appearance of normal spleen. At very high doses, e.g., a single exposure to 900–1000 cGy or fractionated doses to 4450 cGy (Dailey *et al.*, 1980), spleen shrinkage can be sustained and prolonged or permanent, as was described for thymus irradiation. The spleen is also dependent upon extrasplenic cell migration from bone marrow and thymus (Auerbach, 1953) for regeneration (Ford and Micklem, 1956).

If thymectomy preceded TBI, spleens were found to be depleted of lymphoid elements, and no regeneration takes place unless a thymus is regrafted. In irradiated cultured spleens, no lymphoid regeneration can be detected without the addition of other lymphoid tissue grafts. The thymus plays an important role in lymphoid recovery (Miller *et al.*, 1962; Auerbach, 1963; Globerson and Feldman, 1964). Globerson (1966) studied lymphoid recovery in the irradiated mouse spleen cultured *in vitro* and determined that direct interaction occurred between the spleen, thymus, and bone marrow. Isolated spleen explants were incapable of lymphoid regeneration. Thymus had no direct stimulatory effect on spleen lymphoid regeneration; lymph node likewise had no effect. However, bone marrow with or without thymus interacted synergistically with spleen, leading to the reappearance of lymphoid cells and regeneration. No stimulation of lymphopoiesis in bone marrow was conferred by thymus in the absence of spleen. Thus, there was a mutual interaction between irradiated spleen and bone marrow, which was enhanced by the thymus and which led to the appearance and growth of lymphoid cells. Various reports show circulation of bone marrow cells to spleen (Ford *et al.,* 1956; Mekori *et al.,* 1965; Hanks, 1964) and thymus (Ford and Micklem, 1963; Ezine *et al.,* 1983) and with the seeding of spleen by bone marrow (Till and McCulloch, 1961), spleen, and fetal liver/spleen (Siminovitch *et al.,* 1965; Maruyama *et al.,* 1968). The role of cell migration in regeneration of spleen is thus important, but the organ also has great regenerative capacity.

The spleen is an organ that has extensive regenerative capability and can reconstitute normal size, shape, and histological architecture from a small pool of surviving endogenous stem cells. Wolf (1982a,b) ligated the spleen vascular pedicle to produce whole organ necrosis of all except a thin layer of subcapsular cells. Cellular regeneration occurred rapidly from these surviving cells and proceeded inward along reticular fiber tracts. By irradiation with 1000 rad and vessel ligation, repair capacity and stem cell CFU-S number were greatly reduced. A dose of 400 rad had a much more severe effect than 100 rad, but was not as potent at 1000 R plus immediate vessel ligation. Small fragments of human spleen left in the peritoneal cavity can regenerate to form ectopic spleens. Endogenous DNA-synthesizing cells are concentrated around the periarterial sheaths and marginal white pulp zones. When bone marrow cells are transplanted, the cells carrying out active DNA-S are localized under the splenic capsule and adjacent to the trabeculae. Extramedullary hemopoiesis always preceded restoration of the lymphoid components. Although the capacity for self-constitution was effective, it did not fully achieve the original size and was severely compromised by repeated splenic vessel ligation or irradiation and ligation (Wolf, 1982a,b; Wolf and Rosse, 1982). Heavy irradiation thus can disrupt the stem cell number and the reticular network essential for regeneration of the organ. In man, Dailey *et al.* (1980) showed that splenic atrophy and functional asplenia can follow splenic irradiation to 4000 cGy. Shih *et al.* (1984) showed that functional asplenia for radioisotope trapping can also occur after modest doses of spleen irradiation.

Splenic growth in response to antigenic stimulation is also rapid. Spleen response to the presentation of antigen such as *C. parvum* (CP) has been found to be nearly instantaneous, with prompt cell cycle progression evident within a few hours followed by rapid growth of spleen by 5 to 10-fold (Maruyama and Coleman, 1978). CFU-S cells of the spleen progress with entry into DNA-S, evident within a few hours by high specific activity tritiated thymidine suicide assay (Maruyama and Grider, 1981). Spleen growth was extremely rapid, with a cell doubling time of 78 hr. Massive cellular proliferation and spleen hypertrophy occur after antigen administration and the *in situ* organ increases greatly in size in 10 days, with marked increases in DNA-S activity (Maruyama and Coleman, 1978). This was noted by [^{3}H]TdR labeling and uptake and increased levels of DNA-polymerase activity in the organ (Maruyama and Coleman, 1978). Marked

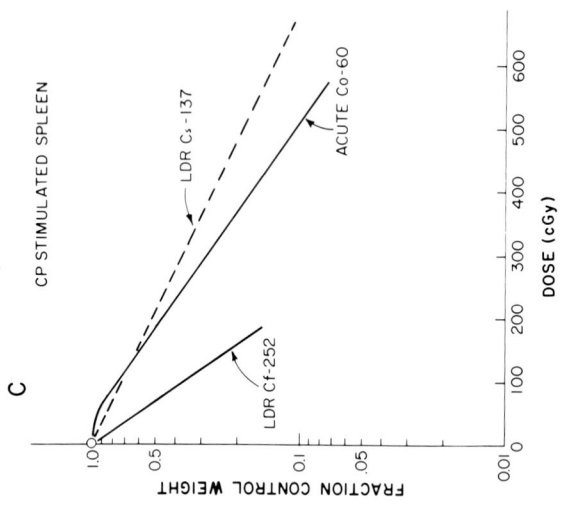

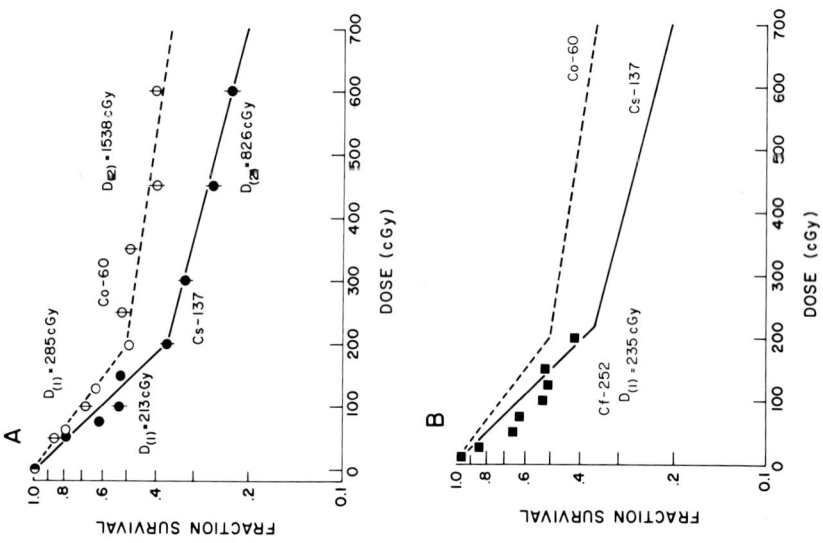

changes in radiosensitivity of CFU-S (Maruyama and Grider, 1981) occurred with antigenic stimulation and was traced by change in radiosensitivity in the 7-hr period after antigen challenge. Suicide assay for cells entering DNA-S using HSATT showed that a large cohort of spleen CFU-S cells entered DNA-S in ~4 hr, then entered a phase of the cell cycle sensitive to the agent vinblastine, which has antimitotic activity, at about 6–7 hr. The results indicated a parasynchronous progression of stem cells from resting to rapidly cycling states, with massive (5–10×) enlargement of spleen occurring in 10 days. Cell cycle progression was sensitive to antihistamine H_2 regulation (Maruyama and Grider, 1981; Byron, 1980). The rapidity of growth response of spleen after CP antigenic stimulation has led to comparative studies of the radiosensitivity and chemoradiosensitivity of spleen for resting (unstimulated) and proliferating (CP stimulated) spleen. Very similar observations were also made by using large doses of HU to stimulate cell progression in the spleen (Maruyama et al., 1979b, 1984a).

Radiosensitivity study of spleen using resting state or nonproliferating (NP) spleen or CP-stimulated proliferating (P) spleen has been the subject of recent investigation using acute ^{60}Co, and LDR ^{137}Cf or ^{252}Cf neutron/γ irradiations. Using weight loss assay done ~2 days after irradiation (Kallman and Kohn, 1955), two component survival curves with exponential decrease in weight were noted after acute ^{60}Co radiation, with D_0 values of 285 and 1850 cGy noted for the first and second slopes (Fig. 8). LDR ^{137}Cs was much more effective by 1.3× for the first slope and 1.9× for the second slope, indicating a more powerful effect of LDR irradiation to produce spleen atrophy than even neutrons. (Maruyama et al., 1984a). The stimulated spleen was studied 10 days after CP stimulation, with irradiation done on day 3 when maximal DNA-S activity and cell proliferation was occurring. Under those conditions, acute ^{60}Co produced single slope but multihit fractional weight loss curves where D_0 was 184 cGy. LDR irradiation with ^{137}Cs γ rays showed a less steep survival curve ($D_0 = 3000$ cGy) which was postulated to be caused by the less rapid accumulation of lethal events in cell targets and much more reparable SLD injury. ^{252}Cf neutron/γ radiations showed a strikingly greater effect in producing a steeper dose–response curve (Fig. 8C) owing to the greater effect of neutrons on DNA-S phase cells and reduced age dependence of neutron effects. Neutrons, thus, appear to

FIG. 8. Typical dose–response curves for irradiated lymphoid organs as measured by weight loss assay. (A) Dose–response spleen weight loss with graded doses of acute ^{60}Co, or low-dose rate (LDR) ^{137}Cs irradiation. Two-component curves were observed, with a marked dose-rate sensitivity. (From Maruyama et al., 1986). (B) Dose–response curve for LDR ^{252}Cf irradiation, with dashed curve for acute ^{60}Co irradiation; solid curve for ^{137}Cs. With LDR ^{252}Cf neutron irradiation, no greater effect was observed in resting spleen. (C) Dose–response curve for spleen by weight loss assay for C. parvum-stimulated animals. LDR ^{137}Cs produced least steep and ^{252}Cf the most steep curve. (Reproduced from Maruyama et al., 1984e, with permission Int. J. Radiat. Biol.)

produce much different biological lesions with greater impairment of pro-liferative capacity than γ irradiation (Maruyama, Feola *et al.*, 1985). Neutrons have a high RBE for spleen injury and for sustained delayed effects on, e.g., spleen weight and thymus cellularity (Ainsworth *et al.*, 1976; Huiskamp *et al.*, 1983). This is probably due to late effects and persistent damage and defects in bone marrow (Ainsworth *et al.*, 1974).

Combination chemoradiotherapy effects were also studied for a number of agents. When acute ^{60}Co radiation was given, its effects dominated the cytotoxic effects completely and no further effects were contributed by the addition of a variety of antimetabolites, alkylating agents, or combinations (Maruyama *et al.*, 1979a, 1981). Cyclophosphamide was found to be an interesting agent which actually stimulated spleen growth rather than having a purely toxic effect. The studies showed that radiation is one of the most powerful agents in producing lymphoid atrophy and organ shrinkage. It did not matter whether CP was given before stimulation, concurrently with stimulation, or after stimulation on the relative effectiveness. However, there were quantitative differences in patterns noted (Maruyama *et al.*, 1981).

Spleen contains hemopoietic CFU-S cells whose radiation sensitivities fol-lowing irradiation have been extensively studied (Siminovitch *et al.*, 1965; Mar-uyama and Eichten, 1968, 1969). The organ is discrete, easily excised, and converted to single cell suspensions for transplantation assay. The density of CFU-S is approximately one-tenth of bone marrow. Dose–response survival curves possess parameters nearly identical to bone marrow CFU-S. Studies of CFU-S sensitivity to thermal neutrons and ^{252}Cf radiation give nearly identical values for RBE of 2.1 (Grahn *et al.*, 1972; Maruyama *et al.*, 1983a,b; Ainsworth *et al.*, 1983). The CFU-S represent a radiosensitive stem cell population. Effects depend upon LET but vary over much narrower ranges than to photon radiation (Ainsworth *et al.*, 1983; Maruyama *et al.*, 1983a). With LDR radiation, single-hit inactivation is observed with photon irradiation whether derived from spleen or bone marrow (Frindel *et al.*, 1972; Puro and Clark, 1972; Krebs and Jones, 1972; Glasgow *et al.*, 1983; Maruyama *et al.*, 1983a). Since spleen growth can be stimulated by antigens, age response of spleen CFU-S has been studied using HSATT suicide assay (Maruyama and Grider, 1981). CFU-S represents an important subpopulation of spleen and probably represents one of many types of lymphohemopoietic precursor cells present. Each has unique and slightly differ-ent radiosensitivities, e.g., T and B lymphocytes, phagocytic cells, and re-ticulum and stromal cells.

The spleen and liver are also targets of lymphoma cell proliferation which is easily measured. Such studies have shown that the kinetics of tumor cell growth in those organs is extraordinarily rapid, with 5–10× increase in mass with intravenous injection of lymphoma cells before host death occurs. Along with this growth are phasic bursts of DNA-S activity, apparently due to periodic relaxation of the capsule, with increase in splenic tumor mass occurring in repeated cycles until the mouse expires (Maruyama and Lee, 1980). The same pattern was seen in the liver and probably also occurs in the bone marrow, lymph nodes, and blood. It has been postulated that the termination of the growth of lymphomatous tumor cell takes place when the cell population reaches a critical or "lethal" cell number. The determination of lethal number of cells has been estimated independently by Maruyama and Brown (1964) and Skipper *et al.* (1964) for the LSA and L1210 tumors, respectively. Based upon these tumors, changes in survival time and frequency of cure in acute lymphocytic leukemia appear to be a result of destroying a fraction of the large, rapidly expanding population of cells. Host death occurs at some critical cell population number at which host physiology is greatly perturbed.

Normal spleen also regenerates rapidly after irradiation. Figure 9A shows a typical shrinkage and regeneration curve for spleen after 400 cGy^{60} irradiation. Bone marrow cells greatly enhance the regeneration of spleen. When stimulated by antigen, it very rapidly enlarges and size increase peaks in 10 days (Fig. 9B). Spleen weight dose–response after single, multiagent, or multimodality treatment produces different patterns of shrinkage (Fig. 9C). Weight changes also depends on resting or proliferating status (Fig. 9D). Radiation is the single most effective agent for producing spleen shrinkage, although the alkylating agent 1,2-bis (2-chloroethyl-nitrosourea (BCNU) and nitrogen mustard (HN_2) have very similar radiomimetic effects (Maruyama *et al.*, 1979a). Actively proliferating spleen exposed to acute ^{60}Co or ^{252}Cf on day 3 after stimulation exhibits biexponential dose–response curves (Fig. 8). There was a high RBE for neutron irradiation and little sensitivity to cell cycle age or to dose rate effect (Maruyama *et al.*, 1984a).

2. Spleen and Cell Sequestration

Lymphocyte migration was studied in detail by Gowans (1964) and circulating lymphocytes were shown to be long-lived cells that circulate continuously through the peripheral lymphoid organs into the lymph and back into the blood in a dual circuit. T and B lymphocytes arrange themselves territorially in lymphoid tissue in T-dependent and B-dependent areas. In Hodgkin's disease, chronic leukemias with hepatosplenomegaly, leprosy, and a number of medical conditions, there are marked changes in lymphocyte circulation. Changes in lymphocyte surfaces apparently lead to change in circulation of cells, and organ accumulation of cells occurs (e.g., in lung or spleen). Experimental modification of lymphoid organs by irradiation, infection, neoplasia, or injection of antigen can lead to transient sequestion of cells in the spleen. Hodgkin's disease patients are staged by splenectomy, and this practice allowed for the study of spleen cells as well as "trapping." This led to the finding that there is T cell accumulation in

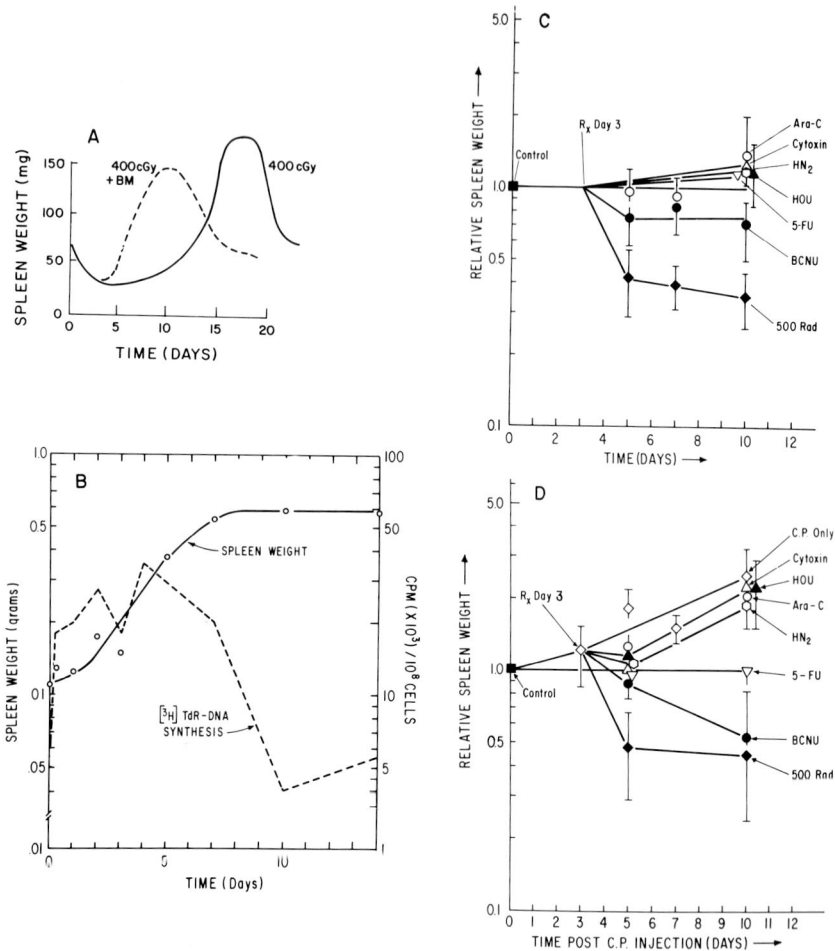

Fig. 9. Spleen effects of radiation and chemotherapy. (A) Spleen weight with time after TBI. A dose of 400 cGy of TBI will produce rapid organ shrinkage, with regeneration beginning on day 10 and peaking on day 15. It decreases in weight thereafter to normal. Bone marrow (BM) injection leads to more rapid restoration of spleen. (Redrawn after Takada *et al.*, 1971.) (B) Spleen weight after antigenic stimulation with *C. parvum* (after Maruyama and Coleman, 1978). Spleen weight increases rapidly for 8–10 days, with increased DNA-S activity for 5 days. (Reproduced with permission of *Cancer Research*.) (C) Alkylating agent, radiation, or antimetabolite effect on weight of normal spleen (after Maruyama *et al.*, 1979a). (Reproduced with permission of *Int. J. Radiat. Oncol. Biol. Phys.*) (D) Effects of alkylating, radiation, antimetabolite agents (see C) on *C. parvum* stimulated proliferating spleen. (Reproduced with permission of *Int. J. Radiat. Oncol. Biol. Phys.*)

the spleen and depletion in the blood (deSousa, 1975). Mycosis fungoides and cutaneous T cell lymphoma (CTCL) are diseases characterized by T lymphocyte trapping in the skin. Certain diseases are characterized by unusual trapping of lymphocytes in the spleen. The surgical removal of the spleen greatly increases the number of lymphocytes circulating. Lymphocyte trapping or sequestration in the spleen represents a potential problem with splenomegaly or with chronic diseases that alter lymphocyte distribution in the body. Lymphocyte trapping can occur rapidly (e.g., after antigen administration). Irradiation of the enlarged spleen with very small doses of irradiation (e.g., 25 R) or doses of 200–2000 cGy can lead to marked and long-lasting changes in remote organ function. These effects are termed abscopal and represent poorly understood remote lymphoid tissue response to irradiation.

C. Lymph Nodes

Sato and Sakka (1969) showed that the overall weight change and dose–effect survival curves of lymphocytes in lymph nodes consisted of two components following biexponential response curves. In lymph nodes, values of D_0 of the lymphocytes were ~200 R for small, medium, and large lymphocytes. Trowell (1952) examined pyknotic changes of lymphocytes in lymph nodes in different parts of the body. Lymphocytes were most radiosensitive in lymph nodes *in vivo,* were less sensitive in lymph nodes *in vitro,* still less sensitive in blood *in vitro,* and radioresistant in the intestinal mucosa. The latter was less responsive by ~10-fold based upon the endpoint of interphase death. Trowell's survival data plotted semilogarithmically follow a one-hit exponential curve for the *in vivo* irradiated cells and were markedly less sensitive in the gut mucosa (Fig. 3B).

Brecher *et al.* (1948) and others (see deBruyn, 1948) have described the histological changes in lymph nodes after irradiation. Sato and Sakka (1969) studied the time course of cell depletion and repopulation of lymph nodes as a function of dose. Lymph node cell depletion was much less than thymus, with much slower recovery. Cell type, i.e., large and medium lymphocytes, reappearance was much slower to recover and the cells were delayed in appearance and recovery pattern. Lymph node and atrophy were observed with local or TBI radiation by numerous workers (Brecher *et al.,* 1948; Kaplan and Brown, 1957; Sato and Sakka, 1969; Watkins *et al.,* 1980). Lymph node recovery was dependent upon proliferation of surviving cells and blood-borne repopulating cells from the thymus and other lymphohemopoietic organs. Cell depletion is a result of cell killing, inhibited cell production, and cell migratory activity. Benninghoff *et al.* (1969) showed that little node shrinkage occurred after local irradiation with 300 cGy, but shrinkage was severe after 300 cGy of TBI. Lymph node cells were thus partly maintained by migratory cells (Hall and Morris, 1964). Repopulation was partly by a long-lived recirculating small lymphocyte which labeled with tritiated thymidine (Miller and Cole, 1967). Repopulation was slowest in lymph nodes compared to thymus and spleen and was greatly promoted by transfused hemopoietic bone marrow cell injection (Congdon, 1966).

Heavy local irradiation also reduces trapping ability of lymph nodes over prolonged observation periods. The barrier function is little affected by low doses of radiation, and therapeutic irradiation to high doses can cause some leakage in the barrier (Dettman *et al.*, 1972). When combined with surgery, lymph node function is greatly reduced, but to irradiation, the barrier function is relatively radioresistant. The impact of irradiation on lymph node function after surgery has been discussed extensively, and some reports have speculated that it can be greatly impaired and can lead to reduced host resistance to cancer and/or enhance the likelihood of spread of cancer (Sternsward *et al.*, 1972,1974). There is little evidence to support this contention. Engeset (1964) showed that high single radiation doses to 3000 R leads to stromal-vascular damage, thus interfering with regenerative capacity. At 1 year, complete lymph node fibrosis and atrophy was observed, or there was only partial restitution. Similar changes can also occur with high-dose fractionated radiation.

Eltringham and Weissman (1970) studied regional lymph node immune function after irradiation with multiple-fraction irradiation. Hall and Morris (1964) studied single-fraction irradiation. Fractionated radiation given before or following immunization depressed both cellular and humoral immunity (Weissman *et al.*, 1973). By timing irradiation at different times after sensitization, it was possible to decrease either humoral, cellular, or both immune functions (Eltringham and Weissman, 1971). Likewise, local or distant immune function (e.g. in the splcen or remote lymph nodes) based upon locally activated migratory immune lymphocytes could be abolished by local lymph node irradiation.

Local lymph node Irradiation can severely alter morphology, cellular and humoral immune function, and memory and messenger migratory lymph node-generated lymphocytes. The ability of the latter to bear information to other remote lymphoid organs can be impaired by local irradiation. With added TBI one can radically impair or ablate local lymph node regenerative capacity and the movement of cells from the central to peripheral lymphoid organs (see Anderson and Warner, 1976). TLI (Strober *et al.*, 1979a,b) or TBI leads to severe depressions of both humoral and cellular immunity, and this tolerant state permits transplantation of homografted tissues or palliates severe autoimmune disorders. This peripheral lymph node function is important in host immune reactivity and TLI can produce a tolerant or immunosuppressed state.

D. Gut-Associated Lymphoid Tissues (GALT/Appendix)

General radiation pathology textbooks described histological changes associated with whole body irradiation. The lymphocyte dose–response in the mesenteric lymph nodes was studied by Trowell (1952) and showed a two-component survival curve with less sensitivity of lymphocytes in the gut environment. Debruyn *et al.* (1965) studied the radiosensitivity of lymphocytes in the appendix by local direct irradiation and indirectly by TBI. A two-component survival curve was observed for X-ray sensitivity of lymphocytes of the appendix. Similar but reduced effects were noted after TBI. The GALT therefore appear to respond exactly like other lymphoid organs and tissues. However, Hale and McCarty (1984) observed that the severity of radiation effects on lymphocytes in Peyer's patches was much more pronounced and protracted than other PLO. They postulated that the microenvironment in the gut was more readily damaged by radiation than in other lymphoid organs. One of the important roles postulated for the GALT is that within the GALT environment may be the bursal equivalent central lymphoid organ (Cooper and Lawton, 1972).

E. Lymphatics

Dettman *et al.* (1972) reviewed the various effects of radiation on lymphatic vessels and transport functions. Lymphatic capillary growth is less radiosensitive than blood capillaries. There is no loss of patency, but only increasing tortuosity and varicosities up to 15,000 R (Engeset, 1964). Unless tissue necrosis or surgery occurs, no blockade of these vessels was noted. All these effects are amplified by surgery, infection, infiltration by malignancy, or tissue necrosis. Radiation lymphedemas are usually caused by tissue and subcutaneous fibrosis which cause interference with vessel regeneration and patency. Radiation combined with radical surgery and removal of numerous regional lymph nodes and lymphatic channels can lead to the greatest frequency of lymphatic blockade and lymphedema. Huge lymphocoeles can appear, e.g., in the pelvis where resected lymph node tissues reform large endothelium-lined spaces in which lymph fluid can accumulate in large cystic sacs (Wharton and Rutledge, 1980) and can be symptomatic.

VI. Therapy of Lymphohemopoietic Tissue and Application to Human Disease

A. Treatment of Acute Childhood Lymphoblastic Leukemia (ALL)

The modern treatment of ALL of childhood is one of the outstanding success stories of modern cancer therapy (Pinkel, 1979). In ALL, cures could not be achieved by single agent chemotherapy. Instead, combination chemotherapy using multiple agents of demonstrated efficacy by different mechanisms and noncompeting limiting toxic side effects led to the use of large doses of multiple drugs to produce remission of disease. One site of relapse was the CNS (Moore *et al.*, 1960) and that was prevented by irradiation (Pinkel, 1979). Previous pharmacologic studies had shown that high levels of drugs could not be achieved in the cerebrospinal fluid of patients owing to the presence of a blood–brain barrier to drugs.

One of the important developments of this period was the recognition that the most important site of relapse was the CNS. Because of the blood–brain barrier, cure could not be achieved by the use of systemic drugs alone. The CNS represented a sanctuary in which leukemia cells could sequester, not be affected by drugs, and could later cause a relapse.

It was at this impasse that the results of experimental studies in mice using chemotherapy plus irradiation of the brain and CNS were reported (Johnson, 1964). Johnson reported cure of L1210 mouse leukemia when whole brain and CNS irradiation were combined with systemic chemotherapy. This led to trials in human leukemia therapy. George and Pinkel (1965) carried out pioneering studies of CNS irradiation in children with ALL. They initially used radiation doses of 500 R to the entire cerebrospinal axis after induction of a complete remission using vincristine and prednisone. CNS irradiation was followed by cyclophosphamide and/or methotrexate. A median hematologic remission of 58.5 weeks was reported and led to extensive studies reported later in detail by Pinkel (1979). Johnson in his experimental studies noted no cures with chemotherapy alone or radiotherapy alone, but this was increased to 70% cures with maximum doses of drug (cyclophosphamide) and radiation (1000 rad) to the CNS. There were 0% cures if the radiation was instead directed to the pelvis. These results were first reported in 1963, and human experience was already being reported in 1965. This shows that in some circumstances very rapid transfer of favorable results from the laboratory to the clinic can occur. The use of routine spinal radiation was later reduced to cranial radiation only, and more recently to reduced doses of radiation (Nesbit et al., 1981). Because of the greater efficacy of current chemotherapy and effective intrathecal chemotherapy, this course of radiation may not be necessary except in more unfavorable presentations of disease with poor prognosis. Nonetheless, sanctuary radiotherapy to drug-inaccessible sites such as the brain, spinal cord, or testis (Sullivan, 1980) contributed to the development of curative therapy of ALL and greatly aided the understanding of therapy.

B. Hodgkin's Disease, Lymphoma and Non-Hodgkin's Lymphoma

In 1832, Thomas Hodgkin described a disease of the lymph nodes in a paper entitled "On Some Morbid Appearances of the Absorbent Glands and Spleen" which first brought medical attention to a disease of the lymph glands and spleen. Wilks recognized an affliction in a series of cases he studied in 1856, with a progressive course, marked glandular enlargement, later induration, and a fatal course. He named the condition after Hodgkin. The leukemias and lymphomas were later described by Virchow and others. Lymphosarcoma was described by Dreschfeld in 1892 and Kundrat in 1893. The follicular or nodular lymphomas were recognized as a separate group of tumors by Brill *et al.* in 1925 and Symmers in 1927 and represent a malignancy of lymph glands of slow evolution and progression. Reticulum cell sarcoma of lymph glands was distinguished from other lymphomas by Roulet in 1930, based upon histological features. Still another distinctive form of lymphoma was discovered by Burkitt (1958) and was a unique form found first in African children, with involvement of the jaw and facial region.

Hodgkin's disease is of lymphoid tissue with unique cellular features and cell types, with the *sine qua non* being a large multinucleated giant cell well described in the literature before the descriptions now attributed to Reed and Sternberg (Kaplan, 1980). The heterogeneity of cellular infiltrates in Hodgkin's disease has led to a long-standing debate on the nature of the condition and is to be distinguished from the non-Hodgkin's lymphomas (NHL) where a single neoplastic lymphoid cell population of uniform histological appearance produces lymph node enlargement, infiltration of other viscera, bone marrow, and may even appear in the blood as leukemia. Controversy about Hodgkin's disease has continued for over a century as to whether it is a malignant neoplastic disease or a form of chronic inflammation of lymphoid tissues (e.g., similar to tuberculosis. It arises usually in lymph nodes, and since the earliest descriptions, it is well known to disseminate to more distant glands and sites, usually by contiguity spread. Widespread discontinuous spread characteristic of NHLs is not commonly seen. The neoplastic precurosr cell is felt to be the giant cell which, however, represents in most tumors only a very small minority of the cells present. Anemia, fever, and intermittent cyclic bouts of fever, anergy to tuberculin, and immunological hypo- or depressed reactivity are characteristics of advanced disease. One of the striking features is immunological deficits and anergy (Schier *et al.*, 1956; Aisenberg, 1964; Good *et al.*, 1962; Hersh and Oppenheim, 1965). Eltringham and Kaplan (1973) and Aisenberg (1962) found that even patients with early or Stage I disease may have significant impairment of immunological reactivity. Sokal *et al.* (1961, 1969) tested reactivity in Hodgkin's disease and found it correlated with disease activity. Some have proposed that Hodgkin's disease affects the immune system and lymphocytes, with autoimmunity being the basis of the interaction between normal and neoplastic lymphocytes, similar to those observed with GVH syndrome (Kaplan and Smithers, 1959). It has also been postulated that there is a viral basis for the disease, with infection of some lymphocytes by a tumor virus leading to neoplastic transformation, with other normal lymphocytes reacting against the transformed cells. This was postulated to lead to chronic GVH reaction, neoplastic reticulum cells, and the complex histological appearance of lymph nodes and tissues (Order and Hellman, 1972). Likewise, the extremely high success of radiotherapy alone, on the order of ~70–80% 5-year survival, and with radiation and chemotherapy of ~90% has continued to feed the controversy of whether it is a true neoplasm. The etiology of the disease and the question of its neoplastic nature will continue. A major observation that continues the controversy is that of Gutensohn and Cole (1968) who found that the disease selectively affected higher socioeconomic groups with small families, fewer playmates, and better educated mothers. They proposed a common viral infection, with the probability of tumorigenesis increasing with age at time of infection. The actual virus has been elusive, but may be Epstein–Barr (EBV), which, however, has been difficult to find (Gallo and Gelman, 1981).

Lymphomas were treated with X rays soon after their discovery in 1902. Lymphoid tumors were found to be extremely responsive (unlike epitheliomas or adenocarcinoma) and to disappear rapidly and dramatically, leading to the descriptive term "radiosensitive." For Hodgkin's disease, it was found that radiotherapy could be curative provided disease was at an early stage. Between 1935 and 1980, sequential evolution of treatment took place concurrently, with stepwise improvement in the types of radiation equipment available. While only 7.7% were 5- year survivors if untreated, low-dose orthovoltage localized radiation therapy led to ~20% 5-year survival (Kaplan, 1980), high-dose local orthovoltage radiotherapy to 39%, and prophylactic extended field therapy to 68% (Peters and Middlemiss, 1958). Megavoltage wide-field radiation with therapeutic doses of ~4000 cGy led to 70% survivals, and with added MOPP chemotherapy the survivals jumped to 80% and more. Beginning with early local radiation therapy directed at apparent disease, therapy has evolved from local to regional to irradiation of all regional lymph nodes using multiple or complex-shaded fields (mantle or inverted Y) or total lymphoid and spleen irradiation (TLI) using megavoltage radiation. The use of radiotherapy of adjacent noninvolved lymphoid tissues was advocated early by Gilbert (1939), effectively applied in low doses by Peters (1966), and extended to high-dose TLI by Johnson et al. (1969, 1970) and Kaplan (1980). Peters (1950) reported that localized versus more widespread disease could be classified in stages. Stage I disease is confined to one anatomic regional site; stage II involves two or more sites on the same side of the diaphragm; and stage III disease presents on both sides of the diaphragm. It could also be classified by the presence of (stage B) or absence of (stage A) systemic symptoms. Stage A patients survived much better than stage B and the major losses in survivorship of stage I and II disease occurred in stage B patients in the 5 to 10-year risk period after radiotherapy (Peters, 1966). Kaplan (1980) and the Stanford group have reviewed the impressive results of radiotherapy and chemotherapy and chemoradiotherapy achieved since the first clinical trials in 1962 for the various stages of disease (NHL Pathological Classification Project, 1982).

Modern staging systems developed after the Peters system represent the current widely used means for staging lymphomatous diseases. The most sensitive test devised for accurate staging of Hodgkin's disease was staging laparotomy which revealed for the first time the hitherto unappreciated frequency of spleen and abdominal-pelvic spread of disease with different disease presentations (Kaplan, 1980). More recently, staging laparotomy is being less frequently used.

The modern strategy of treating Hodgkin's disease directs therapy to multiple adjacent lymph node regions in the body to high-dose, complex-shaped field, i.e., the mantle or inverted Y fields (Fig. 10). Sequential irradiation of all the

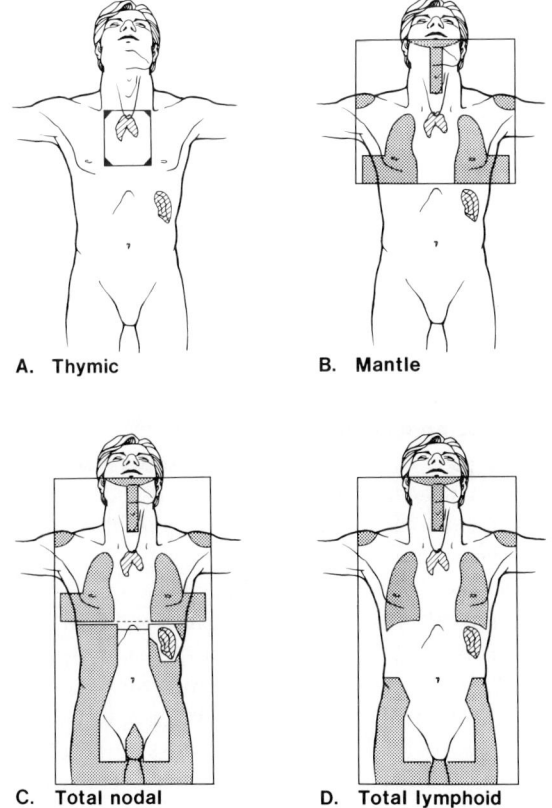

A. Thymic   B. Mantle

C. Total nodal   D. Total lymphoid

Fig. 10. Various irradiation ports used for irradiation of human lymphoid organs. (A) Local port to mediastinum and thymus. (B) Upper mantle port with oral-cervical spine, lung, and humoral head shielding. (C) TNI portals. This combines the upper mantle and the inverted Y and spleen fields. (D) TLI port. This includes mantle and whole abdomen including spleen, liver, and inverted Y port. Treatment is usually done by AP/PA ports. Anterior mantle field only leads to cardiac damage.

major lymph node chains above and below the diaphragm was developed from radiotherapy directed at Stage III disease. The treatment, termed TNI, was surprisingly well tolerated in clinical stage III disease (Johnson et al., 1969; Kaplan, 1980). This led to its wide acceptance and these dramatic advances markedly improved cure and disease-free survival for disease. Split-course radiotherapy was found to be as effective as continuous (Johnson et al., 1976) and to require no increase in total prescribed dose. TNI and extended field radiation combined with multiagent effective chemotherapy, initially MOPP, has displaced the palliative use of single agent chemotherapy, which was once in vogue 10–15 years ago.

For both Hodgkin's disease and NHL of high or advanced stages, therapy now strongly emphasizes systemic chemotherapy combined with extended or localized irradiation fields, whereas early or low-stage disease therapy emphasizes high dose (~1000–5000 cGy) local-regional mantle or inverted Y irradiation therapy, without added adjuvant chemotherapy, and spleen/splenic pedicle and paraaortic lymph node irradiation. For Hodgkin's disease, three excellent and curative chemotherapy regimens have been developed, i.e., MOPP developed by the National Cancer Institute (deVita et al., 1970) (MOPP = nitrogen mustard, oncovin, procarbazine, and prednisone); ABVD developed by Bonnadonna and Santoro (1982) (ABVD = adriamycin, bleomycin, vinblastine, and dihydroimidazole carboximide); and alternating MOPP plus ABVD. These combined agent regimens supplemented by local radiotherapy to areas of originally bulky disease can produce complete remissions in 90–100% of treated patients. However, the best results appear to accrue with alternating radiation, often in reduced dosage and field sizes, with chemotherapy for the treatment of the advanced and systemic presentations of disease.

Immunological responses are profoundly depressed by intensive total lymphoid radiotherapy in patients treated by radiation and/or chemotherapy. Severely depressed immunological reactivity to antigens, T lymphocytopenia, and depressed lymphocyte reactivity is observed. Many patients continue to remain anergic for years following treatment, but nonetheless can remain disease-free indefinitely. As a spin-off of the immune suppression noted after TNI with the absence of delayed leukemia, the Stanford group began studies of TLI in experimental animals and man as a conditioning regimen for organ and bone marrow transplantation and as an immunosuppressive therapy for autoimmune diseases (Strober et al., 1979a,b). Thus, an unanticipated side benefit of TNI learned from radiotherapy of Hodgkin's disease was that a complex immunological disease such as Hodgkin's can be treated by immunosuppressive therapy such as TNI and that such therapy can also be used as a basis to treat other human disease.

For NHL, low-stage and nodular pattern determines outcome, but the tremendous advances made in Hodgkin's disease therapy have not yet altered radiotherapy practice. Localized nodular lymphoma has a very benign course (Quazi *et al.* 1976; Jaffe, 1983). The emphasis is on local radiotherapy to localized disease with/without systemic chemotherapy. Efforts to use TLI and whole abdominal irradiation were not better than more conservative therapeutic measures and have led back again to less radical therapy. Diffuse histiocytic lymphoma (DHL) can respond to CHOP or BCNU combination chemotherapy in about half of the cases (see Rosenberg and Kaplan, 1982). Laparotomy staging is much less accurate in staging NHL than for Hodgkin's. Low-dose TBI to 150 cGy can be effective for producing remission of disease in some patterns of disease (see Rosenberg and Kaplan, 1982; Bush and Gospodarowicz, 1982). For Burkitt's NHL higher dose of radiation with superfractionated schedules appear to be more effective (Norin *et al.*, 1971) with adjuvant chemotherapy. For NHL, future research will hopefully produce new leads for more precise diagnosis, staging, and therapy.

For NHL, research needs to establish the type of lymphocyte (Nadler *et al.*, 1981), stage of the disease, and the frequencies of local versus disseminated presentations by primary site. Some nodular patterns and low-grade forms can be controlled by local irradiation (see, e.g., Bush *et al.*, 1982). However, when of diffuse or high-grade histology, the lymphomas appear to require adjuvant chemotherapy for control and cure.

At present our concepts of Hodgkin's disease and NHL are still far less than clear. For Hodgkin's disease our treatments are effective, but our conceptual understanding is primitive. There is need to conclusively prove the cellular origin of the disease, whether from T, B lymphocytes or the histiocyte, and to establish the causative agent and pathogenesis. The NHLs are undoubtedly neoplastic and reflect arrest of normal lymphocyte development with perverted cellular function, loss of growth regulation, and uncontrolled proliferation. There are a multiplicity of cytological types, behavioral patterns, distribution of cells at presentation, and variation in disease progression so that need for histological diagnosis, classification, and staging is important. For the future, there is a need for improved basic understanding of these diseases.

C. Total Lymphoid Irradiation

The technique of TNI was developed for the treatment of patients with stage III Hodgkin's disease and other malignant lymphomas. Long-term observations are now available on hundreds of patients treated to essentially all the lymphoid tissues of the body using this technique (Kaplan, 1980). The bone marrow was not necessarily totally ablated by 4400 rads of TNI. Instead, TNI's effects depended on age, dose, overall time, and volume treated (Sack *et al.*, 1978).

An outgrowth of the successful therapy of stage III Hodgkin's disease by TNI was TLI. The effort to treat NHL and its tolerance by patients led to its use to condition patients for homografting of organs and bone marrow and palliative treatment of autoimmune disease. It was known that TNI for Hodgkin's disease produced severe immunosuppression of the immune system (Ehringham and Kaplan, 1975) and depressed lymphocyte function (Fuks *et al.*, 1976). One of the major clinically overt immunological effects of TNI was an increase of susceptibility to herpes virus infections (Goffinet *et al.*, 1973). Radiation-induced splenic atrophy has been produced by TNI and has been complicated by fatal pneumococcal sepsis (Dailey *et al.*, 1980). *In vitro* studies have demonstrated that after TNI, depressed lymphocyte reactivity can persist for 10 years or more after successful radiotherapy for Hodgkin's disease (Fuks *et al.*, 1976). The T lymphocytopenia is accompanied by profound abnormalities of T lymphocyte function, including a loss of the capacity to respond in the mixed lymphocyte reaction which can persist for protracted periods. There is also an important impairment of responses to PHA and other lectins which may persist for years. These observations suggested that TLI might be useful as an immunosuppressive regimen for patients undergoing organ and tissue transplantation as well as in patients with severe autoimmune disease.

The ability to produce permanent and specific transplantation tolerance was studied in detail by Slavin *et al.* (1977) and reviewed by Strober *et al.* (1979a,b). Basically, TLI was found to produce tolerance for bone marrow, skin, and heart in inbred and outbred adult animals differing at major histocompatibility loci. Laboratory investigations in mice (Slavin *et al.*, 1977), rats (Slavin *et al.*, 1977), dogs (Gottlieb *et al.*, 1980), and monkeys (Bieber *et al.*, 1979) have confirmed the remarkable efficacy of TLI as a conditioning regimen in organ and tissue transplantation. The cellular mechanisms underlying these immunosuppressive effects of TLI have been extensively investigated and point strongly to the induction of populations of suppressor cells which block both HVG and GVH reactions (Strober *et al.*, 1979a,b; King *et al.*, 1981).

The basic experimental research was a consequence of the success of human therapy. By the clinical or experimental use of 3000–3400 cGy, severe and persistent T lymphocytopenia was produced. Immature T cells appear in the peripheral lymphoid tissue, and there was an absence of GVH disease or homograft immunoreactivity to a variety of homografted tissues. TLI has been tested for BMT in man for aplastic anemia (Ramsey *et al.*, 1980), renal and liver transplantation, and for the treatment of autoimmune disorders of man. Doses used have varied from 750 cGy in a single session to 3500 cGy in multiple fractions 3–4 weeks. TLI has been found by Trentham *et al.* (1981) and Schurman *et al.* (1981) to produce short remissions of refractory rheumatoid or experimental arthritis. By inverted Y therapy alone, remissions may also be obtained (Strober *et al.*, 1981). Many studies are ongoing in augmenting the tolerant state for tissue transplantation. Since TNI has been applied to Hodgkin's disease therapy with few delayed leukemia/lymphomas appearing, it has been suggested that TLI may be a safe method for human therapy. It will be of interest to review in the future its role and possible hazards. Clinical trials are under way for organ and tissue transplantation and for the treatment of autoimmune disorders. Probably TLI will become a form of immunosuppression for conditioning the good to excellent quality matches of homografted tissues (bone marrow, liver, kidney, heart) and TBI, a conditioning regiment for the fair to poor match recipients of homografted organs. For the syngeneic (identical twin) grafts, no barriers to grafts are present. Autologous marrow infusion may become an approach for selected clinical trials in the future. As noted elsewhere, normal BMTs may reduce the later incidence of chemotherapy-induced hematologic neoplasia.

D. Total Body Irradiation—Bone Marrow Transplantation

Observations from TBI, recognition of the radiation syndrome, the atomic bomb effects on man, and reconstitution cellular therapy studies of TBI in man and animals led to the present evaluation of TBI and BMT for certain hematological diseases and malignancies. Intensive chemotherapy is curative in half of childhood ALL patients, but most regimens are limited by lethal bone marrow toxicity as well as limited efficiency. To increase therapeutic efficiency against these diseases, Thomas *et al.* (1975) began studies of lethal TBI and BMT to provide more effective therapy, allow more intensive chemotherapy, and create a means to rescue patients from the toxicities of therapy. The donor may be an identical twin (syngeneic), the patient (autologous), a human leukocyte antigen (HLA) matched sibling (allogeneic), a more remotely related family member (parent), or a nonrelated matched allogeneic donor. BMT allows more intensive chemotherapy and TBI produces a tolerant state to accept the donor marrow graft. BMT with allogeneic grafts exerts an antileukemia effect, expecially with GVH disease, which also is therapeutic in some of the hematologic neoplasms treated. In nonneoplastic disorders such as aplastic anemia and combined system immunodeficiency syndrome, BMT has become an established form of experimental therapy.

Acute nonlymphocytic leukemia (ANLL) has been treated in relapse after first remission produced by chemotherapy. Treatment of the relapsed patient is not effective. Treatment of the ANLL patient in remission was shown to produce 45% actuarial 5-year cures (Thomas *et al.*, 1975, 1978, 1982).

Chronic myelogenous leukemia (CML) also can be successfully treated using TBI and BMT. Patients selected for therapy are generally younger or in an accelerated phase. CML in young people can be a much more aggressive disease than in older adults where it can be a chronic disease of long natural history. The Philadelphia chromosome represents a marker that allows the determination of whether the stem cell (Nowell *et al.*, 1960) has been eliminated (Fefer *et al.*, 1982).

ALL treated in second or subsequent relapse has been studied, but is less effective than treatment of ANLL. A smaller portion of TBI and BMT patients are actuarial survivors at 5 years (25%). With relapse/repeated relapses, survival decreases, and the best results were seen in BMT done after the first remission (Johnson *et al.*, 1981).

GVH, while therapeutic against certain leukemias, is difficult to control in the BMT-treated patient. In an effort to reduce the severity of this condition, use of autologous marrow purged of leukemia cells by cytotoxic antibody and complement therapy has been attempted. Likewise, homologous marrow has been treated with antibodies (e.g. of monoclonal origin) directed against T lymphocytes. This has been used to purge marrow of the cells that cause the GVH syndrome. The latter has been successful in studies carried out at the University of Kentucky. The use of remission or cryopreserved autologous marrow without further tumor removal is ineffective in treating leukemia.

Complications after TBI and BMT have been infections, GVH disease, and, in particular, interstitial pneumonitis with cytomegalic virus (CMV) or radiation. Radiation pneumonitis is reduced in frequency by use of lung shielding from part of the TBI treatment exposure. Shank *et al.* (1981) developed methods for electron boost irradiation of the ribs during a course of multifractionated TBI.

A large variety of conditioning dose regimens has been developed. Thomas *et al.* (1975) used LDR ^{60}Co irradiation to a midline dose of ~1000 cGy. Shank *et al.* (1981) developed a hyperfractionated TBI regimen to 1320 cGy. The University of Kentucky has tested a 3×/day fractionated treatment, but varied doses to 1600 cGy for relapsed leukemia using multiple short acute linear accelerator TBI exposures with partial lung shielding (unpublished data; see Ash *et al.*, 1986). Thomas *et al.* tested single versus fractionated schedules and both were effective in BMT therapy. TBI and BMT are now being tested in a variety of hematological disorders, leukemias of all types, and, more recently, in NHL and other drug- and radiation-sensitive disseminated myelomas and solid tumors.

E. Cutaneous T Cell Lymphoma (CTCL)—Total Skin Irradiation

Malignant T lymphocytes can preferentially infiltrate certain parts of the human body as a characteristic presentation of disease (Edelson *et al.*, 1975). One of these is the clinically and immunologically defined neoplasm known as mycosis fungoides (MF) and Sezary's syndrome, which represents neoplasms caused by T cell lymphomas (Edelson, 1980). In its earliest stages, the disease may be confined to the skin, but with progression widely involves viscera, lymph nodes, liver, and spleen, but with relative sparing of bone marrow (Edelson *et al.*, 1974). Maruyama *et al.* (1972) described an experimental form of cutaneous lymphoma/leukemia cutis of mice.

Whole body electron beam therapy is highly effective for MF and complete remissions can be obtained with low-stage (Bunn and Lamberg, 1979), limited body disease. Remissions are produced by use of low-energy electron beams treating the whole body of low-stage patients to 3000–3600 cGy (Hoppe *et al.*, 1977, 1979). While remissions are frequent, relapse is unfortunately a continuing problem and adjuvant therapy is needed. Because lesions can be quite thick as well as nodular and ulcerative, variable energy electron beam therapy has been advocated (Coffey *et al.*, 1982).

Likewise, added systemic chemotherapy appears necessary for the higher stages of disease (Bunn and Lamberg, 1979). A recent proposal of a workup and staging scheme will allow better stratification of this chronic disease and studies of therapy (Bunn and Lansberg, 1979) in the future.

F. Plasmacytomas

B lymphocytes differentiate into immunoglobulin-synthesizing and antibody-secreting B plasma cells. The pre-B cell has no surface immunoglobulins, but does contain IgM. Early B cells show surface immunoglobulins and complement receptor sites. Actively secreting B cells become plasma cells and respond to pokeweed mitogen, antigens, and T regulatory lymphocytes. Major antibody-forming cells are plasma cells and a single cell and its clonal descendants elaborate a single antibody (Nossal and Aba, 1971).

While multiple myeloma is a widespread systemic and ultimately fatal disease, solitary plasmacytoma represents a localized variant where one anatomic site represents the site of neoplastic plasma cell proliferation. The site can be in bone, i.e., an osseous lesion, or in an extramedullary (nonosseous) site. Plasma cell tumors can be induced by injection of mineral oil intraperitoneally into BALB/c mice (Merwin and Algire, 1959) and have been used (Potter, 1972) for producing antibody-synthesizing tumors. These cells can be fused with other B lymphocytes to produce specific monoclonal antibodies (Kohler and Milstein, 1975). Radiotherapy is frequently given for painful bone lesions with an excellent response using conventional palliative irradiation doses of 2000–3000 cGy.

Solitary plasmacytoma has been a condition of great medical interest and can be associated with long-term survival or cure. It is a rare variant form and 2–10% of plasma cell tumors present with a single skeletal lesion. Extramedullary lesions often present in the upper aerodigestive or gastrointestinal tract. The disease can progress slowly or rapidly to dissemination as multiple myeloma (Wiltshaw, 1976).

The radiotherapeutic response was traced by decline of abnormal serum globulins, which rapidly fell to normal after 4000 rad of local irradiation (Maruyama and Thomson, 1970) in cured patients. Appropriate high-dose radiotherapy produced cure in 67% of treated patients and 85% actuarial survival to

17 years (Mendenhall *et al.,* 1980). About 33% of patients progress to multiple myeloma (Mendenhall *et al.,* 1980; Corwin and Lindberg, 1979). This is more common, i.e., ~50%, for osseous plasmacytoma and less common, i.e., 16%, for extramedullary tumors. While fall of abnormal proteins to normal levels indicates the entire neoplastic plasma cell population has been destroyed, the persistence or reappearance of abnormal levels of serum or urine globulins indicates more disseminated or progressive disease. Extramedullary plasmacytoma differs from bone disease by different patterns of spread and can involve lymph nodes or can even metastasize to lung as discrete nodules (Corwin and Lindberg, 1979).

Dose–response was analyzed in a literature review by Mendenhall *et al.* (1980), and 4000 rad was sufficient to achieve tumor control in 94% of patients. Doses of ≤3000 cGy led to frequent recurrence and progression of disease in 31% of treated patients given less than 4000 cGy. Unusually bulky disease (e.g., in the upper digestive tract and head and neck) appeared to require doses to 5000–6000 cGy. Overall, solitary plasma cell tumors are radiosensitive and radiocurable, and multiple myeloma is readily palliated by light doses of radiation.

VII. Late Effects of Human Therapeutic Irradiation

During the past decade, serious questions arose concerning the hazard and immunosuppressive effects of irradiation, particularly of the mediastinal-thymus region, and of subsequent host resistance to the spread of cancer (Sternsward *et al.,* 1972; see Dubois *et al.,* 1981). The charges were based upon the distribution of B and T lymphocytes in the peripheral blood and their reduced reactivity to stimulation after chest wall irradiation. This was alleged to increase the risk of cancer spread and decrease ability of the host to resist cancer. Breast cancer was chosen as the disease in which thymic irradiation was regularly done and produced the most severe lymphopenia and host vulnerability to later tumor spread. Sternward *et al.* (1972, 1974) reported that radiotherapy of breast cancer resulted in long-term and persistent fall in absolute number of circulating lymphocytes. Meyer (1970) postulated that radiation therapy induced lymphocyte deficiencies, and this led to depressed lymphocyte immune reactivity. He suggested that this could be a factor in later visceral metastases and cancer spread in radiation-treated breast cancer patients.

Stratton *et al.* (1975) carried out an important study of the acute effects of radiation therapy which included or excluded the thymus on lymphocyte sub-populations of cancer patients. They studied patients with breast cancer as well as pelvic carcinomas. Patients were carefully studied serially by total leukocyte and mononuclear cell counts, T and B lymphocyte counts and ratios, and function as measured by *in vitro* response to mitogen stimulation, rosette forming cells, lymphocytes bearing surface, immunoglobulins, etc. All were found to decline to the identical extent following either thymic-mediastinal or pelvic irradiation. There was a rapid decrease in numbers and decrease in function which occurred acutely but was short-lived. Recovery was apparent within 3 weeks after cessation of therapy. Most patients showed only a mild chronic depression in numbers and functional capacities of circulating lymphocytes. T cells were somewhat more sensitive than B cells, but both were affected. Irradiation of the thymus per se had little influence and similar changes occurred whether pelvic (Raben et al., 1976), mediastinal, head and neck (McLaren et al., 1981), or lung (Concannon et al., 1978) irradiation was given.

Sutherland and Maruyama (1974) studied chest radiation effects in mice. Thoracic irradiation with 2000 rad in 5 days caused a similar rapid but reversible lymphopenia which recovered fully by 3 weeks. No suppression of resistance to later syngeneic tumor challenge was found. Other studies in mice have shown that 150 cGy of thymic irradiation was insufficient to cause lymphopenia. Bone marrow shielding or injection leads to rapid repopulation of the irradiated thymus. Only high-dose total lymphoid irradiation leads to severe persistent lymphopenia and immune suppression (Slavin et al., 1977). Heavy irradiation of the thymus with 3000 rads in a single exposure and whole body sublethal irradiation can also produce a completely tolerant state (Maruyama and Barclay, 1967). However, despite severe immunosuppression, Kaplan's (1980) experience with TLI in Hodgkin's disease showed no unusual incidence of delayed tumors in radiation-treated patients despite persistent immunosuppression. When chemotherapy is given with TLI, delayed leukemias are more frequent. Kaplan's success with TLI and the absence of late neoplasms led him to support TLI as a method to treat other benign human diseases (e.g., arthritis: see Strober et al., 1981).

Hematological malignancies (i.e., leukemia/lymphomas) represent a special concern in the study of late radiation effects. Boice and Fraumeni (1984) provided a timely and comprehensive overview of our current knowledge concerning radiation carcinogenesis. In ankylosing spondylitis, late leukemia is established to occur (Court-Brown and Doll, 1957) within a peak 3–5 years after irradiation. Atomic bomb survivors also developed late leukemia 3–5 years following exposure and with a peak incidence at 7–8 years. The most common types were AML, CML in adults and children, and ALL in children. CLL was not radiation related; however, multiple myeloma, a disease of B lymphocytes, is caused by irradiation (Cuzick, 1981). NHL and multiple myeloma have also been observed with greater frequency and the risks for these diseases extends to 40 years or longer. Thereafter, the risk changes to tumors other than leukemia, and this risk is higher in those exposed at a young age versus old age and remains elevated in relation to a nonexposed population (Ishimaru et al., 1981). There is a strong positive correlation between radiotherapy and chemotherapy, particularly with the alkylating agents, in human leukemogenesis. Alkylating agents appear to be a drastically more potent factor to induce oncogenic transformation and delayed leukemia (Greene, 1984), especially for ANLL. Childhood X rays may also increase hematological malignancies (Stewart, 1969). The hematological system is one of the most sensitive for induction of neoplasia to all forms of radiation.

Thymic irradiation, however, is now established to be an ill-conceived and dangerous practice and led to many late effects and neoplasms of the thyroid, depending on dose, field size used, use of anterior and posterior portals, and age at time of treatment (Simpson *et al.*, 1955; Hempelmann *et al.*, 1975). Irradiation of the thymus can lead to chronic and delayed T and B lymphocytopenia and reduced lymphocyte reactivity (Reddy *et al.*, 1976). It is clear that despite poor understanding of thymus function, role in development, immunology, or lymphocyte instruction, regulation, migration, or sensitivity to radiation, many interesting studies have been carried out.

VIII. Summary and Conclusions

Any current review of the effects of radiation on thymolymphatic organs needs to address recent clinical observations and radiation dose–survival curves. Since survival curves were first shown for mammalian and human cells by Puck and Marcus, there have been many studies directed to determine their relevance to murine and human leukemia/lymphomas. Closely related with this was the discovery of survival curves for hemopoietic stem cells and the development of cellular understanding and therapy for the bone marrow syndromes following TBI. Irradiation is known to destroy the hemopoietic system. Research in the past two decades has established that there is an intimate and close interdependent relationship between the hemopoietic and the thymolymphoid system.

Stem cells from one seed the other with an intermediate sojourn for some lymphocyte precursor cells in the thymus or bursal equivalent (bone marrow ?) organ. This has been traced by reconstitution studies, by unique cell markers, and by the enzyme TdT. From these two derivations, two classes of lymphocytes arise (i.e., B and T with their functional repertoires determined and directed to either humoral or cellular immunity. These lymphocytes migrate to and populate the peripheral lymphoid organs. Heavy total body, local, or regional lymphoid, or combinations of these patterns of body irradiation severely damage or impair the lymphohemopoietic systems. These patterns of irradiation applied to the intact mammal or human can alter immunological reactivity, produce a tolerant state and/or allow the homografting of tissues or organs, cure certain neoplastic conditions, and modify the course of autoimmune diseases. The same is true for systemic chemotherapy, particularly with the alkylating agents (''radiomimetic''). The model of the bone marrow, thymus, spleen, lymph nodes, and peripheral lymphoid tissue as a single, unified, and interdependent system allows understanding of local versus whole body irradiation effects in man or mammals and understanding of a variety of tissue responses.

With malignant transformation, many different types of lymphoid tumors arise. The lymphohemopoietic system is very sensitive to tumor induction by radiation. From human experience, lymphoid tumor types appear to be numerous, difficult to characterize, and of widely differing biological behavior. There are only a few animal models developed to date that offer the investigator the ability to study some of the characteristics of lymphoid tumors experimentally. From a few tumor lines, detailed radiation dose–response survival curves have been studied. Interestingly, many show low n values, variable D_0 slopes, and where study for SLD repair has been carried out, some show repair, but many do not and may even show enhancement of radiation sensitivity caused by the first dose. When these tumors are chronically irradiated, SLD repair can be induced or cells exhibiting shouldered survival curves can be selected from the original population.

The apparent limited capacity for SLD repair, and generally greater sensitivity of lymphohemopoietic tissue, and the rapid interphase death process have led to extensive use of radiation for lymphoid tumors. Extensive human studies of lymphocytic leukemia/lymphomas and other lymphoid tumors have led to encouraging observations and responses. However, therapy is complex, since some of these tumors are localized and form masses, others are leukemic and widely disseminated from the start, and still others are restricted to just the lymphatic system. Many novel forms of therapy have emerged, from local to TBI or TNI/TLI irradiation therapy. Usually these are combined with chemotherapy, and for some conditions, with BMT therapy. Neoplasms also show evidence of response and sensitivity to immunological therapy. Thus, lymphoid tumors and radiation response are greatly different from epithelial or similar tumors.

The new understanding and models of the thymo-lymphohemopoietic system have had major impacts on studies and therapy of these systems. It seems likely that no single regimen of therapy will control all the varied presentations of disease manifested by the system. It is important to develop reproducible animal models of the various forms of lymphoid neoplasms and to experimentally characterize them in great detail. There should be a focus on characterizing and accurately developing lymphocyte tumors which can be induced and studied in the primary host as well as in derived transplant tumor models (Schabel *et al.*, 1969; Maruyama *et al.*, 1983b,1985). (Schabel *et al.*, 1969; Maruyama *et al.*, 1983b, 1985). This approach may provide a basis for developing sounder and more productive approaches to therapy than the human clinical trials which are extremely expensive and make only erratic progress. Only when such models are available will it be possible to better assess the influences of radiation on lymphocyte sensitivity, lymphoid organ response, and appropriate lymphoid tumor therapy.

REFERENCES

Abramson, S., Miller, R. G., and Phillips, R. A. (1977). *J. Exp. Med.* **145**, 1567.

Ainsworth, E. J., Fry, R. J. M., Grahn, D., Williamson, F. S., Brennan, P. C., Stearner, S. P., Carrano, A. V., and Rust, J. H. (1974). *In* "Biological Effects of Neutron Irradiation," p. 359. SM-179/1, IAEA, Vienna.

Ainsworth, E. J., Jordon, D. L., Miller, M., Cooke, E. M., and Hulesch, J. S. (1976). *Radiat. Res.* **67**, 30.

Ainsworth, E. J., Kelly, L. S., Mahlmann, L. J., Schooley, J. C., Thomas, R. H., Howard, J., and Alpen, E. L. (1983). *Radiat. Res.* **96**, 180.

Aisenberg, A. C. (1962). *J. Clin. Invest.* **41**, 1964.

Aisenberg, A. C. (1964). *Medicine* **43**, 189.

Aisenberg, A. C. (1981). *In* "Scientific Foundation of Oncology" (T. Symington and R. L. Carter, eds.), p. 537. Yearbook, Chicago.

Aisenberg, A. C., and Bloch, K. J. (1972). *New Engl. J. Med.* **287**, 272.

Alexander, P., and Mikulski, Z. B. (1961). *Biochem. Pharmacol.* **5**, 275.

Altman, K. I., Gerber, G. B., and Okada, S. (1970). "Radiation Biochemistry," Vol. I., p. 247; Vol. II, p. 49. Academic Press, New York.

Anderson, R. E., and Warner, N. L. (1976). *Adv. Immunol.* **24**, 215.

Anderson, R. E., Olson, G. B., Howorth, J. L., Wied, G. L., and Bartels, Ph.H. (1975). Am. J. Pathol. **80**, 21.

Anderson, R. E., Olson, G. B., Autry, R. J., Howarth, J. L., Troup, G. M., and Bartels, P. H. (1977). *J. Immunol.* **118**, 1191.

Anderson, R. E., Lefkovits, I., and Troup, G. M. (1980). *Contemp. Top. Immunobiol.* **11**, 245.

Andreasen, E., and Ottesen, J. (1945). *Acta Physiol. Scand.* **10**, 258.

Arnason B. G., Jankovic, B. D., Waksman, B. H., and Wennestein, L. (1962). *J. Exp. Med.* **116**, 177.

Archer, O., and Pierce, J. L. (1961). *Fed. Proc., Fed. Am. Soc. Exp. Biol.* **20**, 26.

Archer, O. K., Sutherland, D. E. R., and Good, R. A. (1964). *Lab. Invest.* **13**, 259.

Ash, R. C., *et al.* (1986). *Blood* **5**, 159.

Auerbach, R. (1963). *Science* **139**, 1061.

Auerbach, R. (1967). *Dev. Biol.* **3**, 336.

Bacq, Z. M., and Alexander, P. (1961). "Fundamentals of Radiobiology." Pergamon, Oxford.

Bain, B., Vas, M. R., and Lowenstein, L. (1964). *Blood* **23**, 108.

Ballou, J. E. (1976). *ERDA Conf. Ser.* **37**, 740930.

Baral, E., and Blomgren, H. (1976). *Acta Radiol. Ther. Phys. Biol.* **15**, 149.

Barg, M., Mandel, T. E., and Johnson, G. R. (1978). *Aust. J. Exp. Biol. Med. Sci.* **56**, 195.

Barnes, D. W. H., and Loutit, J. F. (1957). *Br. J. Haematol.* **3**, 241.

Barnes, D. W. H., Coup, M. J., Loutit, J. R., and Neal, F. E. (1956). *Br. Med. J.* **2**, 626.

Becker, A. J., McCulloch, E. A., and Till, J. E. (1963). *Nature (London)* **107**, 452.

Becker, A. J., McCulloch, E. A., Siminovitch, L., and Till, J. E. (1965). *Blood* **26**, 296.

Beer, J. Z. (1973). *Stud. Biophys.* **36**, 175.

Belli, J., and Andrews, J. R. (1963). *J. Natl. Cancer Inst.* **31**, 689.

Belli, J., Dicus, G. J., and Bonte, F. J. (1967). *J. Natl. Cancer Inst.* **38**, 673.

Belli, J. A., Dicus, G. J., and Nayle, W. (1970). *Front. Radiat. Ther. Oncol.* **5**, 40.

Benninghoff, D. L., Tyler, R. W., and Everett, N. B. (1969). *Radiat. Res.* **37**, 381.

Bergsagel, D. E., and Valeriote, F. A. (1968). *Cancer Res.* **28,** 2187.

Bergsagel, D. E., Alison, R. E., Bean, H. A., Brown, T. C., Bush, R. S., Clark, R. M., Chua, T., Dalley, D., Deboer, G., Gospodarowicz, H., Hasselback, R., Perrault, D., and Rideout, D. F. (1982). *Cancer Treat. Rep.* **66,** 717.

Berry, R. J. (1967). *Br. J. Radiol.* **401,** 285.

Berry, R. J., and Andrews, J. R. (1961). *Radiology (Easton, Pa)* **77,** 824.

Bieber, C. P., Jamieson, S., Raney, A., Burton, N., Bogarty, S., Hoppe, R., Kaplan, H. S., Strober, S., and Stinson, E. B. (1979). *Transplantation* **28,** 347.

Blackett, N. M. (1965). *Int. J. Radiat. Biol.* **9,** 323.

Blomgren, H. (1969). *Exp. Cell Res.* **58,** 353.

Blomgren, H. (1971). *Cell Tissue Kinet.* **4,** 443.

Blomgren, H., Edsmyr, F., Naslund, I., Petrini, B., and Wasserman, J. (1983). *Clin. Oncol.* **9,** 289.

Bloom, W. (1949). "Histopathology of Irradiation." McGraw-Hill, New York.

Boersma, W. J. A. (1983). *Exp. Hematol.* **11,** 922.

Boersma, W., Betel, I., Daculsi, R., and van der Wester, G. (1981). *Prog. Cancer Res. Ther.* **26,** 181.

Boice, J. D., and Fraumeni, J. F. (1984). "Radiation Carcinogenesis." Raven, New York.

Boivin, J. F., and Hutchison, G. B. (1984). *In* "Radiation Carcinogenesis Epidemiology and Biological Significance" (J. D. Boice and J. F. Fraumeni, eds.), p. 181. Raven, New York.

Bollum, F. J. (1974). *Enzymes* **10,** 145.

Bollum, F. J. (1981). *Adv. Exp. Med Biol.* **145,** 1.

Bond, V. P., Fliedner, T. M., and Archambeau, J. O. (1965). "Mammalian Radiation Lethality." Academic Press, New York.

Bonnadonna, G., and Santoro, A. (1982). *Adv. Cancer Res.* **36,** 257.

Boyum, A. (1968). *Scand. J. Lab. Invest.* **21** (Suppl. 97), 9.

Boyum, A., Carstens, A. L., Chikkappa, C., Cook, L., Bullis, J., Honikel, L., and Cronkite, E. P. (1978). *Int. J. Radiat. Biol.* **34,** 201.

Bradley, T. R., and Metcalf, D. (1966). *Aust. J. Exp. Biol. Med.* **44,** 287.

Bradley, T. R., Metcalf, D., and Robinson, W. (1967). *Nature (London)* **213,** 926.

Brecher, G., Endicott, K. M., Gump, H., and Brawner, H. P. (1948). *Blood* **3,** 1259.

Bruce, W. R., and van der Gaag, H. (1963). *Nature (London)* **199,** 79.

Bruce, W. R., Meeker, B. E., and Valeriote, F. A. (1966). *J. Natl. Cancer Inst.* **37,** 233.

Buckton, K. E., Smith, P. G., and Court-Brown, W. M. (1967). "Human Radiation Cytogenetics." North Holland Publ., Amsterdam.

Bunn, P. A., and Lamberg, S. (1979). *Cancer Treat. Rep.* **63,** 725.

Burg, G., and Braun-Falgo, O. (1983). "Cutaneous Lymphomas." Springer-Verlag, Berlin.

Burkitt, D. (1958). *Br. J. Surg.* **46,** 218.

Burnet, F. M. (1959). "The Clonal Selection Therory of Acquired Immunity." Cambridge Univ. Press, London.

Burton, R. C., Grail, D., and Warner, N. L. (1978). *Br. J. Cancer* **37,** 806.

Bush, R. S., and Bruce, W. R. (1964). *Radiat. Res.* **21,** 612.

Bush, R. S., and Gospodarowicz, M. (1982). *In* "Malignant Lymphomas" (S. A. Rosenberg and H. S. Kaplan, eds.), p. 485. Academic Press, New York.

Byron, J. W. (1980). *In* "Biological Basis of Immunodeficiency" (E. W. Gelfand and H. M. Dosch, eds.), p. 143. Raven, New York.

Caldwell, W. L., Lamerton, L. F. A., and Bewley, D. K. (1965). *Nature (London)* **208,** 168.

Cantor, H., and Boyse, E. A. (1977). *Contemp. Top. Immunobiol.* **7,** 47.

Carabell, C., Chaffey, J. T., Rosenthal, D. S., Maloney, W. C., and Hellman, S. (1979). *Cancer* **43,** 994.

Chang, L. M. S. (1987). *Biochem. Biophys. Res. Commun.* **44,** 124.

Claman, H. N., and Chaperon, E. A. (1969). *Transplant. Rev.* **1,** 92.

Claman, H. N., and Mosier, D. E. (1972). *Prog. Allergy* **16,** 40.

Coffey, C. W., Maruyama, Y., Stewart, B. L., and White, G. A. (1982). *J. Ky. Med. Assoc.* **80,** 398.

Cole, L. J. (1967). *Brookhaven Symp. Biol.* **20,** 263.

Coleman, M. S., Hutton, J. J., DeSimone, P., and Bollum, F. J. (1974). *Proc. Natl. Acad. Sci. U.S.A.* **71,** 4404.

Concannon, J. P., Dalbow, M. H., Davis, W., Hodgson, S. E., Mitchell, J., and Markopoulos, E. (1978). *Int. J. Radiat. Oncol. Biol. Phys.* **4,** 255.

Cooper, M. D., and Lawton, A. R. (1972). *Contemp. Top. Immunobiol.* **1,** 49.

Cooper, M. D., Peterson, R. D. A., South, M. A., and Good, R. A. (1966a). *J. Exp. Med.* **123,** 75.

Cooper, M. D., Peterson, R. D., Gabrielsen, A., and Good, R. A. (1966b). *Cancer Res.* **26,** 1165.

Corwin, J., and Lindberg, R. D. (1979). *Cancer* **43,** 1007.

Court-Brown, W., and Doll, R. (1957). *MRC Special Rep. Ser.* (295).

Courtenay, V. D. (1965). *Int. J. Radiat. Biol.* **9,** 581.

Covelli, V. (1981). *Tumori* **67,** 1.

Cronkite, E. P., Fliedner, T. M., Bond, V. P., and Rubini, J. R. (1959). *Ann. N.Y. Acad. Sci.* **77,** 803.

Cross, A. M., Davies, A. J. S., Doe, B., and Leuchars, B. (1964). *Nature (London)* **203,** 1239.

Cunliffe, P. N., Mann, J. R., Cameron, A. H., Roberts, K. D., and Ward, H. W. C. (1975). *Br. J. Radiol.* **48,** 374.

Curry, J. L., and Trentin, J. J. (1967). *Dev. Biol.* **15,** 395.

Cuzick, J. (1981). *N. Engl. J. Med.* **304,** 204.

Dailey, M. O., Coleman, C. N., and Kaplan, H. S. (1980). *N. Engl. J. Med.* **302,** 215.

Dalculsi, R., Aslier, T., Legrand, E., and Duplan, J. F. (1985). *Thymus* **4,** 45.

Dardenne, M., and Bach, J. F. (1981). *In* ''The Thymus Gland'' M. D. Kendall, ed., p. 113. Academic Press, London.

Debruyn, P. P. H. (1948). *In* ''Histopathology of Irradiation From External and Internal Sources'' (W. Bloom, ed.), p. 348. McGraw-Hill, New York.

Debruyn, P. P. H., Tornova-Svehlik, M. M., and Venter, J. H. (1965). *Radiat. Res.* **24,** 15.

DeSousa, M. A. E. (1975). *Contemp. Top. Immunobiol.* **2,** 119.

Dettman, P. M., Demis, A. J., and Storaasli, J. P. (1972). *Front. Radiat. Ther. Oncol.* **6,** 1428.

DeVita, V. T., Serpick, A., and Carbone, P. P. (1970). *Ann. Intern. Med.* **73,** 881.

DeVries, M. J., and Vos, O. (1958). *J. Natl. Cancer Inst.* **21,** 1117.

Dewey, W. C., Humphrey, R. M., and Cork, A. (1963). *Int. J. Radiat. Biol.* **6,** 463.

Doria, G., Agarossi, G., and Adorini, C. (1982). *Immunol. Rev.* **65,** 23.

Dougherty, I. F., Berliner, M. L., Schneefie, G. L., and Berliner, D. L. (1964). *Ann. N.Y. Acad. Sci.* **113,** 825.

Drewinko, B., and Humphrey, R. M. (1971). *Int. J. Radiat. Biol.* **20,** 169.

Drewinko, B., and Humphrey, R. M. (1973). *Int. J. Radiat. Biol.* **23,** 1.

Dubois, J. B., Serrou, B., and Rosenfeld, C., eds. (1981). ''Immunopharmacologic Effects of Radiation Therapy.'' Raven, New York.

Dukor, P., Miller, J. F. A. P., House, W., and Allman, V. (1965). *Transplantation* **3,** 639.

Dunlap, C. E., and Warren, S. (1942). *Arch. Pathol.* **34,** 562.

Edelson, R. L. (1980). *J. Am. Acad. Dermatol.* **2,** 89.

Edelson, R. L., Kirkpatrick, E. H., Shevath, E. M., Schein, P. S., Smith, R. W., Green, J., and Lutzner, M. (1974). *Ann. Intern. Med.* **80,** 685.

Edelson, R., Schein, P., Green, I., Kirkpatrick, C., and Ahmed, A. (1975). *Ann. Intern. Med.* **83,** 534.

Edwards, S. E., Miller, R. S., and Phillips, R. A. (1970). *J. Immunol.* **105,** 719.

Eltringham, J. R. and Kaplan, H. S. (1973). *Natl. Cancer Inst. Monogr.* **36,** 107.
Eltringham, J. R., and Kaplan, H. S. (1975). *In* "Immunologic Deficiencies in Man and Animals" (D. Bergsma and R. A., Good, eds.), p. 278. J. Finstad, Natl. Found. March of Dimes.
Eltringham, J. R., and Weissman, I. (1970). *Radiology* **94,** 438.
Eltringham, J. R., and Weissman, I. L. (1971). *J. Immunol.* **106,** 1185.
Elves, M. W., Roath, S., and Israels, M. C. G. (1963). *Lancet* **1,** 806.
Engeset, A. (1964). *Acta Radiol. Suppl.* **229,** 1.
Evans, R. G., and Norman, A. (1968). *Nature (London)* **217,** 455.
Everett, N. B., Coffrey, R. W., and Ricke, W. O. (1964). *Ann. N.Y. Acad. Sci.* **113,** 887.
Ezine, S., Weissman, I. L., and Rouse, R. V. (1984). *Nature (London)* **309,** 629.
Fabrikant, J. (1969). *Radiology* **93,** 887.
Fabrikant, J. (1970). *Am. J. Roentgenol.* **108,** 729.
Fabrikant, J. (1975). *ERDA Conf. Ser.* 731005, p. 504.
Fefer, A., Cheeves, M. A., Greenberg, P. D., Appelbaum, F. R., Boyd, C. D., Buckner, C. D., Kaplan, H. G., Ramberg, R., Sanders, J. E., Storb, R., and Thomas, E. D. (1982). *New Engl. J. Med.* **306,** 63.
Feola, J. M. (1977). *Rad. Res.* **70,** 118.
Feola, J., and Maruyama, Y. (1979). *Cancer Treat. Rep.* **63,** 1409.
Feola, J. M., Lawrence, J. H., and Welch, G. P. (1969). Rad. Res. **40,** 400.
Feola, J., Nava, C. A., and Maruyama, Y. (1982). *Int. J. Radiat. Biol.* **41,** 33.
Feola, J., Maruyama, Y., and Todd, P. W. (1983). *Proc. Int. Radiat. Res. Congr, 7th* p. 402.
Feola, J. M., Hwang, H. N., Beach, J. L., and Maruyama, Y. (1985). *J. Radiat. Res.* **26,** 140.
Fibach, E., Gerassi, E., and Sachs, L. (1976). *Nature (London)* **259,** 127.
Filipovich, A. H., Zerbe, D., Spector, B. D., and Kersey, J. H. (1984). *In* "Pathogenesis of Leukemias and Lymphomas: Environmental Influences" (Magrath, I. T. *et al.,* eds.), p. 225., Raven, New York.
Fliedner, T. M., Kesse, M., Cronkite, E. P., and Robertson, J. S. (1964). *Ann. N.Y. Acad. Sci.* **113,** 578.
Ford, C. E., and Micklem, H. S. (1963). *Lancet* **1,** 359.
Ford, C. E., Lamerton, J. L., Barnes, D. W. H., and Loutit, J. F. (1956). *Nature (London)* **177,** 452.
Ford, C. E., Micklem, H. S., Evans, E. P., Gray, J. G., and Ogden, D. A. (1966). *Ann. N.Y. Acad. Sci.* **129,** 283.
Ford, W. L. (1975). *Prog. Allergy* **19,** 1.
Fox, M. (1979). *Int. J. Radiat. Biol.* **36,** 335.
Fox, M., and Fox, B. W. (1973). *Int. J. Radiat. Biol.* **23,** 333.
Fox, M., and Gilbert, C. W. (1966). *Int. J. Radiat. Biol.* **11,** 339.
Fox, M., and Lajtha, L. G. (1967). *Int. J. Radiat. Biol.* **12,** 251.
Frindel, E., Hahn, G. M., Robaglin, D., and Tubiana, M. (1972). *Cancer Res.* **32,** 2096.
Fu, K. K., Phillips, T. L., Kane, L. J., and Smith, V. (1975). *Radiology* **114,** 709.
Fuks, Z., Strober, S., Bobrove, A. M., Sasazuki, T., McMichael, A., and Kaplan, H. S. (1976). *J. Clin. Invest.* **58,** 803.
Furth, J. J. (1946). *J. Gerontol.* **1,** 46.
Furth, J., and Kahn, M. C. (1937). *Am. J. Cancer* **31,** 276.
Gallatin, W. M., Weissman, I. L., and Butcher, E. C. (1983). *Nature (London)* **304,** 30.
Gallo, R. C., and Gelman, E. P. (1981). *New Engl. J. Med.* **304,** 169.
Gatti, R. H., and Good, R. A. (1971). *Cancer* **28,** 89.
George, P., and Pinkel, D. (1965). *Proc. Am. Assoc. Cancer Res.* **6,** 22.
Geraci, J. P., Thower, P. D., Jackson, K. L., Christensen, G. M., and Fox, M. S. (1974). *Int. J. Radiat. Biol.* **25,** 403.

Geraci, J. P., Christensen, G. M., and Jackson, K. L. (1975). *Radiat. Res.* **61,** 158.

Gerber, M. (1984). *Radiat. Res.* **100,** 365.

Gilbert, R. (1939). *Am. J. Roentgenol. Radium Ther.* **41,** 198.

Glasgow, G. P., Beetham, K. L., and Mill, W. B. (1983). *Int. J. Radiat. Oncol. Biol. Phys.* **9,** 557.

Glick, B. (1964). *In* "The Thymus in Immunobiology" (R. A. Good and A. E. Gabrielsen, eds.), p. 343. Hoeber, New York.

Glick, B., Chang, T. S., and Jaap, R. G. (1956). *Poultry Sci.* **35,** 224.

Globerson, A. (1966). *J. Exp. Med.* **123,** 25.

Globerson, A., and Feldman, M. (1964). *Transplantation* **2,** 212.

Goffinet, D. R., Glatstein, E., and Kaplan, H. S. (1973). *Natl. Cancer Inst. Monogr.* **36,** 463.

Goldstein, A. L., Bananjei, S., Schneebel, G. L., Dougherty, T. F., and White, A. (1970). *Radiat. Res.* **41,** 579.

Good, R. A., Kelly, W. D., Rotstein, J., and Varco, R. C. (1962). *Prog. Allergy* **6,** 187.

Gotoff, S. P., Amirmokri, E., and Liebner, E. J. (1967). *Am. J. Dis. Child.* **114,** 617.

Gottlieb, M., Strober, S., Hoppe, R. T., Grumet, C., and Kaplan, H. S. (1980). *Transplantation* **29,** 487.

Gowans, J. L., and Knight, E. J. (1964). *Proc. R. Soc. Ser. B* **159,** 257.

Gowans, J. L., and McGregor, D. G. (1965). *Prog. Allergy* **9,** 1.

Gowans, J. L., McGregor, D. D., Cowen, D. M., and Ford, C. E. (1962). *Nature (London)* **196,** 651.

Grahn, D., Ainsworth, E. J., Williamson, F. S., and Fry. R. J. M. (1972). *In* "Radiobiological Applications of Neutron Irradiation," p. 211. IAEA, Vienna.

Greenberger, J. S., Weichelbaum, R. R., Botnick, L. E., Sakakenny, M., and Maloney, W. C. (1979). *Exp. Hematol.* **7,** 279.

Greene, M. H. (1984). *In* "Radiation Carcinogenesis: Epidemiology and Biol. Significance" (J. D. Boice and J. E. Fraumeni, eds.), p. 199. Raven, New York.

Gutenson, N., and Cole, P. (1968). *New Engl. J. Med.* **304,** 135.

Hale, M. L., and McCarty, K. F., (1984). *Radiat Res.* **99,** 151, (1984).

Hall, J. G., and Morris, B. (1964). *Lancet* **1,** 1077.

Hamilton, L. D. (1959). *In* "The Kinetics of Cellular Proliferation" (F. Stohlman, ed.), p. 151.

Hanks, G. E. (1964). *Nature (London)* **203,** 1393.

Hanks, G. E., and Kaplan, H. S. (1965). *Radiat Res.* **26,** 84.

Hanna, M. G. (1964). *Lab. Invest.* **13,** 95.

Haran-Ghera, N. (1965). *Radiat. Res.* **26,** 442.

Harcourt, S. A., Lehmann, A. R., Stevens, S., and Bridges, B. A. (1975). *Nature* **258,** 427.

Harris, J. E., and Ford, C. E. (1964). *Nature (London)* **201,** 884.

Hedges, M. J., and Hornsey, S. (1978). *Int. J. Radiat. Biol.* **33,** 291.

Heineke, H. (1904). *Munch. Med. Woehenschr.* **51,** 785; (1905). *Dtsch. Z. Chir.* **78,** 196.

Hempelmann, L. H., Hall, W. J., Phillips, M., Cooper, R. A., and Ames, W. R. (1975). *J. Natl. Cancer Inst.* **55,** 519.

Hendry, J. H. (1972). *Br. J. Radiol.* **45,** 923.

Hersh, E. M., and Oppenheim, J. J. (1965). *New Engl. J. Med.* **273,** 1006.

Herzenberg, L. A., and Herzenberg, L. A. (1978). *Handb. Exp. Immunol.* **3,** 22.

Heublein, A. C. (1932). *Radiology* **18,** 1051.

Hewitt, H. B., and Wilson, C. W. (1959). *Br. J. Cancer* **13,** 69.

Hirschhorn, K., Bach, F., Kolodny, R. L., Firschein, I. L., and Hashem, N. (1963). *Science* **142,** 1185.

Hollcroft, J., Lorentz, E., Congdon, C. C., and Jacobson, L. O. (1953). *Rev. Mex. Radiat.* **7,** 115.

Homo-Delarche, R. (1984). *Cancer Res.* **44,** 431.

Hoppe, R. T., Fuks, Z., and Bagshaw, M. A. (1977). *Int. J. Radiat. Oncol. Biol. Phys.* **2,** 843.

Hoppe, R. T., Cox, R. S., Fuks, Z., Price, N. M., Bagshaw, M. A., and Farber, E. M. (1979). *Cancer Treat. Rep.* **63**, 691.
Hoppe, R. T., Cox, R. S., Rosenberg, S. A., and Kaplan, H. S. (1982). *Cancer Treat. Rep.* **66**, 743.
Howard, A., and Pelc, S. R. (1953). *Hereditas Suppl.* **6**, 261.
Huiskamp, R., Davids, J. A. G., and Vos, O. (1983). *Radiat. Res.* **95**, 370.
Hume, D., and Weldemann, M. J. (1980). "Mitogenic Lymphocyte Transformation." Elsevier, Amsterdam.
Ihle, J. N. (1978). *Semin. Hematol.* **15**, 95.
Inada, T., Kasuga, T., Noriji, I., Furuse, T., and Hiraoka, T. (1976). *Gann* **67**, 451.
Ishimaru, T., Ichimaru, M., Mikami, M., Yamada, Y., and Tomonasa, Y. (1981). *Radiat. Effects Res. Found. Tech. Rep.* 12–81.
Jackson, K. L., Christensen, G. M., and Forkey, D. J. (1968). *Radiat. Res.* **34**, 366.
Jackson, K. L., Christensen, G. M., and Bustline, R. N. (1969). *Radiat. Res.* **38**, 560.
Jacobson, L. O., Marks, E. K., Robson, M. J., Gaston, E. O., and Zirkle, R. F. (1949). *J. Lab. Clin. Med.* **34**, 1538.
Jacobson, L. O., Simmons, E. L., and Marks, E. K. (1950). *J. Lab. Clin. Med.* **35**, 746.
Jacobson, L. O., Marks, E. K., and Gaston, E. O. (1954). In *Radiobiol. Symp. Proc.* Butterworth, London.
Jacobson, L. O., Goldwasser, E., and Gurney, C. W. (1960). *CIBA Found. Symp. Haemopoiesis* p. 423.
Jaffe, E. (1983). *J. Natl. Cancer Inst.* **70**, 401.
Johanson, K. J., Wlodek, D., and Szumiel, I. (1982). *Int. J. Radiat. Biol.* **41**, 261.
Johansson, L., Carlsson, J., and Nilsson, K. (1982). *Int. J. Radiat. Biol.* **41**, 411.
Johnson, R. E. (1964). *J. Natl. Cancer Inst.* **32**, 1333.
Johnson, R. E. (1975). *Br. J. Cancer* **31** (Suppl. 2), 450.
Johnson, R. E., Thomas, L. B., and Hardt, J. (1969). *Proc. Am. Assoc. Cancer Res.* **10**, 43.
Johnson, R. E., Thomas, L. B., Schneiderman, M., Glenn, D. W., Faw, F., and Hafermann, M. (1970). *Radiology* **96**, 603.
Johnson, R. E., Ruhl, R., Johnsson, S. K., and Glover, M. (1976). *Cancer* **37**, 1713.
Johnson, F. L., Thomas, E. D., Clark, B. S., Chard, R. L., Hartmann, J. R., and Storb, R. (1981). *N. Engl. J. Med.* **305**, 846.
Juraskova, V. (1975). *Folia Biol.* **21**, 340.
Juraskova, V. (1978). *Folia Biol.* **24**, 253.
Juraskova, V., and Drasil, V. (1984). *Br. J. Cancer Suppl.* **49**, 217.
Kadish, J. C., and Basch, R. S. (1975). *J. Immunol.* **114**, 452.
Kallman, R. F., and Kohn, H. I. (1955). *Radiat. Res.* **2**, 280.
Kallman, R. F., and Kohn, H. I. (1956). *Radiat. Res.* **3**, 77.
Kann, H. E., Blumenstein, B. A., Petkas, A., and Schott, M. A. (1980). *Cancer Res.* **40**, 771.
Kaplan, H. S. (1980). "Hodgkin's Disease." Harvard Univ. Press, Cambridge, Massachusetts.
Kaplan, H. S., and Brown, M. B. (1952). *J. Natl. Cancer Inst.* **13**, 185.
Kaplan, H. S., and Smithers, D. W. (1959). *Lancet* **2**, 1.
Kaplan, H. S., Brown, M. B., and Paul, J. (1953). *J. Natl. Cancer Inst.* **14**, 303.
Karlsberg, I., Edstrom, S., Ekmann, L., Johansson, S., Scherstein, T., and Lundholm, K. (1981). *Cancer Res.* **41**, 4154.
Kataoka, T. Oh-Hashi, F., Tsukagoshi, S., and Sakurai,Y. (1977). *Cancer Res.* **37**, 964.
Kataoka, Y., and Sato, T. (1975). *Immunology* **29**, 121.
Kelly, L. S. (1961). *Brookhaven Symp. Biol.* **14**, 32, (1961).
Kennedy, J. C., Till, J. E., Siminovitch, L., and McCulloch, E. A. (1965). *J. Immunol.* **94**, 715.
Kimler, B. F., Park, C. H., Yakar, D., and Mies, R. M. (1984). *Br. J. Cancer* **49**, 221.
King, D. P., Strober, S., and Kaplan, H. S. (1981). *J. Immunol.* **127**, 1085.

Knapp, W. (1981). "Leukemia Markers." Academic Press, New York.
Knox, S. J., Shifrine, M., Wilson, F. D., and Rosenblatt, L. S. (1981). *Exp. Hematol.* **4,** 926.
Knox, S. J., Shifrine, M., and Rosenblatt, L. S. (1982). *Radiat. Res.* **89,** 575.
Kohler, G., and Milstein, C. (1975). *Nature (London)* **256,** 495.
Konings, A. W. T. (1981). *J. Radiat. Res.* **22,** 282.
Kraal, G., Geldof, A. A., and Boden, D. (1980). *Cell. Immunol.* **49,** 110.
Krebs, J. S., and Jones, D. C. L. (1972). *Radiat Res.* **51,** 374.
Kwan, D. K., and Norman, A. (1977). *Radiat. Res.* **69,** 143.
Lacassagne, A., and Gricouroff, G. (1958). "An Introduction to Radiotherapy." Grune & Stratton, New York.
Lajtha, L. G. (1965). *Curr. Top. Radiat. Res.* **1,** 141.
Lajtha, L. G., Gilbert, C. W., and Guzman, G. (1971). *Br. J. Haematol.* **20,** 343.
Lawrence, J. H., and Tennant, R. (1937). *J. Exp. Med.* **66,** 667.
Leone, C. A. (1962). "Effects of Ionizing Radiation on Immune Processes." Gordon & Breach, New York.
Lett, J. T., Caldwell, I., Dean, C. J., and Alexander, P. (1967). Nature (*London*) **214,** 790.
Little, J. R., Brecher, G., Bradley, T. R., and Rose, S. (1962). *Blood* **19,** 236.
Lorenz, E., Uphoff, D., Reed, T. R., and Shelton, E. (1951). *J. Natl. Cancer Inst.* **12,** 197.
Ly, I., and Mishell, R. I. (1974). *J. Immunol. Methods* **5,** 239.
McCaffrey, R., Harrison, T. A., Parkmoy, R., and Baltimore, L., (1975). *New Engl. J. Med.* **292,** 775.
McCulloch, E. A. (1970). *In* "Regulation of Hematopoiesis" (A. S. Gordon, ed.) Vol. 1, p. 133. Appleton, New York.
McCulloch, E. A., and Till, J. E. (1964). *Radiat. Res.* **22,** 383.
McLaren, J. R., Olkowski, Z. L., Skeen, M. J., McConnell, F. M. S., Beniyo, B., Mansory, K., Nixon, D. N., Bells, R., and Shah, N. K. (1981). *In* "Immunopharm. Effects Rad. Ther." (Dudois *et al.,* eds.), p. 253. Raven, New York.
McNally, N. J. (1972). *Int. J. Radiat. Biol.* **22,** 407.
MacKinney, A. A., Jr., Stohlman, F., and Brecher, G. (1962). *Blood* **19,** 349.
Madhvanath, Y., Raju, M., and Kelly, L. S. (1976). *ERDA Symp.* **37,** 74093, p. 125.
Malaise, E. P., Chavaudra, C., Charbit, A., and Tubiana, M. (1974). *Eur. J. Cancer* **10,** 451.
Mariani, T., Maruyama, Y., and Good, R. A. (1972). *Natl. Cancer Inst. Monogr.* **35,** 309.
Martinez, C., Kersey, J., Papermaster, B. W., and Good, R. A. (1962). *Proc. Soc. Exp. Biol. Med.* **109,** 193.
Maruyama, Y. (1963). *Nature (London)* **198,** 1181.
Maruyama, Y. (1964). *Nature (London)* **201,** 93.
Maruyama, Y. (1967). *Int. J. Radiat. Biol.* **12,** 277.
Maruyama, Y. (1968a). *Int. J. Cancer* **3,** 593.
Maruyama, Y. (1968b). *Int. J. Cancer* **3,** 788.
Maruyama, Y. (1968c). *Radiology* **91,** 657.
Maruyama, Y. (1969). *Radiology* **92,** 630.
Maruyama, Y. (1971). *J. Natl. Cancer Inst.* **46,** 963.
Maruyama, Y. (1973). *Adv. Radiat. Res. Biol. Med.* **3,** 1233.
Maruyama, Y. (1976). *Int. J. Radiat. Oncol. Biol. Phys.* **1,** 1159.
Maruyama, Y. (1978). *Cancer Biochem. Biophys.* **2,** 173.
Maruyama, Y. (1979). *Int. J. Radiat. Oncol. Biol. Phys.* **5,** 215.
Maruyama, Y., and Barclay, T. J. (1967). *Am. J. Roentgenol.* **100,** 944.
Maruyama, Y., and Brown, B. W. (1964). *Int. J. Radiat. Biol.* **8,** 59.
Maruyama, Y., and Ceman, C. (1970). *Radiat. Res.* **41,** 552.
Maruyama, Y., and Coleman, M. S. (1978). *Cancer Res.* **38,** 1617.

Maruyama, Y., and Deland, M. M. (1977). *Cancer Biochem. Biophys.* **2**, 19.
Maruyama, Y., and Eichten, J. G. (1968). *Am. J. Roentgenol. Radium Ther. Nucl. Med.* **102**, 46.
Maruyama, Y., and Eichten, J. G. (1969). *J. Minn. Acad. Sci.* **35**, 104.
Maruyama, Y., and Feola, J. M. (1973). *Oncology* **28**, 52.
Maruyama, Y., and Feola, J. (1976). *Rev. Inter. Am. Radiol.* **1**, 1.
Maruyama, Y., and Feola, J. (1980). *Leuk. Res.* **4**, 427.
Maruyama, Y., and Grider, C. (1981). *Stem Cells* **1**, 81.
Maruyama, Y., and Johnson, E. A. (1969). *Cancer* **23**, 309.
Maruyama, Y., and Knuth, P. (1972). *Growth* **30**, 453.
Maruyama, Y., and Lee, T. C. (1972). *Cancer* **30**, 84.
Maruyama, Y., and Lee, T. C. (1980). *Cancer Biochem. Biophys.* **4**, 257.
Maruyama, Y., and Raju, M. R. (1974). *J. Natl. Cancer Inst.* **53**, 285.
Maruyama, Y., and Thomson, J. (1970). *Cancer* **26**, 110.
Maruyama, Y., Briese, F. W., and Brown, Jr., B. W. (1967). *Radiat. Res.* **30**, 96.
Maruyama, Y., Ackerman, E., and Khan, F. M. (1969). *Int. J. Cancer* **4**, 793.
Maruyama, Y., Ceman, C., and McHugh, R. B. (1970). *Cancer Res.* **30**, 2245.
Maruyama, Y., Engles, E. P., Mariani, T., and Good, R. A. (1973). *Cancer* **31**, 1106.
Maruyama, Y., Lee, T. C., and Sullivan, S. (1973). *Radiology* **109**, 213.
Maruyama, Y., Gedgaudas, R. K., Lee, T. C., and Yung, W. K. (1974). *Cancer Res.* **34**, 2580.
Maruyama, Y., Lee, T. C., and McMillin, R. D. (1975). *ERDA Conf.* 731205, p. 239.
Maruyama, Y., Krolikiewicz, H., DeLand, F. H., Beihn, R., and Hafner, T. (1977). *In* "New
    Methods in Tumor Localization" (Y. Maruyama, ed.), p. 138. Univ. of Ky., Lexington.
Maruyama, Y., Grider, C., and Feola, J. (1979a). *Int. J. Radiat. Oncol. Biol. Phys.* **5**, 1691.
Maruyama, Y., Magura, C., and Feola, J. (1979b). *Acta Radiol. Oncol.* **18**, 136.
Maruyama, Y., Grider, C., and Williams, A. C. (1981). *Int. J. Radiat. Biol.* **40**, 475.
Maruyama, Y., Coleman, M. S., and Feola, J. (1982b). *Leuk. Res.* **6**, 845.
Maruyama, Y., Williams, A., Feola, J., and Nava, C. (1982b). *J. Cancer Res. Clin. Oncol.* **103**,
    107.
Maruyama, Y., Feola, J. M., Hwang, H. N., Williams, A., and Beach, J. L. (1983a), Radiat.
    Environ. Biophys. **21**, 265.
Maruyama, Y., Feola, J., and Muir, W. (1983b). *Gann* **74**, 426.
Maruyama, Y., Nava, C., Feola, J., Beach, J. L., Hwang, H. N., and Williams, A. (1983c). *Int. J.
    Radiat. Oncol. Biol. Phys.* **9**, 1049.
Maruyama, Y., Feola, J. M., and Magura, C. (1984a). *Int. J. Radiat. Biol.* **46**, 779.
Maruyama, Y., Williams, A. C., and Feola, J. M. (1984b). *Leuk. Res.* **8**, 629.
Maruyama, Y., Feola, J. M., Magura, C., and Beach, J. L. (1986). *J. Rad. Res.* **27**, 112.
Maruyama, Y., Feola, J. M., and Kryscio, R. (1985). *Jpn. J. Cancer Res.*, **26**, 1236.
Mekori, T., Chieco-Bianchi, L., and Feldman, M. (1965). *Nature (London)* **206**, 367.
Mendenhall, C. M., Thar, T. L., and Million, R. R. (1980). *Int. J. Radiat. Oncol. Biol. Phys.* **6**,
    1497.
Merwin, R. M., and Algire, G. H. (1959). *Proc. Soc. Exp. Biol. Med.* **101**, 437.
Metcalf, D. (1960). *Br. J. Haematol.* **6**, 324.
Metcalf, D. (1966). *Recent Results Cancer Res.* **5**, 1.
Metcalf, D. (1970). *In* "Regulation of Hematopoiesis" (A. S. Gordon, ed), p. 187. Appleton, New
    York.
Metcalf, D. (1977). *Recent Results Cancer Res.* **6**, 1.
Metcalf, D., and Wiadrowski, M. (1966). *Cancer Res.* **26**, 483.
Meyer, K. K. (1970). *Arch. Surg.* **101**, 114.
Micklem, H. S., Ford, C. E., Evans, E. P., and Gray, J. (1966). *Proc. R. Soc. London Ser. B* **165**,
    78.

Miller, J. J., and Cole, L. J. (1967). *J. Immunol.* **98,** 982.

Miller, J. F. A. P. (1961). *Lancet* **2,** 748.

Miller, J. F. A. P., and Mitchell, G. F. (1968a). *J. Exp. Med.* **128,** 801.

Miller, J. F. A. P., and Mitchell, G. F. (1968b). *Proc. Natl. Acad. Sci. U.S.A.,* **59,** 296.

Miller, J. F. A. P., Marshall, A. H. E., and White, R. G. (1962). *Adv. Immunol.* **2,** 111.

Miller, J. F. A. P., Doak, S. M. A., and Cross, A. M. (1963). *Proc. Soc. Exp. Biol. Med.* **112,** 785.

Miller, J. F. A. P., Leuchars, E., Cross, A. M., and Dukor, P. (1964). *Ann. N.Y. Acad. Sci.* **120,** 205.

Miller, R. G. (1976). *In* "In Vitro Methods in Cell-Mediated and Tumor Immunity" (B. R. Bloom and J. R. David, eds.), Vol. 2, p. 283. Academic Press, New York.

Miller, R. G., and Phillips, R. A. (1969). *J. Cell. Physiol.* **73,** 191.

Miller, R. G., Gorezynski, R. M., Lafleur, L., MacDonald, H. R., and Phillips, R. A. (1975). *Transplant. Rev.* **25,** 59.

Mitchell, G. F. (1974). *Top. Immunobiol.* **3,** 97.

Mitchell, G. F., and Miller, J. F. A. P. (1968). *J. Exp. Med.* **128,** 821.

Montour, J. L., Wilson, J. D., Rogers, C. C., Theus, R. B., and Atttix, F. H. (1974). *Cancer* **34,** 54.

Moore, E. W., Thomas, L. B., Shaw, R. K., and Freireich, E. J. (1960). *A.M.A. Arch. Intern. Med.* **105,** 451.

Moorehead, P. S., Nowell, P. C., Mellman, W. J., Battips, D. M., and Hungerford, D. A. (1960). *Exp. Cell Res.* **20,** 613.

Mosier, D. E. (1967). *Science* **158,** 1573.

Mosier, D. E., and Coopleton, L. W. (1970). *Proc. Natl. Acad. Sci. U.S.A.* **61,** 542.

Murray, R. G. (1958). *In* "Histopathology of Irradiation" (W. Bloom, ed.) pp. 243, 348. McGraw-Hill, New York.

Nadler, L. M., Ritz, J., Griffin, J. D., Todd, R. T., Reinherz, E. L., and Schlossman, S. F. (1981). *Prog. Hematol.* **12,** 187.

Nesbit, M. E., Sather, H. N., Robison, L. L., Ortega, J., Littman, P. S., D'Angio, G. J., and Hammond, G. D. (1981). *Lancet* **1,** 461.

Nias, A. H. W., and Fox, M. (1968). *Br. J. Radiol.* **41,** 468.

Non-Hodgkin's Lymphoma Pathological Classification Project (1982). *Cancer* **49,** 2112.

Norin, T., and Onyango, J. (1977). *Int. J. Radiat. Oncol. Biol. Phys.* **2,** 399.

Norin, T., Clifford, P., Einhorn, J., Einhorn, N., Johansson, B., Klein, G., Onyango, J., de-Schryver, A., and Nalstam, R. (1971). *Acta Radiol. Ther.* **10,** 545.

Nossal, G. J., and Ada, A. (1971). "Antigens, Lymphoid Cells and the Immune Responses." Academic Press, New York.

Nowell, P. C. (1960). *Cancer Res.* **20,** 462.

Olson, G. B., Anderson, R. E., and Bartels, P. H. (1979). *Hum. Pathol.* **10,** 179.

Order, S. E., and Hellman, S. (1972). *Lancet* **1,** 571.

Osgood, E. E., Aebersold, P. C., Erf, L. A., and Packham, E. A. (1942). *Am. J. Med. Sci.* **204,** 372.

Osoba, D. (1970). *J. Exp. Med.* **132,** 368.

Oughterson, A. W., and Warren, S. (1956). "Medical Effects of the Atomic Bomb in Japan." McGraw-Hill, New York.

Parrott, D. M. V., and deSousa, M. A. B. (1966). *Adv. Med. Biol.* **5,** 293.

Paterson, M. C., and Smith, P. J. (1979). *Annu. Rev. Genet.* **13,** 291.

Paterson, M. C., Bech-Hanson, N. T., Smith, P. J., and Mulvihill, J. J. (1984). *In* "Radiation Carcinogenesis" (J. D. Boice and J. F. Fraumeni, eds.), p. 319. Raven, New York.

Patt, H. M., Tyree, E. B., Straube, R. L., and Smith, D. E. (1949). *Science* **110,** 213.

Patt, H. M., Blackford, M. E., and Straube, R. L. (1952). *Proc. Soc. Expl. Biol. Med.* **80,** 92.

Pazdernik, T. L., and Nishimura, T. (1978). *Agents Actions* **8**, 229.

Pearmain, G., Lycette, R. R., and Fitzgerald, P. H. (1963). *Lancet* **1**, 637.

Pernis, B., Forni, L., and Amante, L. (1970). *J. Exp. Med.* **132**, 1001.

Peters, M. V. (1950). *Am. J. Roentgenol. Rad. Ther.* **63**, 299.

Peters, M. V. (1966). *Cancer Res.* **26**, 1232.

Peters, M. V., and Middlemiss, K. C. H. (1958). *Am. J. Roentgenol. Rad. Ther.* **79**, 114.

Peters, M. V., Alison, R. E., and Bush, R. S. (1966). *Cancer* **19**, 308.

Peterson, E. A., and Evans, W. H. (1967). *Nature (London)* **214**, 824.

Pinkel, D. (1979). *Cancer* **43**, 1128.

Playfair, J. H. L., and Cole, L. J. (1965). *J. Cell. Comp. Physiol.* **65**, 7, (1965).

Porteous, D. D., and Munro, T. R. (1972). *Int. J. Cancer* **10**, 112.

Porteous, D. D., Porteous, K. M., and Hughes, M. J. (1979). *Br. J. Cancer* **39**, 603.

Porter, K. A., and Cooper, E. H. (1962). *Lancet* **2**, 317.

Potter, M. (1972). *Physiol. Rev.* **52**, 631.

Powers, W. E., and Tolmach, L. J. (1963). *Nature (London)* **197**, 710.

Puck, T. T. (1966). *Radiat. Res.* **27**, 272.

Puck, T. T., and Marcus, H. M. (1956). *J. Exp. Med.* **103**, 653.

Puro, E. A., and Clark, G. M. (1972). *Radiat. Res.* **52**, 115.

Quazi, R., Aisenberg, A. C., and Long, J. C. (1976). *Cancer* **37**, 1923.

Raben, M., Walach, N., Galih, U., and Schlesinger, M. (1976). *Cancer* **37**, 1417.

Raff, M. C. (1969). *Nature (London)* **224**, 378.

Raff, M. C. (1970). *Immunology* **19**, 637.

Ramsay, N. K. C., Kim, T., Nesbit, M. E., Krivit, W., Coccia, P. F., Levitt, S. H., Woods, W. G., and Kersey, J. H. (1980). *Blood* **55**, 344.

Rappaport, H. (1966). *Atlas Tumor Pathol.* **8** (Sect. III). AFIP, Washington, D.C.

Reddy, M. M., Cooly, K. O., and Hempelmann, L. H. (1976). *ERDA Symp.* **37**, 740930, p. 192.

Reif, A. E., and Allen, J. M. V. (1964). *J. Exp. Med.* **120**, 413.

Reif, A. E., and Allen, J. M. (1966). *Cancer Res.* **26**, 123.

Richards, F., Spurr, C., Pajak, T. F., Blake, D. D., and Raben, M. (1974). *Am. J. Med.* **57**, 862.

Rickinson, A. B. and Ilbery, P. L. T. (1971). *Cell Tissue Kinet.* **4**, 549.

Rixon, R. H. (1967). *Radiat. Res.* **32**, 42.

Rocha, B., Freitas, A. A., and Continho, A. A. (1983). *J. Immunol.* **131**, 2158.

Rosenberg, S. A., and Kaplan, H. S. (1982). "Malignant Lymphomas: Etiology, Immunology, Pathology and Treatment." Academic Press, New York.

Rozenszajn, L. A., Shoham, D., and Kalechman, I. (1975). *Immunology* **29**, 1041.

Sack, E. L., Goris, M. L., Glatstein, E., Gilbert, E., and Kaplan, H. S. (1978). *Cancer* **42**, 1057.

Sainte-Marie, G., and Leblond, C. P. (1958). *Proc. Soc. Exp. Biol. Med.* **97**, 263.

Sainte-Marie, G., and Leblond, C. P. (1964). *In* The Thymus in Immunobiology" (R. A. Good and A. Gabrielsen, eds.), p. 207. Harper New York.

Sainte-Marie, G., and Sin, Y. M. (1970). *In* "Regulation of Hematopoiesis" (A. S. Gordon, ed.), Vol. I, p. 1339. Appleton, New York.

Sasaki, M. S., and Norman, A. (1966). *Nature (London)* **210**, 913.

Sato, K., and Hieda, N. (1979). *Rad. Res.* **78**, 167.

Sato, C., and Sakka, M. (1969). *Radiat. Res.* **38**, 204.

Sato, C. Kojima, K., Orozawa, M., and Matsuzawa, T. (1972). *Int. J. Radiat. Biol.* **22**, 479.

Sato, C., Kojima, K., Matsuzawa, T., Sairenji, T., and Hinuma, Y. (1974). *J. Radiat Res.* **15**, 25.

Scaife, N., and Broher, (1967). *Int. J. Radiat. Biol.* **13**, 111.

Schabel, F. M., Skipper H. E., Trader, M. W., Laster, W. R., and Simpson-Herren, L. (1969). *Cancer Chemother. Rep.* **53**, 329.

Schier, W. W., Roth, A., Ostroff, G., and Schrift, M. H. (1956). *Am. J. Med.* **20**, 94.

Schooley, J. L., Bryant, B. J., and Kelly, L. S. (1964). *In* "The Thymus in Immunobiology" (R. A. Good and A. E. Gabrielsen, eds.), p. 208. Harper, New York.

Schrek, R. (1946). *Radiology* **46**, 395.

Schrek, R. (1947). *J. Cell. Comp. Physiol.* **30**, 203.

Schrek, R. (1961). *Ann. N.Y. Acad. Sci.* **95**, 839.

Schrek, R., and Stefani, S. (1964). *J. Natl. Cancer Inst.* **32**, 507.

Schlesinger, M., and Yron, I. (1969). *Science* **164**, 1412.

Schurman, D. J., Hirshman, H. P., and Strober, S. (1981). *Arthritis Rheum.* **24**, 38.

Schwartz, E. E. (1958). *Proc. Am. Assoc. Cancer Res.* **2**, 343.

Scollay, R. G., Butcher, E. C., and Weissman, I. L. (1980). *Eur. J. Immunol.* **10**, 210.

Scott, D., Fox, M., and Fox, B. W. (1974). *Mutat. Res.* **22**, 207.

Scott, K. G., and Lawrence, J. H. (1941). *Proc. Soc. Exp. Biol. Med.* **48**, 155.

Shank, B., Hopfan, S., Kim, J. H., Grossband, E., Kapoor, L., Reid, A., Chui, C., Mohani, C., Finigan, D., and O'Reilly, R. J. (1981). *Int. J. Radiat. Oncol. Biol. Phys.* **7**, 1109.

Sharp, J. G., and Crouse, D. A. (1981). *In* "Experimental Hematology Today," p. 63. Karger, Basel.

Sharp, J. G., and Thomas, D. B. (1975). *Radiat. Res.* **64**, 293.

Shen, J., Wilson, F., Shifine, M., and Gershwin, M. (1977). *J. Immunol.* **119**, 1299.

Shih, W. J., Domstad, P. A., Deland, F. H., and Maruyama, Y. (1985). *Clin. Nucl. Med.* **10**, 184.

Shortman, K. (1972). *Annu. Rev. Biophys. Bioeng.* **1**, 93.

Silini, G., and Maruyama, Y. (1965). *J. Natl. Cancer Inst.* **35**, 841.

Silini, G., Maruyama, Y., and Pozzi, L. V. (1965). *Radiol. Med.* **51**, 515.

Silini, G., Ardreozzi, U., and Pozzi, L. V. (1976). *Cell Tissue Kinet.* **9**, 341.

Siminovitch, L., Till, J. E., and McCulloch, E. A. (1965). *Radiat Res.* **24**, 840.

Simonsen, M., Engelbreth-Holm, J., Jensen, E., and Poulsen, H. (1960). *Acta Unio Int Cancrum* **16**, 1179.

Simpson, C. L., Hempelmann, L. H., and Fuller, L. M. (1955). *Radiology* **64**, 482.

Simonsen, M., Engelbreth-Holm, J., Jensen, E., and Poulsen, H. (1960). *Acta Unio Int. Cancrum* **16**, 1179.

Simpson, C. L., Hempelmann, L. H., and Fuller, L. M. (1955). *Radiology* **64**, 840.

Skipper, H. E., Schabel, F. M., and Wilcox, W. S. (1964). *Cancer Chemother. Rep.* **35**, 1.

Skipper, H. E., Schabel, F. M., Trader, M. W., and Laster, W. R. (1969). *Cancer Chemother. Rep.* **53**, 345.

Slavin, S., Strober, S., Fuks, Z., and Kaplan, H. S. (1977). *J. Exp. Med.* **146**, 34.

Smith, C., and Kieffer, D. A. (1957). *Proc. Soc. Exp. Biol. Med.* **94**, 601.

Smith, E. B., White, D. C., Hartsock, R. J., and Dixon, A. C. (1967). *Am. J. Pathol.* **50**, 159.

Smith, P. G. (1984). *Prog. Cancer Res. Ther.* **26**, 107.

Sokal, J. E., and Aungst, C. W. (1969). *Cancer* **24**, 128.

Sokal, J. E., and Primikirios, H. (1961). *Cancer* **14**, 597.

Spiegler, P., and Norman, A. (1969). *Radiat. Res.* **39**, 400.

Sredni, B., Kalechman, Y., Michlin, H., and Rozenszajn, L. A. (1976). *Nature (London)* **259**, 130.

Stefani, S., and Schrek, R. (1964). *Radiat. Res.* **123**, 126.

Sternsward, J. (1974). *Lancet* **2**, 1285.

Sternsward, J., Jondahl, M., Vanky, F., Wigzell, H., and Sealy, E. (1972). *Lancet* **1**, 1352.

Stewart, A. M. (1969). *USAEC Symp. Ser.* **7**, 690501, p. 681.

Stewart, C. C., and Perez, C. A. (1976). *Radiology* **118**, 201.

Storer, J. B., Harris, P. S., Furchner, J. E., and Langham, W. H. (1957). *Radiat. Res.* **6**, 188.

Stratton, J. A., Byfield, P. E., Byfield, J. E., Small, R. C., Benfield, J. E., and Pilch, R. (1975). *J. Clin. Invest.* **56**, 88.

Strober, S., Slavin, S., Gottlieb, M., Zan-Bar, J., King, D. P., Hoppe, R. T., Fuks, Z., Grumet, F. C., and Kaplan, H. S. (1979a). *Immunol. Rev.* **46,** 87.

Strober, S., Kotzin, B. L., Hoppe, R. T., Slavin, S., Gottlieb, M., Calin, A., Fuks, Z., and Kaplan, H. S. (1979b). *Int. J. Radiat. Oncol. Biol. Phys.* **7,** 1.

Strober, S., Kotzin, B. L., Hoppe, R. T., Slavin, S., Gottlieb, M., Calin, A., Fuks, Z., and Kaplan, H. S. (1981). *Int. J. Rad. Oncol. Biol. Phys.* **7,** 1.

Stutman, O. (1978). *Immunol. Rev.* **42,** 138.

Suit, H. D., Shalek, R. J., and Wette, R. (1965). *In* "Cellular Radiation Biology," p. 514. Wilkins, Baltimore.

Sullivan, M. P., Chen, T., Dynent, P. G., Huizdala, E., and Staubes, C. P. (1982). *Blood* **60,** 948.

Sullivan, M. P., Perez, C. A., Henson, J., Silva-Sousa, M., Land, V., Dyment, P. G., Chen, R., and Ayala, A. G. (1980). *Cancer* **46,** 508.

Sundaran, K., Bains, G. S., and Moorthy, P. S. (1967). *Int. J Radiat. Biol.* **13,** 335.

Sutherland, J., and Maruyama, Y. (1974). *Oncology* **30,** 85.

Szumiel, I., Budzicka, E., Neipokojczycka, E., Wlodek, D., and Beer, J. Z. (1981). *Radiat. Res.* **87,** 592.

Takada, A., and Takada, Y. (1972). *Transplantation* **13,** 276.

Takada, A., Takada, Y., Huang, C. C., and Ambrus, J. L. (1969). *J. Exp. Med.* **129,** 445.

Takada, A., Takada, Y., Kim, U., and Ambrus, J. L. (1971). *Radiat. Res.* **45,** 522.

Taliaferro, W. H., and Taliaferro, L. G. (1970). *J. Immunol.* **104,** 1364.

Taliaferro, W. H., Taliaferro, L. G., and Jaroslow, B. N. (1964). "Radiation and Immune Mechanisms." Academic Press, New York.

Talmadge, D. W. (1955). *Annu. Rev. Microbiol.* **9,** 335.

Tanaka, T., and Lajtha, L. G. (1970). *Eur. J. Clin. Biol. Res.* **15,** 2187.

Thomas, E. D., Storb, R., Clift, R. A., Fefer, A., Johnson, F. L., Neiman, P. E., Lerner, K., G., Glucksberg, H., and Buckner, C. D. (1975). *New Engl. J. Med.* **292,** 837.

Thomas, E. D., Buckner, D., Fefer, A., Neiman, P. E., and Strob, R. (1978). *Adv. Cancer Res.* **27,** 269.

Thomas, E. D., Clift, R. A., Hersman, J., Sanders, J. E., Stewart, P., Buckner, C. D., Fefer, A., McGuffin, R., Smith, J. W., and Storb, R. (1982). *Int. J. Radiat. Oncol. Biol. Phys.* **8,** 817.

Thomas, F., and Gould, M. N. (1982). *Radiat. Environ. Biophys.* **20,** 89.

Till, J. E., and McCulloch, E. A. (1961). *Radiat. Res.* **14,** 213.

Till, J. E., and McCulloch, E. A. (1963). *Radiat. Res.* **18,** 96.

Till, J. E., and McCulloch, E. A. (1964). *Ann. N.Y. Acad. Sci.* **114,** 115.

Trentham, D. E., Belli, J. A., Anderson, R. J., Buckley, J. A., Goetze, E. J., David, J. A., and Austen, K. F. (1981). *New Engl. J. Med.* **305,** 976.

Trentin, J. J. (1957). *Am. Assoc. Cancer Res.* **2,** 256.

Trentin, J. J. (1970). *In* "Regulation of Hematopoiesis" (A. S. Gordon, ed.), Vol. 1, p. 159. Appleton, New York.

Trowell, O. A. (1952). *J. Pathol. Bacteriol.* **64,** 687.

Trowell, O. A. (1961). *Ann. N.Y. Acad. Sci.* **95,** 849.

Vallera, D. A., Soderling, C. C. B., Carlson, G. J., and Kersey, J. H. (1981). *Transplantation* **31,** 218.

Van Bekkum, D. W., and DeVries, M. J. (1967). "Radiation Chimaeras," Logos Press, Academic Press, London.

Vassort, F., Winterholer, M., Frindel, E., and Tubiana, M. (1973). *Blood* **41,** 789.

Vos, O. (1967). *Int. J. Radiat. Biol.* **13,** 317.

Warner, N. L., Szenberg, A., and Burret, I. M. (1962). *Aust. J. Exp. Biol. Med.* **40,** 373.

Warren, S., MacMillan, J. C., and Dixon, F. J. (1950). *Radiology* **55,** 375.

Watkins, E. B.,and Sharp, J. G. (1979). *J. Reticuloendothel. Soc.* **26,** 209.

Watkins, E. B., Sharp, J. G., Crouse, D. A., and Metcalf, W. K. (1980). *Exp. Hematol. Today* p. 81.

Weichselbaum, R. R., Greenberger, J. S., Schmidt, A., Karpas, A., Maloney, W. C., and Little, J. B. (1981). *Radiology* **139,** 485.

Weisman, I. L., Peacock, M., and Eltringham, J. R. (1973). *J. Immunol.* **110,** 1300.

Wendling, F., Jullien, and Tambourin, P. (1973). *Radiat. Res.* **55,** 177.

Wharton, J. T., and Rutledge, F. N. (1980). *In* "Textbook of Radiotherapy" (G. H. Fletcher, ed.), p. 774. Lea & Febiger, Philadelphia.

Wilcox, W. S. (1969). *Natl. Cancer Inst. Monogr.* **24,** 257.

Wilcox, W. S., Griswold, D. P., Laster, W. R., Schabel, F. M., and Skipper, H. E. (1965). *Cancer Chemother. Rep.* **47,** 27.

Wilson, F. D., Whalley, C. B., Shifrine, M. *et al.* (1980). *Exp. Hematol.* **8,** 802.

Wilson, J. D., and Dalton, G. (1976). *Aust. J. Exp. Biol. Med. Sci.* **54,** 27.

Wiltshaw, E. (1976). *Medicine* **55,** 217.

Wodinsky, I., Swiniarski, J., and Kensler, C. J. (1967). *Cancer Chemother. Rep.* **51,** 415.

Wolf, N. S. (1982a). *Exp. Hematol.* **10,** 98.

Wolf, N. S. (1982b). *Exp. Hematol.* **10,** 108.

Wolf, N. S., and Rosse, (1982). *Am. J. Anat.* **163,** 131.

Wolf, N. S., and Trentin, J. J. (1968). *J. Exp. Med.* **1278,** 205.

Wu, A. M., Till, J. E., Siminovitch, L., and McCulloch, E. A. (1967). *J. Cell. Physiol.* **69,** 177.

Wu, A. M., Till, J. E., Siminovitch, L., and McCulloch, E. A. (1968). *J. Exp. Med.* **127,** 455.

Yau, T. M., Kim, S. C., Gregg, E. C., and Nygaard, O. F. (1979). *Int. J. Radiat. Biol.* **35,** 577.

Yoffey, J. M. (1970). *In* "Regulation of Hematopoiesis" (A. S. Gordon, ed.), Vol. 1, p. 1421. Appleton, New York.

Yoffey, J. M., Reinhardt, W. O., and Everett, N. B. (1961). *J. Anat.* **95,** 293.

Relative Radiosensitivities of the Small and Large Intestine

ALDO BECCIOLINI

LABORATORY OF RADIATION BIOLOGY
DEPARTMENT OF CLINICAL PHYSIOPATHOLOGY
UNIVERSITY OF FLORENCE
FLORENCE 50134, ITALY

I. Introduction

The alimentary tract includes the esophagus, stomach, small and large intestines, and rectum. The structures of these tissues are similar, and from the innermost to outermost layers each consists of (1) the mucosa, made up of epithelial cells resting on a lamina formed by connective tissue, (2) the submucosa, a thin layer of stromal tissue, and (3) the muscularis, smooth muscle which ends at the muscularis externa.

The functions of the mucosa require complex cellular organization along the length of the gastrointestinal tract. Nutrients that have been minced in the mouth and homogenized by saliva are mixed with other oral secretions before entering the alimentary canal. External secretions cause macromolecules to be hydrolyzed, and the products of those reactions are absorbed by the mucosa. Once digestion is completed, wastes are discarded by the alimentary tract.

The mucosa undergoes continuous stress due to passage of food, microtraumas, and the action of intestinal flora. In order to remain unimpaired in its functions, both digestive and protective, the mucosa must renew itself rapidly to replace cells exposed to the passage of food. Its fast turnover is supported by marked mitotic activity. Because of this high turnover rate, the alimentary tract is very radiosensitive. The relative radiosensitivities of different regions of the gut, from highest to lowest, may be listed as follows: duodenum, jejunum, ileum, esophagus, stomach, colon, and rectum (Rubin and Casarett, 1968a,b: Roswit *et al.*, 1972a; Casarett, 1980).

II. Stomach

Radiobiological studies of the human stomach have not been as extensive as have studies of the small intestine. The data collected so far, however (Regaud *et al.*, 1912; Hamilton, 1947; Ricketts *et al.*, 1948; Roswit *et al.*, 1972a,b), are very important and reveal both acute and delayed tissue injury, particularly late damage.

A. Histological and Morphological Aspects

The stomach wall consists of mucosa, submucosa, muscularis, and adventitia. There are three types of glands which are defined according to their locations in the stomach: the cardiac, fundic, and pyloric glands.

The surface epithelium produces mucigen which lubricates the surface of the mucosa. Under physiological conditions, the epithelial cells are desquamated continuously with a 3- to 4-day turnover time (Stevens and Leblond, 1953; Leblond and Walker, 1956; Potten and Hendry, 1983). Proliferative activity occurs in the isthmus between the gastric pits and the glands as well as in the neck regions of the pits. After an [^{3}H]thymidine pulse, the cells in this region are labeled, and later zymogen and parietal cells contain [^{3}H]TdR labels. The renewal of zymogen and parietal cells is much slower than that of the surface epithelium, and it seems to take months in animals and even longer in man (Cameron, 1971; Lipkin, 1973).

Gastric glands are the most important structures in the stomach as they secrete the majority of the gastric juices. They consist of four types of cells: (1) chief or zymogen-secreting cells, which secrete pepsinogen and intrinsic factor and are located in the lower half of the glandular tubules; (2) parietal cells, which apparently are associated with hydrochloric acid or production of its precursors; (3) mucous neck cells, which secrete mucus; and (4) argentaffin cells, which are found in the fundic glands.

Whether the various types of cells originate from one single stem cell or each type has its own maturation line is still to be determined. The lamina propria consists of connective tissue which fills in the spaces between the glands and the muscularis mucosa. Together with collagenous and reticular fibrils, it contains blood cells and fibroblasts. It also includes the vascular network.

B. Effects of Radiation: Histopathological and Clinical Aspects

Radiation effects in the stomach are similar to those observed in the small intestine, although the stomach has been considered less radiosensitive. The

$TD_{5/5}$ minimal tolerance dose (i.e., the dose that can cause 5% injury in 5 years) and $TD_{50/5}$ maximum dose (i.e., the dose that can cause 50% injury in 5 years) have been calculated in the alimentary canal by assessing the incidence of ulcers and strictures following irradiation. The $TD_{5/5}$ and $TD_{50/5}$ are, respectively, esophagus, 60 and 75 Gy; stomach, 45 and 50 Gy; small intestine and colon, 45 and 65 Gy; and rectum, 55 and 80 Gy (Rubin and Casarett, 1972).

Among the different structures in the stomach, cardiac and pyloric glands appear to be less radiosensitive than fundic glands. The cells of the surface luminal epithelium, parietal cells, and argentaffin cells as well as zymogen-secreting cells exhibit low radiosensitivity whereas the proliferative cells in the deeper regions of the gastric pits and necks of the glands are highly sensitive. A few hours after irradiation, mitotic activity may be lost (Bloom, 1948; Friedman, 1952; Rubin and Casarett, 1968a; Casarett, 1980), and in proliferative cells, marked degenerative features may be observed. In connective tissue, edema and vasodilation may be present.

One day after a single dose of 15 Gy, the cells of the gastric pit and surface epithelium show marked alterations and are reduced in number. After 2 days, the cells in the gland neck show partial recovery whereas a reduction of cells is apparent in the glandular epithelium. Occlusive processes are observed in the vessels. Edema and inflammatory responses persist in the stroma.

The arrested proliferative activity leads to a decrease in differentiated cells at different times, depending on their turnover rate. The onset of damage is revealed by glandular atrophy and a reduction in the thickness of the mucosa which may cause ulcers. If the dose of low linear energy transfer (LET) radiation is less than 10 Gy, proliferation resumes and the epithelium returns to control levels. If, on the other hand, the dose is high, i.e., 20 Gy (Casarett, 1980) in a single treatment, the arrest in proliferative activity lasts longer, and the number of dead or desquamated cells is so high that there may be ulceration 1–2 weeks after irradiation. This alteration may affect only the mucosa or may involve the underlying stroma and the muscular layer as well. In this case, there may be hemorrhage, generally in the necrotic areas, inflammation, and fibrosis. These ulcers may cicatrize over a period of several months or years. If ulcers do not occur within 2–3 weeks, the glands return to normal and the appearance of the surface epithelium, after a period of temporary hyperplasia, becomes similar to controls. Incomplete repair, mainly at the stromal level, may cause sclerosis of blood vessels accompanied by progressive glandular atrophy and fibrosis of connective tissue.

Friedman (1952) evaluated the incidence and severity of alterations after radiotherapy in relation to dose. Doses lower than 20 Gy cause histological and functional modifications which do not appear to be particularly severe. After doses of 25–34 Gy, 20% of the patients present with dyspepsia and ulcers. Doses of about 44 Gy cause radiation gastritis in ~20% of the subjects, and after higher doses patients suffer from radiation-induced ulcers and gastritis. In all cases the lesions are very severe for a few months to a year after irradiation.

C. Physiological and Biochemical Modifications

Biochemical alterations are induced by radiation. A decrease in HCl production (Rider et al., 1957), and sometimes its disappearance after 15–17 Gy of X rays given over 12 days, occurs before morphological modifications are evident; this deficiency will continue as long as the parietal cells are altered. Even after stimulation with histamine, the acidity is markedly lowered and remains so even when the mucosa returns to normal. Production of pepsin is reduced to a lesser extent and not as rapidly; its return to normal also is slower. The zymogen-secreting cells recover more slowly than do the parietal cells.

Ionizing radiation was used in the treatment of peptic ulcers for several years starting in 1939 (Palmer and Templeton, 1939). The low doses used were successful in reducing gastric acidity soon after radiotherapy was completed. In most patients, however, the effect diminished after about a year. The doses employed were 16 Gy for 10 days (Goldgraber et al., 1954) and 15 Gy over 21 days (Doig et al., 1951).

Biopsies taken at different times have indicated that there were no marked modifications in the structure of parietal and zymogen cells during the first days of irradiation. Morphological alterations appeared 5 days after irradiation and later involved the basal part of the glands in which decrease in the number of cells and marked structural disorganization were evident. In the surface epithelium, modifications were slight. The mucosa was thinner, but with these doses proliferative activity resumed and repopulation occurred. In the second to third weeks, there were alterations in the connective tissue with edema and inflammatory cell infiltration. Recovery was slow and damage to the mucosa was still evident 7 weeks after irradiation. After treatment for duodenal ulcers, no fibrosis was observed, but such a condition may appear after higher doses. In the latter case, progressive fibrosis may be observed months after irradiation and may develop into chronic atrophic gastritis and stenosis. Delayed effects can include impairment of gastric motility.

Nausea and vomiting occur in many mammals after whole body irradiation with doses of 1 Gy or more. Recent studies (Dubois et al., 1984) on dogs have demonstrated that within 2 hr after irradiation gastric passage of solid and liquid substances is blocked and a delayed gastric emptying time is evident. A similar result was observed in rats (Conard, 1953; Hulse, 1969). The mechanism underlying this alteration is not yet clear, but some antiemetics (Dubois et al., 1984) can prevent the symptoms.

III. Small Intestine

A. Experimental Studies

1. Histological and Morphological Aspects

Evaginating structures in the small intestine facilitate absorption. The system of *valvulae conniventes,* crypt-villus, and microvilli in villus cells increase the surface for absorption by 600 times. Because of its morphology, the epithelium of the small intestine is a valuable model for studies of ionizing radiation effects on tissues. The same type of cells prevails in a single layer. High mitotic activity and a rapid renewal rate for the epithelial cells allow studies of damage and recovery within a few days after irradiation. The proliferative compartment is well separated from the differentiated one. Differentiation is characterized by, among other phenomena, the synthesis of specific enzymes which will be present in the mature cell when it reaches the tip of the villus.

The cell population which lines the crypt-villus formation is constituted of subpopulations having different morphological and functional characteristics (Leblond and Cheng, 1976; Potten and Hendry, 1983). These cells seem to originate from a single type of stem cell located at the base of the crypt. When they divide, these cells produce new proliferative and differentiating cells. From the base of the crypt to the villus junction, the proliferative compartment is composed of the following cell types: (1) totipotent stem cells, usually in G_0, and differentiated cells such as Paneth and enteroendocrine cells; (2) partially differentiated cells which retain the ability to divide; and (3) cells which have lost the capacity for division and are differentiating.

At the tips of the villi the differentiated cells are shed continuously. Columnar cells, which constitute more than 90% of the epithelial cells, are characterized by the absence of secretion products and by a well-developed brush border. Of the cells in the jejunum, 6% are mucous cells and originate from the oligomucous cells of the crypts which remain capable of dividing (Leblond and Cheng, 1976; Cheng, 1974a). Mucous cells produce neutral or acid polysaccharides which line the glycocalix, thus providing a protective coat on the intestinal mucosa against the mechanical and chemical microtraumas caused by bolus, action of enzymes, and bacterial flora. Enteroendocrine and Paneth cells are very few in number (Trier, 1968; Cheng, 1974b; Cheng and Leblond, 1974b; Schultze and Kellerer, 1979).

2. Organization of the Proliferative Cellular Compartment

Totipotent stem cells, present in the first 4–5 positions at the base of each crypt, divide to produce daughter cells which, by the time they reach positions

6–9, have undergone partial differentiation. Three mitoses are necessary to produce differentiated columnar cells: two for mucous cells and one for enteroendocrine cells; Paneth cells derive from direct differentiation (Leblond and Cheng, 1976; Cheng and Leblond, 1974b). Cell cycle time in rats is 13.3 hr: 3.7 hr in G_1, 7.8 hr in S, 1.3 hr in G_2, and 0.5 hr in M (Schultze and Kellerer, 1979).

Totipotent stem cells do not divide unless required to do so and the turnover of the epithelium is supported by proliferative cells. Only under special circumstances do stem cells leave G_0 and enter the cycle, as happens when physical or chemical agents cause severe damage to the epithelium. According to Hayflick (1965), a stem cell can undergo only a limited number of divisions during the life of the organism. Under normal conditions, the epithelial cell number remains constant as a result of cellular equilibrium. In rats and mice the turnover lasts about 3 days (Leblond and Messier, 1958; Trier, 1968; Cheng and Leblond, 1974a), and it lasts about 5 days in man (Cameron, 1971; Lipkin, 1973).

The distribution of cells which have taken up [^{3}H]thymidine along the crypt 1 hr after administration has been studied in histological sections on the longitudinal axis of the villus. A low frequency of cells labeled or in mitosis is seen at the base of the crypt (Cairnie et al., 1965; Al Dewachi et al., 1979; Becciolini et al., 1983a). The increase in frequency occurring in the positions immediately above this level is so rapid that it can rise above 50%. Halfway up the crypt, values decrease and labeled cells are absent in the outer third. Cells in mitosis present a different distribution (Cairnie et al., 1965; Al Dewachi et al., 1979; Becciolini et al., 1983a), with the frequency following that of the labeled cells (Fig. 1). The distribution of labeled and mitotic cells along the small intestine is similar, but in the duodenum and jejunum, S-phase cells are more frequent than in the ileum.

Differentiation occurs in the upper third of the crypt where columnar cells lose their proliferative capacity and acquire specific functions. Even the composition of the plasma membrane changes during differentiation (Quaroni et al., 1980). A class of glycoproteins, absent in the crypt, is synthesized and appears in the luminal region of cell membranes in the villus (Danielsen et al., 1981; Montgomery et al., 1981). Tri- and disaccharases and tri- and dipeptidases turn over rapidly (Seethram et al., 1980; Berteloot and Hugon, 1982) and are also found outside the epithelium, probably due to the action of pancreatic enzymes (Alpers and Tedesco, 1975); Alpers, 1977) and perhaps to the action of lysosomal enzymes which seem to act on the membrane carriers.

3. Physiological Aspects

Digestion of food is the function of the small intestine. A partial hydrolysis of starch and protein takes place in the mouth and in the stomach, respectively. In the first part of the small intestine, macromolecules in the bolus, mixed with biliary and pancreatic secretions, undergo further hydrolysis. Simple molecules

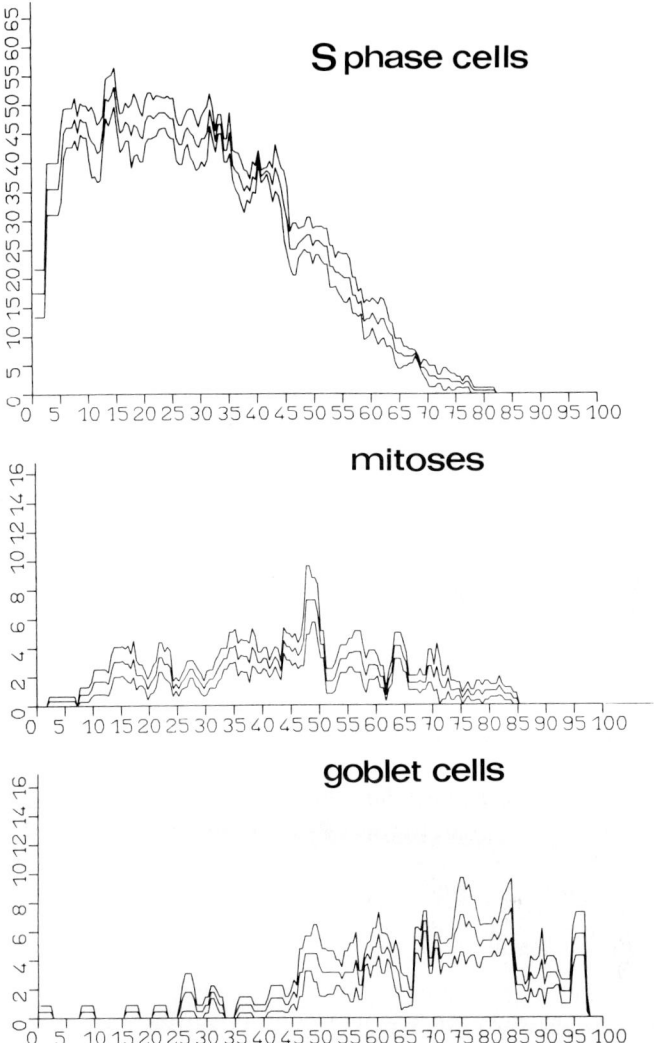

FIG. 1. Frequency of distribution of S-phase cells, mitoses, and goblet cells in different positions along the crypt of control animals is reported. Total height of the crypt is represented as 100%. The curves represent the mean value ±SE. (From Becciolini *et al.,* 1983a, 1984.)

can be absorbed as such, whereas larger molecules require further hydrolysis before they may be absorbed. For many years this process was assumed to have been carried out in the intestinal lumen by enzymes secreted directly by the intestinal cells in the succus entericus. More recently (Newey and Smith, 1959; Dahlquist, 1962; Crane, 1966), the participation of the membrane in hydrolysis of certain molecules has been demonstrated, and the process has been called "membrane digestion" (Ugolev, 1968; Ugolev and De Laey, 1973). Smaller fragments, arising from the partial digestion of polysaccharides and proteins, come into contact with the glycocalyx in which pancreatic enzymes are caught and are degraded further into simpler forms. Once they have reached the outer protein layer of the plasma membrane, di- and trisaccharides and di- and tripeptides are hydrolyzed to monomers by the action of di- and trisaccharases and by di- and tripeptidases located in the membrane.

The above-mentioned breakdown products are absorbed preferentially via specific carriers which are located near the enzymes. This circumstance might explain the more rapid absorption of dimers with respect to the correspondent monomers (Berteloot et al. 1981). It is worth noting that the final stage of digestion occurs, protected by the glycocalyx, under sterile conditions so that the bacterial flora cannot utilize the monomers produced by membrane digestion. During the turnover of epithelial cells, these membrane proteins are renewed several times (Alpers and Tedesco, 1975; Alpers, 1977; Seethram et al., 1980).

Along the gastrointestinal tract, the disaccharases are absent in the stomach, gradually increase from the duodenum to the jejunum, then decrease in the ileum until they reach very low levels of activity in the terminal portion of the ileum. These enzymes are not present in the colon and in the rectum. The importance of the brush border enzymes in the digestive process was demonstrated in patients with congenital or acquired deficiency of specific enzymes. In these cases, disaccharides are hydrolyzed by the microflora and cause diarrhea because of toxic products (Semenza, 1968).

4. Effects of Irradiation

When a mammal is exposed to whole body or to abdominal irradiation, the gastrointestinal tract undergoes a series of alterations which can cause death via a well-defined gastrointestinal syndrome. Quastler (Quastler, 1956; Quastler and Sherman, 1959) distinguished four levels of organization which demonstrate damage successively: cell, tissue, organ, and organism. The radiation-induced damage to DNA in the proliferative cells of the crypt leads to the death of some cells and an arrest of mitotic activity of others. Inhibited proliferation in the crypts and the natural loss of cells from the apex of the villus due to turnover results in the shortening of the crypt-villus unit. Reduction in the number of cells in the proliferative compartment produces a subsequent decrease in differentiating cells and later in differentiated ones. If the dose is high enough to protract blockage of proliferative activity, epithelial cells are lost and ulcerations may follow. The disappearance of the intestinal barrier leads to the death of the organism. When a sublethal dose has been received, some cells repair their damage, others are recruited from those in the G_0 phase, and replacement of the epithelial cells occurs.

In the alimentary tract, cell turnover is rapid and manifestation of acute damage occurs in a few days; the epithelium can return to normal equally rapidly, however. Late damage is tied to alterations in the connective and vascular networks, and, as a secondary effect, can lead to late damage to epithelial structures.

a. Changes following a Single Radiation Exposure of the Abdomen.

i. Morphological Modifications. The structures of the small intestine in man and in the rat are remarkably similar, and findings in laboratory animals are comparable to those observed in irradiated patients. One must remember that the temporal evolution of radiation effects is different, since cell turnover times are longer for patients and treatment schedules may be quite different.

The temporal sequence of epithelial modifications has been investigated accurately, histologically as well as ultrastructurally, for a number of years (Quastler, 1956; Quastler and Sherman, 1959; Trier, 1962; Hugon *et al.,* 1965; Trier and Browning, 1966; Wiernick, 1966). Animals irradiated in the abdominal region with 6 Gy and sacrificed starting 30 min later showed at the level of the crypts that as early as 1–2 hr after exposure, alterations such as increased cell size appear along with other degenerative phenomena such as pycnosis, karyolysis, karyorhexy, and inhibition of mitosis (Becciolini *et al.,* 1976). Because of turnover, the altered cells progressively involve the whole crypt and villus while inflammatory cells increase in the stroma. Forty-eight hours after irradiation altered cells are apparent throughout the whole organ. At the same time, renewed proliferation is in evidence, and some cells with normal morphology are observed in the crypts. The number of inflammatory cells then diminishes. At 96 hr the epithelium returns to its normal appearance.

ii. Ultrastructural Modifications. After a dose of 6 Gy, ultrastructural modifications which concern crypt cells almost exclusively are quite moderate. Within 4 hr, regression and irreversible phenomena appear (Quastler and Hampton, 1962; Detrick *et al.,* 1963; Hugon and Borgers, 1965; Trier *et al.,* 1968). In villus columnar cells, modifications are slight, e.g., engulfing of cells, moderately shortened villi, expansion of the Golgi complex, increase in ribosomes, and enlarged mitochondria. A large number of ribosomes, free or aggregated, in the rough endoplasmic reticulum (Tedde *et al.,* 1972, 1975; Becciolini *et al.,* 1976) appear in the differentiating cells of the crypt-villus junction. Later, regenerative phenomena appear in the crypt while marked alterations such as swelling, fragmentation, and dense and packed organelles move to the intestinal lumen. After 72 hr, the alterations involve mainly the tips of the villi. After longer intervals, the columnar cells recover and seem to return to a normal appearance.

iii Quantitative Morphological Modifications. Changes in the crypt cell numbers and in mitotic and labeling indices in rats at different times after exposure of the abdomen to 8 Gy are shown in Table I. Reduction in the parameters listed above progresses up to 48 hr after irradiation; then, due to a resurgence of proliferative activity, the irradiated epithelium appears to be composed of a significantly larger number of mitotic cells than the unirradiated controls. Only after many days does the proliferative level return to normal. The distribution of labeled cells along the side of the crypt shows an inhibition of [^{3}H]thymidine uptake, particularly in the lower half of the crypt. A progressive extension of the proliferative compartment follows, and at 72 hr, labeled cells are present in the entire crypt, with a marked reduction in the differentiation zone. After that there is a gradual return to normal at 20 days after irradiation.

iv. Biochemical Modifications. In order to assess radiation-induced biochemical changes, (1) the activity of the brush border enzymes, whose presence reflects the intactness of the differentiation apparatus and the integrity of the structure in which these enzymes are located, was determined, and (2) the activity of lysosomal enzymes—an index of cell degeneration during the early postirradiation period and, later on, of inflammatory infiltration—was measured.

TABLE I

EFFECT OF IRRADIATION OF THE ABDOMEN WITH 8 GY[a]

Time after irradiation	Crypt epithelial cells	Mitotic index	Labeling index
0 hr	100.0 ± 0.60	100.0 ± 7.01	100.0 ± 2.71
6 hr	90.03 ± 2.01	16.67 ± 9.26	32.92 ± 3.69
12 hr	70.50 ± 4.23	42.29 ± 6.63	26.02 ± 1.59
20 hr	63.30 ± 1.54	21.61 ± 6.45	44.42 ± 4.88
36 hr	50.18 ± 1.75	26.95 ± 3.73	49.87 ± 13.85
48 hr	49.00 ± 0.85	9.47 ± 9.47	89.28 ± 9.83
72 hr	55.30 ± 2.0	20.36 ± 11.36	164.58 ± 3.08
96 hr	115.00 ± 1.93	103.69 ± 8.50	144.87 ± 28.31
126 hr	119.00 ± 3.44	87.89 ± 7.05	144.00 ± 11.59
150 hr	121.50 ± 1.17	89.51 ± 15.95	141.60 ± 19.90
11 days	111.14 ± 1.75	44.74 ± 9.04	125.12 ± 5.00
20 days	109.00 ± 5.20	57.76 ± 8.99	97.32 ± 5.86

[a]This table shows changes in the number of epithelial crypt cells as well as of mitotic and labeling indices expressed as percentage of control values.

Single acute doses of 2, 3, 5, 6, 6.5, 8, 12, and 20 Gy were used to irradiate either the abdomen or the whole body (Becciolini *et al.*, 1970; 1972, 1974, 1976; 1977a, 1982a–c, 1983d,e). In all cases, three separate phases could be distinguished: (1) Within a few hours after irradiation there was an increase in the activity of brush border enzymes which was statistically significant after the highest doses. The increase in enzyme activity did not involve an increase in the functional capacity of these cells; in fact, disaccharide absorption was lower than that in controls (Becciolini *et al.*, 1977b). Morphological modifications at this time involved crypt cells only. Lysosomal enzyme activity showed an increase at the last intervals of this phase at a time when protein content also decreased. (2) Over the period spanning 40–72 hr, cellular alterations became more severe and involved the entire organ. In this phase, the activity of brush border enzymes was very low, whereas that of lysosomal enzymes was at its highest. Protein content was lower. At these intervals, labeled cells occupied the whole crypt and cells in differentiation were low in numbers. (3) Subsequently, recovery affecting morphologically normal cells lining the crypt, and later the villus, occurred. Lysosomal enzyme activity and protein content returned to control values, while the activities of brush border enzymes *continued at* a statistically significant lower level. Only after longer postirradiation intervals did this parameter return to normal.

The severity of the damage and its duration depended on the radiation dose. After doses less than 5 Gy, alterations were limited and the intestine returned to normal rapidly. Intermediate doses caused more severe damage and concomitantly slower recovery. After doses as high as 12 and 20 Gy, the activity of brush border enzymes was almost undetectable during the acute phase of damage.

b. Changes after Fractionation of the Radiation Dose Including Multiple Daily Dose Fractionation (MDF). A morphological and biochemical study after 2 × 6 Gy doses to the abdomen, separated by 6- to 120-hr intervals, has demonstrated that when times up to 24 hr elapsed between fractions, results were similar to those obtained after a single dose of 12 Gy with the three phases described above (Fig. 2) (Becciolini *et al.*, 1973, 1975). When the elapsed time between fractions was 48 or 72 hr, at early intervals, disaccharase and dipeptidase activities were significantly lower than in controls inasmuch as the second fraction was given at a time of maximum damage to the epithelium. The effect was severe, but there was a return to control values as early as 5 days.

When 120 hr elapsed between fractions, the three phases were observed again, and they were similar in duration and extent to those seen after single doses of 6 Gy. At the time of the second fraction, the epithelium had recovered, and it showed a response similar to that elicited by the first dose fraction.

i. Morphological Modifications. The effects of MDF with a total dose of 12 Gy to the abdomen have been studied according to the following schedules: 2

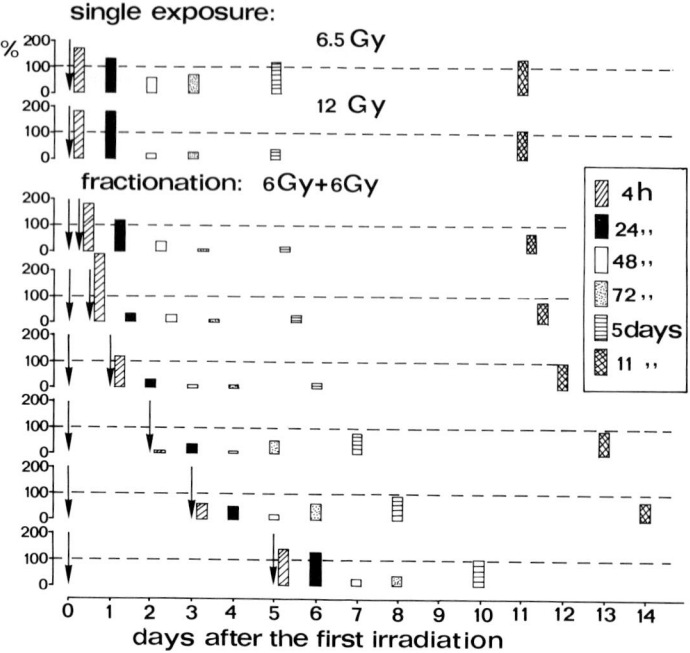

FIG. 2. Modifications of maltase activity in animals exposed to 6 Gy and 6 Gy administered at intervals between 6 and 120 hr. In the upper part of the figure, the modifications after 6.5 and 12 Gy are shown. The values are expressed as percentage of control.

Gy × 3 fractions/day × 2 days (MDF A), 2 Gy × 3 fractions/day every 4 hr followed by a 16-hr split and later by a second cycle (MDF B), and 3 Gy × 2 fractions/day × 2 days (MDF C) (Becciolini, 1980; Becciolini et al., 1983d,e 1986). The numbers of cells in the crypts after the different MDFs are reported in Table II. In all cases, the numbers of epithelial cells were significantly fewer than those for controls up to 36 hr. After that time, values returned rapidly to or were higher than control levels. Multiple daily fractionation scheme A presented a lower initial reduction in epithelial cells than did the others.

The labeling index (Table III) decreased significantly at early time intervals after irradiation; this reduction appeared greatest and lasted longest in MDF B. Subsequently, there was a return to control values with a temporary overshoot, particularly in MDF B and C. The mitotic index confirmed the initial inhibition of proliferative activity and behavior similar to that resulting from a single dose of 8 Gy observed in MDF A and MDF B. At 5 days values of all MDFs were similar. The results indicate the tolerance of the small intestine for schedules of irradiation using these MDF schedules.

TABLE II

EFFECT OF THREE MULTIPLE-DOSE FRACTIONS TO THE ABDOMEN
ON EPITHELIAL CRYPT CELLS[a,b]

Sacrifices	2 Gy × 3 × 2 (A)	2 Gy × 3 × 2 (B)	3 Gy × 2 × 2 (C)
1 hr	59.38 ± 1.89	47.20 ± 0.83	49.42 ± 0.71
24 hr	57.27 ± 3.06	60.12 ± 1.98	47.17 ± 2.12
36 hr	67.18 ± 2.90	56.18 ± 6.32	54.35 ± 1.80
72 hr	95.62 ± 4.68	106.5 ± 3.11	107.70 ± 5.09
5 days	98.84 ± 5.53	116.0 ± 4.12	111.40 ± 1.70
11 days	119.19 ± 3.06	99.90 ± 2.48	115.10 ± 1.51
19 days	120.29 ± 2.14	102.2 ± 2.89	111.30 ± 2.69

[a]Results are expressed as percentage of control values.
[b]Intervals among fractions: (A), $t = 8$ hr; (B), $t_1 = 4$ hr and $t_2 = 16$ hr; (C), $t = 12$ hr.

Regarding the localization of labeled cells along the crypt, the distribution among the MDFs presents substantial differences which do not seem to be justified by the different durations of treatment. At early intervals, MDF B causes an arrest in the uptake of [^{3}H]thymidine which is much higher than that seen after the other MDFs and which also has a distribution similar to that found at early intervals after a dose of 8 Gy. A smaller reduction was observed after MDF A (Fig. 3). Later, differences became smaller; reduction in the frequencies of labeled cells at the bases of the crypts was still evident at 36 hr, whereas S-phase cells remained numerous in the higher parts of the crypts (Fig. 4).

TABLE III

CHANGES IN THE LABELING INDEX OF EPITHELIAL CRYPT CELLS
AFTER MULTIPLE DAILY FRACTIONS[a,b]

Sacrifices	2 Gy × 3 × 2 (A)	2 Gy × 3 × 2 (B)	2 Gy × 2 × 2 (C)
1 hr	81.29 ± 1.98	22.56 ± 4.03	66.72 ± 3.58
24 hr	102.60 ± 9.22	83.94 ± 3.32	124.70 ± 11.52
36 hr	109.10 ± 7.95	130.00 ± 7.25	139.65 ± 4.58
72 hr	104.36 ± 3.91	135.60 ± 9.89	145.17 ± 15.31
5 days	108.89 ± 1.65	107.37 ± 3.76	111.43 ± 7.48
11 days	103.33 ± 2.59	116.27 ± 4.13	96.63 ± 15.01
19 days	111.20 ± 17.26	94.21 ± 11.43	98.56 ± 8.89

[a]Results are expressed as percentage of control values.
[b]Intervals among fractions: (A), $t = 8$ hr; (B), $t_1 = 4$ hr and $t_2 = 16$ hr; (C), $t = 12$ hr.

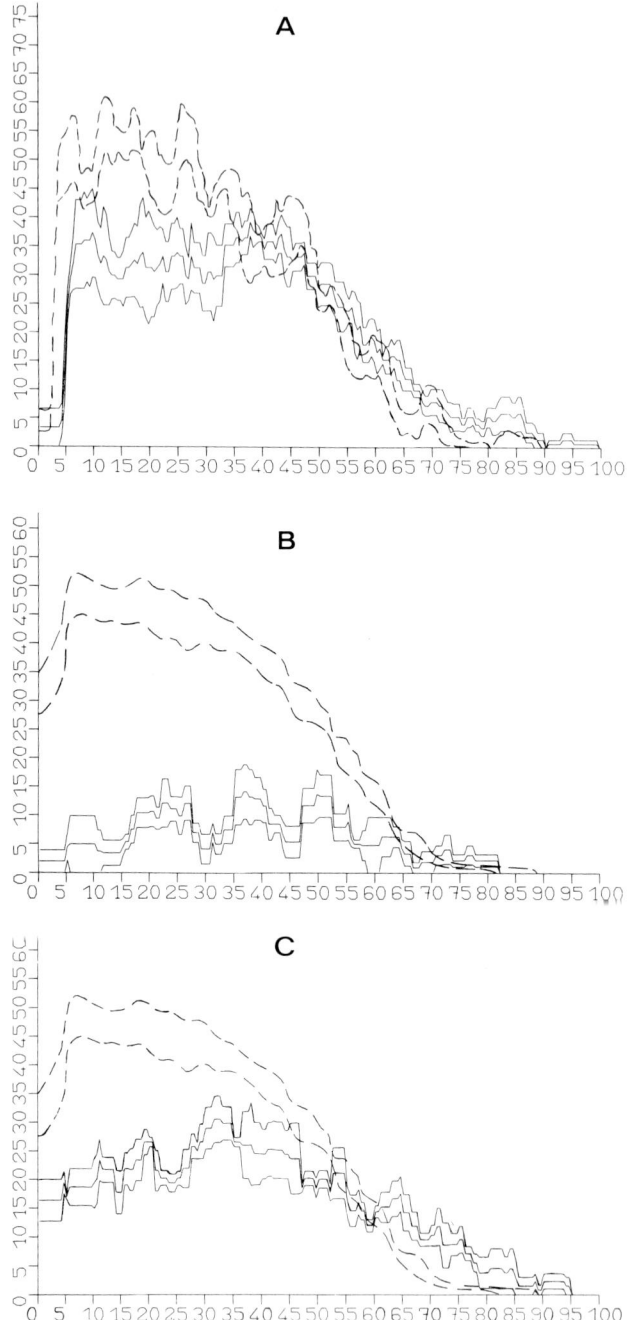

FIG. 3. Frequency of S-phase cells along crypt in rats exposed to MDFs and sacrificed 1 hr after the last fraction. A, MDF 2 Gy × 3 × 2 with 8-hr split. B, MDF 2 Gy × 3 × 2 with a 4-hr split among the first three fractions and a 16-hr split before the second three fractions. C, MDF 3 Gy × 2 × 2 with 12-hr intervals among fractons. Mean values ±SE in each position are reported. Curves for data from control rats sacrificed at the same time of the day are represented as a dashed line. (B, From Becciolini *et al.*, 1986c; C, from Becciolini *et al.*, 1983e.)

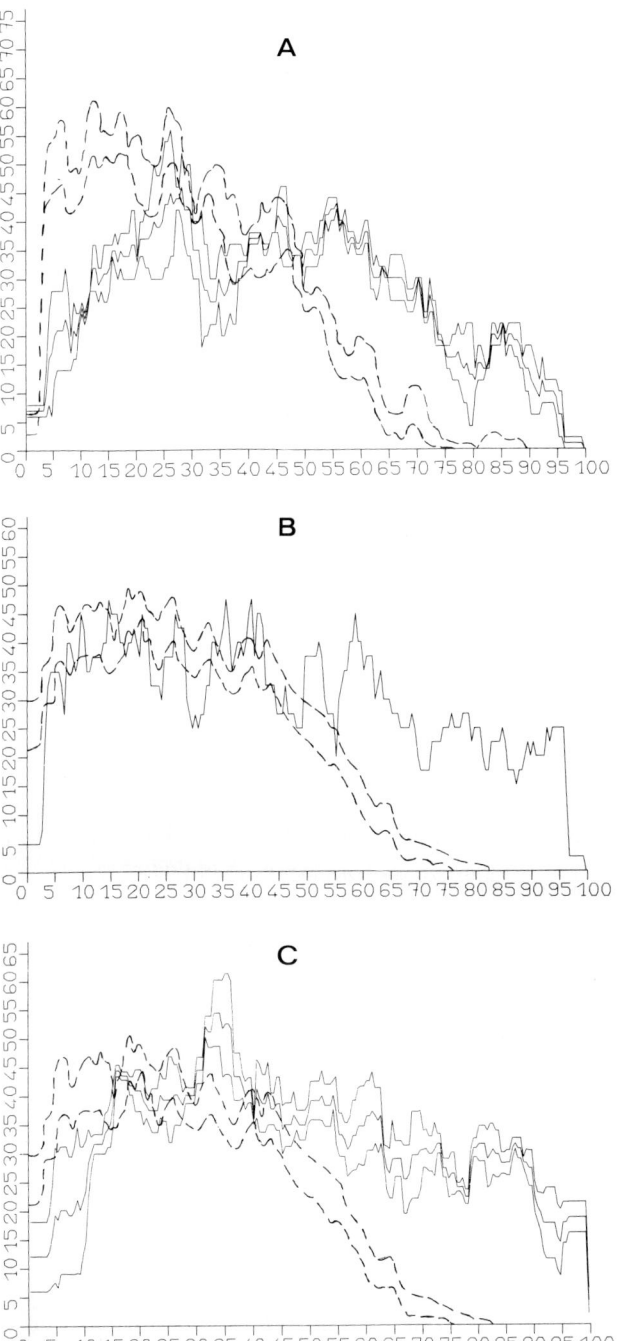

FIG. 4. Frequency of S-phase cells along the crypt in rats exposed to different MDF (see Fig. 3 for explanation) and sacrificed 36 hr after last fraction. (B, From Beccioloni *et al.*, 1986c; C, from Becciolini *et al.*, 1983e.)

At 72 hr, the labeled cell frequency decreased and the curves were similar to those observed 96 and 120 hr after a single dose. Later, the distribution resembled controls even when a temporary extension of the proliferative compartment was still present.

Some interesting modifications were observed in the numbers of goblet cells. After a single dose, goblet cells increased their numbers significantly at early intervals, were markedly reduced during acute damage, and showed a tendency to return to control levels when the epithelium had recovered. The effect was more severe when the dose was increased (Becciolini *et al.*, 1985). The use of MDFs produced similar effects as compared with a single dose except for the fact that return to a "normal" state occurred earlier. Among MDFs, MDF B caused a more marked initial increase in number, whereas with MDF A, return to normality was more rapid than that after the other two schemes.

ii. Biochemical Modifications. Modifications in the activity of brush border enzymes after MDF occur and follow the three phases described already for the results of a single dose of radiation. The most important differences, with respect to a single 8-Gy dose, are (1) the presence of activities higher than or similar to those of controls over much longer times and consequent later attainment of minimum values, (2) lower or equal reduction, and (3) values which approach control levels more rapidly during recovery. The general behavior of the different activities was the same even though there were differences among the different enzymes. Invertase (Fig. 5) presented a slight initial increase and a more moderate reduction. Generally, MDF A showed less severe enzyme damage and a more rapid return to normal. Lysosomal enzymes showed the same behavior as that observed after single doses (Becciolini, 1980; Becciolini *et al.*, 1983e). Return to normal values was slower than that observed after 8 Gy.

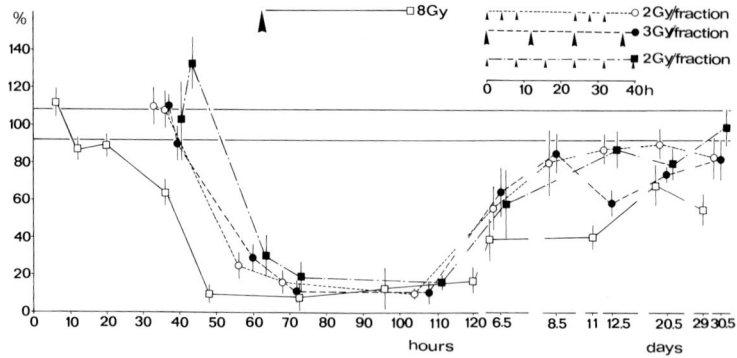

Fig. 5. Invertase activity after different MDFs. Mean value ±SE of controls is expressed as 100%. The values are compared with those of animals irradiated with 8 Gy and sacrificed at the same time of the day.

In conclusion, the results indicate that (1) MDFs are tolerated better than are single doses; (2) among MDFs, an 8-hr interval between 2-Gy fractions induced less acute damage and allowed rapid recovery; (3) MDF B showed initial modifications similar to those resulting from a single dose of 8 Gy (4- and 16-hr intervals between fractions do not appear to be sufficient to repair the damage in any effective way); and (4) tolerance to 3 Gy $\times$ 2 $\times$ 2 MDF has been good.

Increase in brush border enzyme activity, blockage of [³H]thymidine uptake, diminished numbers of cells capable of taking up labeled nucleoside, and increases in the numbers of goblet cells observed at early posttreatment intervals confirm the conclusion that ionizing radiations can induce early differentiation. Through this process, radiation-altered proliferative cells would continue to differentiate, although they may not divide. MDFs were more effective than were single doses in eliciting this process, and 2 Gy/4 hr was the most efficient scheme.

5. Radiation Effects on Bioperiodic Phenomena

Recently, variations in certain functional and cellular kinetic parameters depending on the times of day they were observed have been reported by a number of investigators (e.g., Sigdestad and Lesher, 1970; Hendry, 1975; Al Dewachi et al., 1976; Saito et al., 1976; Stevenson and Fierstein 1976; Becciolini et al., 1977c; Potten et al., 1977; Aschoff, 1979; Kaufman et al., 1980). In animals kept on a 12-hr (06:30–18:30) constant light/darkness cycle, the activity of brush border enzymes was at its highest in darkness—the time of highest activity in rats—and at its lowest during daylight (Fig. 6). No circadian

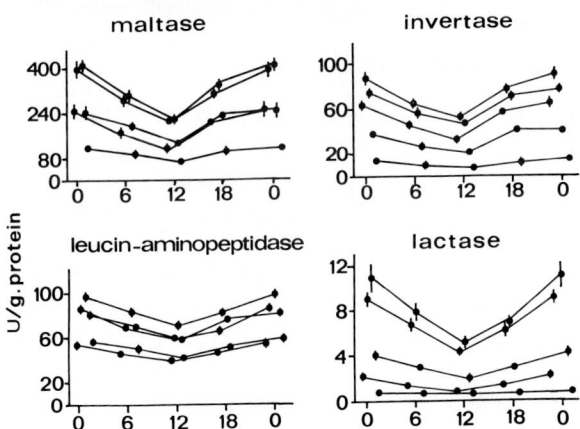

FIG. 6. Brush border enzyme activities (maltase, invertase, LAP, and lactase) in control rats sacrificed at different times of dark/light cycle. The values ($\pm$SE) in the five segments into which the small intestine was cut are reported.

modifications have been observed in lysosomal enzyme activity and protein content (Fig. 6) (Becciolini *et al.*, 1977c, 1982b), whereas protein synthesis seems to oscillate (Benucci *et al.*, 1981).

Regarding the determination of quantitative morphological parameters, some authors (Sigdestad and Lesher, 1970; Hendry, 1975; Potten *et al.*, 1977; Klein, 1980) have observed that the mitotic index in rats exhibits circadian fluctuations with peak values at the midpoint of the dark period and minimum values during the light period. Other authors have reported different fluctuation periods (Alov, 1963; Al Dewachi *et al.*, 1976; Al Nafuss and Wright, 1982).

At night and at midday we observed mitotic indices higher than at other times of the day (Becciolini *et al.*, 1983a). In the attempt to find irradiation conditions capable of inducing different effects on normal tissues after equal doses, animals were irradiated at different hours of the day. A single 3-Gy whole body dose of 8 Gy to the abdomen alone was administered at 0:00, 6:00, 12:00, or 18:00 hr. The results demonstrated that qualitative morphological modifications did not show substantial differences in all the groups except for an earlier return to control numbers of cells in the group irradiated at 6:00 hr (Becciolini *et al.*, 1983a). The labeling index returned to control levels earlier in those groups treated during the light period.

Groups of animals which received 8 Gy at different times of the day did not show significant differences up to 72 hr after treatment as injury and recovery were similar in all groups at that time. The subsequent shrinking of the proliferative compartment occurred earlier in the group irradiated at 6:00 hr (Becciolini *et al.*, 1983b,c). Early return to a distribution of labeled cells similar to controls was paralleled by measured values for levels of brush border enzymes which became similar to those for controls (Becciolini *et al.*, 1982b).

6. Radiation Effect on Intestinal Weight

Lack of appetite, vomiting, and diarrhea characterize the intestinal syndrome in mammals and lead to loss of body weight. The decrease is not immediate, but it becomes significant when the intestinal syndrome is more severe. After 12 Gy, the decrease may be as high as 40% (Becciolini *et al.*, 1982c), and after lower doses (e.g., 6–8 Gy), a decrease of ~20% may be seen; the lowest level is reached after between 4 and 6 days, after which body weight slowly returns to control levels. A similar weight loss occurs after MDF with a total dose of 12 Gy. When a 2×6 Gy fractionation scheme with about 24 hr between fractions is used, loss of body weight is more rapid than it is after a 12-Gy single dose. When the time between fractions exceeds 48 hr, weight reduction is less.

The decrease in the weight of the small intestine is rapid and minimum values are reached between 2 and 3 days postirradiation. After an MDF either of 6 Gy or of 12 Gy, these decreases fluctuate between 65 and 75% of control weights. Four to five days after irradiation, normal values are attained which later rise to 130–140% of control weights. After about 30 days, the weights of irradiated small intestines are similar to those of controls.

A 2 × 6 Gy fractionation scheme causes greater intestinal weight reductions in animals exposed twice within 24 hr than does a 12-Gy single dose. Return to normal values occurs later, as it does after a single 12-Gy exposure. When the interval between fractions is longer, weight reduction is less extreme and restoration to normal weight is more rapid.

It should be noted that the term "intestinal weight" refers to the whole intestinal wall, i.e., the mucosa and the muscularis. The mucosa, however, is made up of radiosensitive cells and is the major component which undergoes variations in weight.

7. Radiation Effects on DNA and Other Macromolecules as well as Electrolyte Loss and Protein Leakage

Damage to the epithelial component of the small intestine leads to a progressive decrease in DNA content (Nygaard and Potter, 1959, 1962; Gerber and Altman, 1970). The higher the dose (Nygaard and Potter, 1959; Kay and Entenman, 1959) and dose rate, the higher and more persistent the decrease (Looney, 1966). RNA synthesis does not seem as impaired as does that of DNA, but both show minimum values within 24 hr after irradiation (Gerber et al., 1963; Maisin, 1966).

Tritiated thymidine given to whole body irradiated rats 1 hr before sacrifice revealed an inhibition of proliferative capacity as early as 8 hr after exposure (Nygaard and Potter, 1962). Autoradiography showed that few molecules of the tracer were taken up by the rare cells synthesizing DNA. After doses of up to 2 Gy, return to control levels occurred as early as 20 hr after exposure; after higher doses, recovery was delayed and occurred after a period of marked increase (Nygaard and Potter, 1959; Gerber et al., 1963; Maisin, 1966).

In the animals irradiated at different times of the day and sacrificed 1 hr after [³H]thymidine injection, marked differences were observed only in the groups exposed to 8 Gy (Fig. 7). The group irradiated at 6.00 hr showed levels of nucleic acid labeling lower than those of other groups during recovery and a more rapid return to normal levels. After 12-Gy MDF, labeling of nucleic acids did not differ substantially from that after an 8-Gy single dose when the first fraction given was taken as the starting point (Becciolini and Lanini, 1986).

Protein synthesis decreases more slowly and to a lesser degree than does labeling of DNA (Edwards et al., 1964). After doses of about 5 Gy, the lowest level of uptake of [¹⁴C]glycine occurs 48 hr after irradiation (Lipkin et al., 1963; Looney, 1966; Maisin, 1966). The rate of disappearance of [¹⁴C]leucine 2 hr after exposure to 8 Gy appears more rapid than does that in controls (Casati et

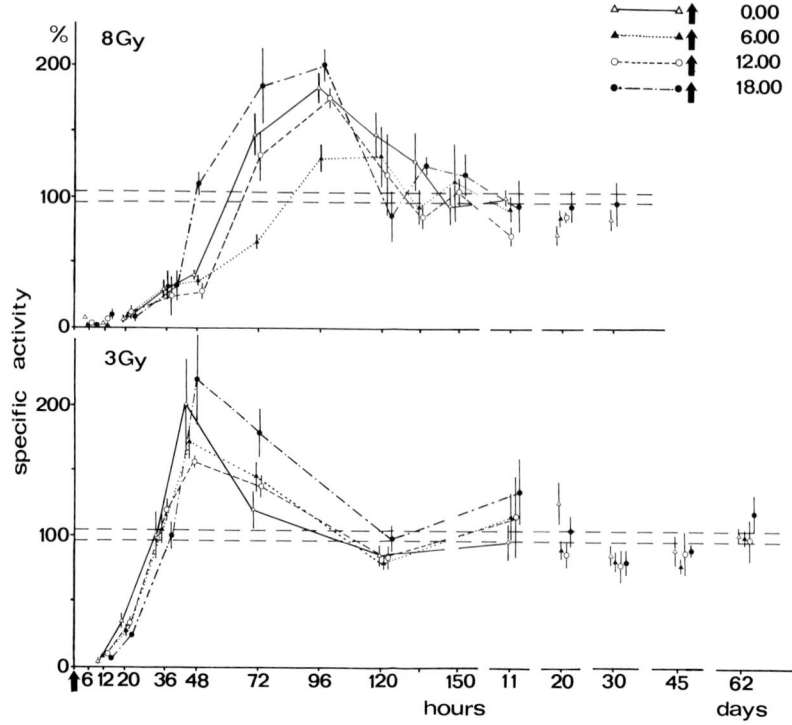

FIG. 7. Radioactivity incorporated into nucleic acids in rats exposed to 3 or 8 Gy at different times of the day. The mean values ±SE in the small intestine are expressed as percentage of control animals.

al., 1979). When irradiated rats are injected instead with [¹⁴C]leucine 4 hr before sacrifice, the uptake increases. Incorporation is about 2.5 times as high as that in controls when cells in mitosis increase during recovery (Benucci et al., 1981).

B. Clinical Aspects

1. Clinical Manifestation of Radiation Injury

Animal studies have provided information on the basis of which the syndrome of malabsorption following irradiation could be delineated and defined. Investigations carried out in man until the 1950s dealt mainly with qualitative evaluations of late damage and concerned dogs or surgical patients and patients deceased at various times after irradiation. The results of those studies provided an incentive to further investigations of animals and man. Subsequent availability of radioactively labeled compounds allowed their utilization in absorption tests in animals and patients irradiated in the abdominal region. Often, however, the few patients studied at the beginning of and throughout treatment were not the same, and their clinical conditions were not identical. This made interpretation of results difficult. Moreover, data were not easily comparable because radiation doses, treatment schedules, and absorption tests were different in each case. Reports on more than one parameter in the same patient are infrequent (Dalla Palma, 1968; Goldberg, 1972; Tarpila, 1971) and some of these are questionable (Ratzowski and Hochman, 1968).

Choice of patients is an important factor in the evaluation of malabsorption. In fact, the malabsorption syndrome has been described in patients with malignant disease not involving the intestine. Lower absorption of xylose and fats as well as metabolic disorders of folic acid which can contribute to anemic conditions were found in such patients. Jejunal biopsies showed abnormal histology (Sleisenger et al., 1953; Creamer, 1964; Dymock, 1969; Dymock et al., 1967).

In an accurate study, a number of investigations involving assays of disaccharases and other enzymes in biopsies taken from the mucosa before and at the end of radiotherapeutic treatment to volumes including parts of the abdomen were performed with the light and electron microscopes by Tarpila (1971). At the end of treatment, partial villus atrophy was observed in some cases when the biopsy site received 30 Gy or more, but no statistically significant changes in the mitotic index were observed. Brush border enzyme activity was reduced at the end of treatment, but a statistically significant decrease appeared only when villus atrophy was observed; 1–5 months following the end of radiotherapy, morphology and enzyme activities returned to normal.

The first research done in a fairly rigorous way on a relatively large number of patients was carried out at the Institute of Radiology of the University of Florence (De Giuli, 1963, 1965, 1975; De Dominicis and Grechi, 1965; De Marzi, 1965; Giannardi et al., 1965, 1967; De Giuli et al., 1967; Dalla Palma, 1968; Cionini et al., 1971, 1976; Becciolini et al., 1972, 1974, 1976, 1979). This work continued for some years and, as far as we know, does not seem to have been repeated elsewhere (although we have considered data from other authors in this review). Several radiological and functional parameters were analyzed in subjects who presented no gastrointestinal abnormalities prior to radiotherapy. Observations of each patient were repeated halfway through and at the end of treatment.

A telecobalt unit (dose rate = 60–80 cGy/min) was used as the radiation source in all cases and the daily dose was calculated at the midline. Fractions at 2 Gy were given 5 days per week up to a maximum total dose of 45–55 Gy. Three irradiation fields were used (Fig. 8): (1) abdominopelvic; (2) lower hemiabdomen; and (3) epigastric. In many patients it was possible to perform the majority of routine tests. Only the results from those cases for which the treat-

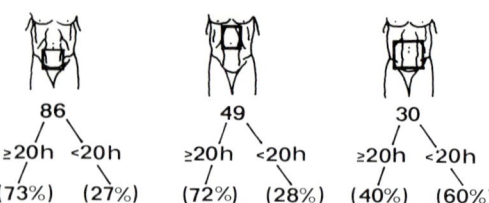

FIG. 8. Transit time (in hours) in patients irradiated on abdominopelvic, lower hemiabdomen and epigastric region, respectively, is presented.

ment scheme was not modified were reported; this limited the number of cases. The number of patients in the first group was much higher than the numbers in other groups.

The following tests were performed: (1) bioptical examination of the mucosa; (2) radiological investigation; (3) assays of blood, urine, and feces for the absorbed quantities of substances administered orally; (4) assessment of the loss of intravenously administered substances through the intestinal mucosa; and (5) aspiration by probe of intestinal contents to determine the components thereof and to analyze microbial flora.

Investigations revealed radiation damage such as (1) altered motility, as that can affect contact time between intestinal contents and the mucosa and consequently affect absorption; (2) morphological modifications; (3) altered membrane properties, active and passive transport through the cell membrane; (4) altered gastric, biliary, and pancreatic secretions; and (5) changes in bacterial flora.

In these patients, radiation-induced intestinal injury was rather moderate, and only in some cases did treatment have to be discontinued for a few days. Nausea and vomiting generally appeared after the first sessions. These symptoms were highly variable in severity and lasted for about a week. Such manifestations were more frequent and intense when either the upper half of the abdomen or the epigastric region had been irradiated. Diarrhea appeared during the third week when the total accumulated dose reached 25–30 Gy, particularly in women in whom the abdominopelvic field included the lower abdomen (Fig. 8). Gastric pain was observed in men irradiated in the epigastric region, but only rarely did diarrhea occur in those who received irradiation of the lower midabdomen and pelvis. Weakness and lack of appetite also occurred at this time.

a. Transit Time. In order to evaluate transit time, radiological investigations were carried out with $BaSO_4$ alone or mixed with substances whose absorption had to be assessed. Radiographs were taken every 30 min up to 5 hr and then again at 7 and 8 hr (DeMarzi, 1965; Pelù *et al.*, 1965; DeMarzi and Aulisi, 1966). Gastric emptying did not show any extraordinary changes after irradiation, but partial acceleration was observed. Transit time was markedly affected by the exposed volume. In fact, acceleration was prevalent in most patients irradiated in the region of the lower hemiabdomen whereas in the rest, values were closer to normal (Fig. 9).

In most patients who had diarrhea, accelerated transit was found in the small intestine and colon. In some patients, accelerated transit occurred only in the colon.

The most striking alterations appeared midway through the treatment, but there was no worsening later due to the administration of alteration-limiting drugs. Hypermotility was confirmed by other authors (Reeves *et al.*, 1965) in 42% of the patients after completion of treatment. Nonetheless, transit time after experimental irradiation has been noted to increase. A marked increase in propulsive activity was observed within a few hours after irradiation, and after that there was delayed gastric emptying and a delay in intestinal transit due to mucosal damage (Conard, 1956; Kurstin, 1963; Bond, 1963; Summers *et al.*, 1970).

b. Morphology. Morphological investigation of human intestinal mucosa showed modifications similar to those observed in laboratory animals. In patients, absorption tests revealed a progressive decrease in the mitotic index after a dose of 2 Gy, with the lowest indices following doses of ~40 Gy. The number of cells in the villus-crypt formation showed a similar reduction, which remained constant until the treatment ended (Bond, 1963; De Dominicis *et al.*, 1967, 1968). During this phase, inflammatory cells infiltrated the lamina propria. Epithelial morphology returned to normal within 1–4 months after the termination of treatment.

It should be noted that in the intestinal loops outside of the irradiation field, changes were similar to those observed in the loops hit directly by the beam of radiation. The extent of the damage was substantially smaller, but it was out of proportion to the very low dose of scattered irradiation received (De Dominicis *et al.* 1969). In patients receiving chest irradiation, the intestinal mucosa was not altered.

2. Malabsorption

After radiotherapy involving the entire abdomen, malabsorption of several dietary components is observed in most patients. This condition can occur as early as halfway through treatment and generally increases in severity when therapy has been completed. Very often, reactions are dependent only in part on the dose of radiation received, however, and the response of the individual patient may play a major role in the syndrome. There are cases, in fact, in which malabsorption does not occur at all or diminishes immediately at the end of

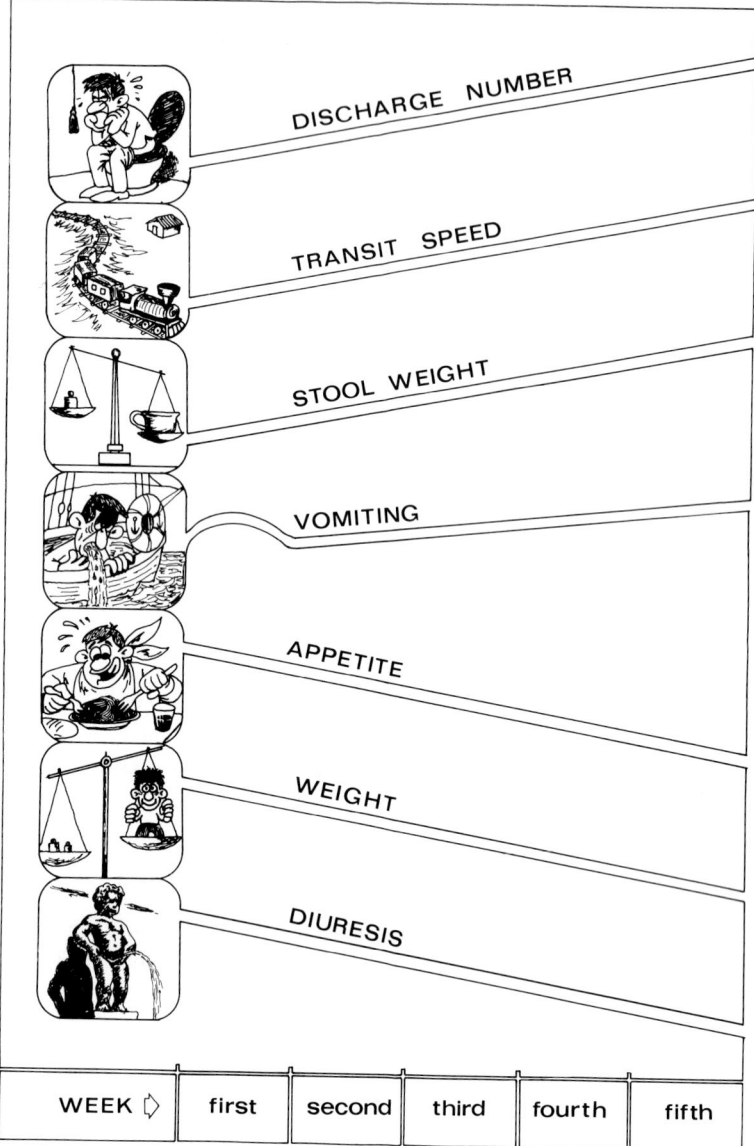

FIG. 9. Schematic representation of clinical symptoms in patients during radiotherapy.

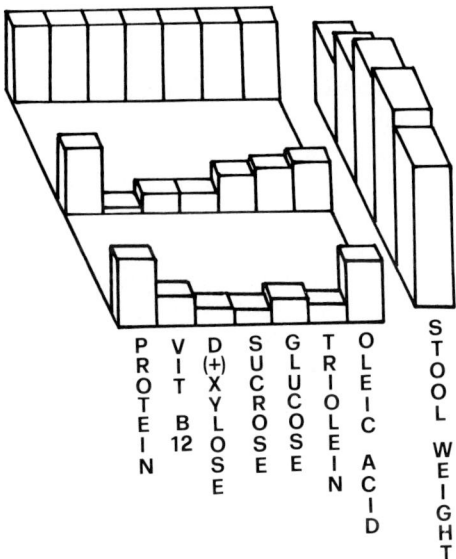

P R O T E I N   V I T · B 12   D (+) X Y L O S E   S U C R O S E   G L U C O S E   T R I O L E I N   O L E I C A C I D   S T O O L W E I G H T

FIG. 10. Results of absorption tests and stool weights in patients before, halfway, and at the end of radiotherapy. The plateau represents the 50% value.

treatment. Malabsorption test results also may differ depending upon which compounds are used in the tests. Pretreatment absorption values were generally comparable to those found in tests of healthy subjects (Fig. 10).

   a. Carbohydrate Absorption.   In patients receiving radiation in the region of the pelvis or lower hemiabdomen, carbohydrate absorption was evaluated by means of blood sugar curves after a glucose of sucrose oral load and by urinary excretion after a load of $D(+)$-xylose. Halfway through and at the end of treatment, the maximum blood level for glucose was 75.1 ±5.6% and 67.5 ±7.5%, respectively, when compared with control levels (Giannardi *et al.*, 1965), 63.4 ±6.8% and 59.8 ±6.0% for sucrose (Giannardi *et al.*, 1966, 1967), and 68.2 ±4.5% and 66.7 ±6.1% for xylose (Dalla Palma *et al.*, 1964; Dalla Palma, 1968). The greater reduction of sucrose absorption compared with that of glucose is related to a reduction in the activity of relevant brush border enzymes. Other authors also have observed a reduction in blood sugar levels (Dickinson, 1955; Sullivan, 1961; Bond, 1963; Perris, 1968; Sobo and Johnston, 1970). After an oral load of lactose, a substance capable of inducing diarrhea in adults by itself, a statistically significant reduction in absorption was observed between the first and the fifth week of treatment (Stryker *et al.*, 1978).

b. Amino Acid Absorption.   In the case of proteins and certain amino acids, absorption did not decrease to the same extent as it did in the case of carbohydrates, although the differences were statistically significant (Cionini *et al.*, 1971) with respect to control values. At the middle and end of treatment, the percentages of compounds absorbed were 93.8 ±0.1% and 95.1 ±0.1%, respectively, for albumin, and similar values were observed for monoiodotyrosine. For diiodotyrosine, an amino acid absorbed by passive transport, these values were 86.2 ±2.4% and 78.6 ±5.1%.

Moderate modifications in fecal nitrogen excretion were also demonstrated by other authors (Nathans *et al.*, 1960; Pelu *et al.*, 1964; Wiseman, 1968). Decreased protein absorption was also found in animals (Bennett *et al.*, 1951; Kiseley and Okulov, 1962; Gerber and Altman, 1970).

c. Fat Absorption.   Fat absorption was investigated using an oral load of fat to which had been added ^{131}I-labeled oleic acid or ^{131}I-labeled triolein. Pretreatment absorption of these molecules was 6–7% less in patients than it was in healthy subjects. After irradiation, oleic acid absorption was 91.6 ±2.6% and 88.0 ±2.5% of control values halfway through and at the end of treatment, respectively. Triolein values were 78.5 ±2.2% and 72.1 ±5.4% of control values. Tests performed 1–2 months later showed normal values (Dalla Palma *et al.*, 1962). As with other substances, the reduction in absorption was greater when the subjects had received abdominopelvic irradiation. No marked variations in the absorption of oleic acid during treatment either at 8 weeks or at the end of treatment have been observed by others (Goodrich and Hickman, 1962; Reeves *et al.*, 1965).

Data from animals are contradictory. Some authors have reported a decrease in absorption (Schwartz and Shapiro, 1961), while others did not observe significant changes after X-irradiation (Mead *et al.*, 1951; Maisin *et al.*, 1968).

d. ^{59}Fe Absorption.   Preirradiation absorption values of ^{59}Fe were higher in patients than in healthy subjects, perhaps because of disease. Absorption was 27.9 ±5.6% and 72.1 ±6.4%, respectively, halfway through and at the end of treatment when compared with pretreatment values (Dalla Palma, 1968). In irradiated animals, decreased ^{59}Fe absorption was observed only when the intestinal epithelium showed marked alteration (Donati *et al.*, 1969).

e. Vitamin B_{12} Absorption.   The function of the ileum has been studied by means of the vitamin B_{12} absorption test. This molecule is absorbed by the ileum only. Absorption in irradiated subjects was 64.9 ± 7.2% at the middle and 70.4 ±10.2% at the end of treatment (Dalla Palma and Cionini, 1964; Dalla Palma, 1968). Reduced capacity of absorption was noted also in patients who developed diarrhea (Yeoh *et al.*, 1984) and at the end of treatment (Goldberg, 1972).

Modification of intestinal bacterial flora might be responsible for vitamin B_{12} malabsorption, and tests performed on patients treated with antibiotics showed similar alterations (Dalla Palma and Cionini, 1964). Accelerated transit might also influence absorption (Pelu et al., 1964; Reeves et al., 1965; Yeoh et al., 1984).

A recent study of late damage in surviving patients who received external or intracavitary irradiation with radium applicators in 1966–1968 showed that serum folate levels were low in 3/23 patients and that vitamin B_{12} malabsorption was present in 5/23 (Lantz and Einhorn, 1984).

3. Other Factors: Water and Electrolytes; Intestinal Fluids

a. Water and Electrolytes. The loss of water and electrolytes due to diarrhea is one of the most important factors in acute radiation syndrome. In experimental animals, a postirradiation increase in the quantity of Na^+ present in the gastrointestinal tract and in the feces has been demonstrated. The loss is 10–15% of the Na^+ in the body (Sullivan, 1968; Mathieu and Gerber, 1970; Gits and Gerber, 1973), and it is concomitant with a decrease in absorption (Goodner et al., 1955; Curran et al., 1960; Zsebok and Petranyi, 1964; Perris et al., 1966). The results suggest that electrolyte loss is not a decisive factor in causing death by the gastrointestinal syndrome. In patients (Cionini et al., 1976), the loss of water and electrolytes did not cause important changes even when diarrhea was present, and diet and suitable drugs appeared to compensate for it. Some of the drugs used in these patients include opium, bismuth, and papaverine to act on the intestinal muscles or atropine and others to block the cholinergic effectors (De-Giuli et al., 1967). Determinations of plasma and blood volumes and total body water showed no significant modifications either halfway through or at the end of treatment (Cionini et al., 1976).

b. Intestinal Fluids. Changes in biliary and pancreatic secretions have been studied to elucidate the mechanisms responsible for malabsorption in patients. After a standard test meal containing carbohydrates, fats, and proteins, duodenal contents were analyzed for pH, pancreatic enzymes, and biliary secretion and presented no marked radiation-induced changes beyond the normal range of physiological variation (Dalla Palma et al., 1965; Becciolini et al., 1979). Patients who received epigastric radiation exhibited reduction in acidity and in the amounts of fluid collected, probably due to delayed gastric emptying time (Becciolini et al., 1979). The permeability of the mucosa remained within normal limits (Cionini et al., 1971) as demonstrated by a study of the quantity of macromolecules lost via the intestinal mucosa during treatment. In irradiated animals, this loss can be severe (Vatistas and Hornsey, 1966). The daily quantity of feces during the second week of treatment was 1.5–2 times as that seen initially (Pelù et al., 1965; Dalla Palma, 1968). In animals, intestinal contents enhance radiation injury in the small intestine (Hiatt and Warner, 1967; Morgenstern and Hiatt, 1967). The effects of different factors on intestinal injury were evaluated by means of suitable surgical techniques in dogs which received irradiation and parenteral feeding. Pancreatic secretions proved to be the most important factor in the production of intestinal damage, and trypsin was the most harmful constituent of those secretions (Rauch and Stenstrom, 1972; Yolk *et al.*, 1966). The intestinal segment in which trypsin was present showed the typical radiation-induced damage, whereas those segments in which only biliary secretions were present exhibited less severe injury.

4. Biliary Secretion after Abdominal Irradiation

The action of bile salts was studied in order to gain insight into the cause of radiation-induced diarrhea. Excessive amounts of bile salts in the colon increase water secretion and motility, and this causes biliary enteropathy (Hoffman, 1967; Archambeau *et al.*, 1967). It has been demonstrated that biliary duct canulation before irradiation prevents diarrhea (Jackson and Entenman, 1959; Sullivan, 1962, 1965; Hoffman, 1967). If physiological quantities of bile salts are injected into the duodenum of such animals, diarrhea appears again. Absorption of bile salts occurs in the terminal ileum; irradiation of this segment of the ileum leads to malabsorption with subsequent secretion of water and electrolytes. Reabsorption of bile acids decreases around the fifth week of treatment when the daily stool frequency increases considerably (Stryker *et al.*, 1977; Anderson *et al.*, 1978; Stryker and Demers, 1979). The late damage observed in these patients was contamination of the ileum by bacteria and malabsorption of bile salts and deconjugation of nonabsorbed bile salts (Newman *et al.*, 1973). Treatment with an ion-exchange resin, such as cholestyramine, capable of binding biliary salts has made radiation-induced diarrhea less severe (Condon *et al.*, 1978; Heusinkweld *et al.*, 1978). A low-fat diet also has been used successfully (Bosaeus *et al.*, 1979).

5. Radiation Effects on Intestinal Microflora

Under normal conditions, apart from the area beyond the ileocecal valve where they are abundant, bacteria normally are few in number. Coliform bacteria and enterococci in the midsection of the small intestine and cecum increase by factors of 100–1000 in irradiated animals. Numbers of lactobacilli, which are normally more numerous, either decrease or remain unchanged. These results indicate that there is a change in the number of constitutive microorganisms rather than an introduction of new species.

In patients irradiated for treatment of uterine carcinoma, coliform organisms increased in 61% of the cases observed, and staphylococci increased in 72% of the cases. Whenever intestinal motility is slowed, contents remain in the lumen longer, and this may lead to an increase in bacterial flora and to the production of toxic compounds. The role of infection in radiation death is important along with concomitant damage to the bone marrow (Hammond *et al.*, 1954; Congdon *et al.*, 1955; Rosoff, 1963). Bacteria that originated in the intestine have been found in the blood 1–2 weeks after a lethal dose of radiation was received (Kent *et al.*, 1968). Administration of antibiotics before or immediately after irradiation for a period of a few weeks results in a substantial increase in survival (Hammond *et al.*, 1954; Rosoff, 1963; Webster, 1967), probably by allowing the recovery of the intestinal epithelium. After doses above 10 Gy, no important differences are observed.

Studies on germ-free animals showed a higher postirradiation survival when such animals were compared with conventional ones. This result may also be explained based on differences in turnover rates of intestinal epithelium which, in germ-free animals, is longer. When higher doses are employed, death occurs in both types of animals when diffuse ulcerations appear (Matsuzawa and Wilson, 1965).

IV. Large Intestine

The main feature of this part of the alimentary tract is the lack of villi starting at the ileocecal valve. Compared with the small intestine, the crypts are higher in the colon and twice as high in the rectum. In these two structures, goblet cells are present in great numbers while enteroendocrine cells are rare and Paneth cells are absent. Proliferation is limited to the lower part of the crypt. Cell cycle time in the rat colon is 21.7 hr (G_1 = 13.5 hr; S = 6.3 hr; G_2 = 1.2 hr; M = 0.7 hr) (Galand, *et al.*, 1968), while in man it lasts about 24 hr (Lipkin, 1973). Cell turnover in the colon and rectum is slower than that in the small intestine, i.e., 3 days in the rat and about 6 days in man (Cole and McKalen, 1961; MacDonald *et al.*, 1964; Lipkin, 1965; Cameron, 1971). During radiotherapeutic treatment, the high sensitivity of the duodenum, the first part of the jejunum, and the terminal ileum could be due to the relative fixation of these segments. The relatively high radiotolerance of the colon could be attributed to the longer turnover time of its cells and to the presence of cells in prolonged interphase.

A. Acute and Late Radiation Injury

The sequence of events following a single dose of radiation is the same as that observed in the small intestine. The death of some epithelial cells and the inhibition of mitosis lead to a decrease in cell number as well as to alterations in cell morphology. In the submucosal and serosal layers, edema and vascular changes are noted, and lymphatic tissue disappears in the irradiated area. Damage to the mucosa is repaired in a few weeks. During treatment for pelvic tumors, hypermotility and tenesmus are observed, and the rectal mucosa appears edematous and hyperemic. When the period of treatment ends, most patients no longer show these effects, but late lesions, far more severe than the acute damage, may appear as early as during the first year. Progressive degeneration of connective tissue and blood vessels and stenosis may also occur. In the most severe manifestations of late damage, there may be ulceration or fistulae between the rectum and the vagina (Anderson *et al.*, 1955; Kottmeier, 1964; Rubin and Casarett, 1968b; Joelsson, 1970; Joelsson and Raf, 1973).

After high doses, severe alterations in the cellular organization of the mucosa and degenerative phenomena in the blood vessels can be seen, and these may lead to ulcerations. In all cases, the damage to connective tissue and small vessels can be very considerable and progressive: Edema, local inflammatory processes, thrombosis, and hyaline thickening are the most frequent features of this damage. Ulcers may occur in any part of the intestinal tract.

During the chronic phase of radiation damage, endarteritis and fibrinoid degeneration of blood vessels are seen in the intestinal wall. Ulceration of the mucosa can lead to inflammation in the deeper layers. Hyalinization and extensive fibrosis can cause intestinal obstruction.

The time that may elapse between radiotherapy and the onset of delayed symptoms varies from less than 6 months to as long as 22 years before the appearance of late effects (Fabrikant *et al.*, 1959; Joelsson and Raf, 1973; Van Nagell *et al.*, 1974). Five years can be regarded as the period of risk. Late injuries are less frequent in the large intestine than they are in other parts of the alimentary tract, and a small percentage of patients treated for gynecological cancer develop such injury. Generally, this sort of damage is ascribed to the motility of this part of the intestinal tract. It is seen most commonly in the (immobile) rectum.

During intracavitary or external irradiation for carcinoma of the uterus, the dose to the small intestine is low when adhesions are absent. When adhesions are present, a high radiation dose is given to limited segments of the small intestine, and the patients may be predisposed to radiation injury. In such cases, the radiotherapeutic regimen must be modified to reduce intestinal injury (Green *et al.*, 1975; Maruyama *et al.*, 1974). An early report on patients treated before 1945 did not mention intestinal complications after the application of radium with the old Stockholm method because the serous surface of the uterus was exposed to relatively low doses (Kottmeir, 1953). Subsequent technical advances and changes have brought about a significant increase in the survival of patients, but show an increased frequency of radiation-induced complications.

The effects of treatment with one fraction of 5.8 Gy/week for 5 weeks in patients with stage IIIB carcinoma of the cervix were compared with conventional fractionation (10 Gy in 5 daily fractions). Both treatments were followed by an intracavitary radium insertion with a dose of 40 Gy to point A. A few days after the end of the treatment, 5.8 or 6.7 Gy in a single session was given to the first group, and 10 Gy in five daily fractions were given to the second (Singh, 1978). Doses were calculated so as to have the same time/dose/fractionation (TDF); in addition, the nominal standard dose (NSD) and cumulative radiation effect (CRE) were similar. Early normal tissue reactions and early tumor regression were similar in the two groups. Late normal tissue reactions in the second group were more severe than in the first group and all patients showed late reactions or complications such as proctitis, perforation, and vesicovaginal or rectovaginal fistulae. There were more deaths in the second group (7/24) than there were in the first one (5/24).

Evaluation of the results indicates that radiotherapists should be advised against the use of large doses per fraction with NSD and CRE isoeffective as the standard 2-Gy dose in five daily fractions per week. Such large doses were not tolerated by the large intestine and normal tissues.

No relationships were observed between early reactions arising during and immediately after radiotherapy and late complications. Buchler et al. (1971) observed that early bowel reactions were present in only 7 out of 16 patients (from a total of 410 individuals) who developed late gastrointestinal or genitourinary fistulae. Moreover, 14 out of 33 patients who developed a stricture of the rectosigmoid showed moderate reactions during radiotherapy. Cowell (1953) reported that after radium therapy (45–55 Gy to point A) followed by 40 Gy of X-ray therapy over a period of 4 weeks, 59.5% of the patients developed complications. Frick et al. (1960) observed 4 deaths and 6 instances of severe complications in 65 cases after they had received 15–20 daily fractions of 2 Gy, whereas Deeley (1954) noted complications in about 55% of patients subsequent to a rectal dose of 54 Gy.

The incidence of delayed effects in the small intestine varies between 0.6 and 5% (Peckham et al., 1969; Dickson, 1972; Wellwood and Jackson, 1973; Maruyama et al., 1974; Schmitz et al., 1974; Green et al., 1975; Loiudice et al., 1977; Lederman et al., 1982; Meoz et al., 1983). In 7% of patients with malignant tumors of the ovary, small bowel obstructions have been observed in the first year after treatment (Potish, 1980).

Late alterations also occur in cases of epithelial carcinoma of the ovary treated by the intraperitoneal application of radiocolloids alone or in tandem with external radiotherapy to the pelvis (Pezner et al., 1978). In patients who had radiocolloid applications only, bowel complications at 5 years were 2.2%, but the frequency of such sequelae rose to 24% in patients treated with external radiotherapy as well; many of the latter patients subsequently required surgery.

In a series which primarily included patients who received pelvic irradiation for tumors of the cervix, uterus, or bladder and who had external radiotherapy, it was observed that 71% of the cases showed strictures and 54% showed chronic proctocolitis, and only some patients showed fistulae and necrosis (Jackson,

1976). These remarkable figures show that radiotherapy must be done carefully if severe bowel damage is to be avoided.

The rates of intestinal complications rose in a stepwise fashion at the 1900 to 2000 ret level following radiotherapeutic treatment of bladder cancer (Morrison, 1975). In fact, severe complications such as intestinal obstruction, fistula formation, bladder contracture, and hemorrhage rose from a level of 5% after 1700 ret to 10% after 1700 ret and to 40% after 2000 ret (25 fractions over a period of 35 days); respective tumor control rates were 53%, 61%, and 75%. Other authors reported similar complication rates (Liegner et al., 1962; Miller and Jones, 1962; Goodman and Balfour, 1964; Laing and Dickinson, 1965; van der Werf-Messing, 1965).

Typical symptoms in patients with large bowel complications are rectal bleeding and changes in bowel habits. At the time of diagnosis, clinical manifestations of late damage often are attributed initially to a recurrent tumor. In fact, it may be difficult to distinguish a rectal lesion from a recurrent tumor because of ulcerated and necrotic areas. Intestinal complications can progress and the results may be fatal. Colonic or rectal symptoms in patients who have been treated with radiotherapy should be investigated by sigmoidoscopy, rectal biopsy, and X-ray examinations such as barium enemas. This type of examination should reduce fatalities caused by intestinal obstruction and perforation (Kaplan et al., 1965; Green et al., 1978; Schraub et al., 1978). In addition, tests for malabsorption and CAT scanning can aid in the diagnosis.

It should be noted that surgical treatment may be technically difficult and hazardous for the patient if there is extreme fibrosis in the irradiated tissue or if there are infections which can lead to delayed healing and to peritonitis and fistula formation (Joelsson and Raf, 1973; Loiudice et al., 1977; Galland and Spencer, 1979; Rotman et al., 1979; Cochrane et al., 1981; Schmitt and Symmonds, 1981). Sometimes damage to the lower end of the ureters is also so severe that necrosis and ureteral fistulae can occur a few days after an operation. Sclerotic vessels a few centimeters away from the radiation field can be observed at times, and this makes it difficult to reestablish normal conditions in the intestine after surgery.

In some cases, surgery may be necessary and may resolve chronic radiation disease accompanied by anorexia, weight loss, abdominal pain, diarrhea, intermittent intestinal obstruction, fistulae, etc.

B. Predisposing Factors in the Production of Radiation Injury

There have been many studies of the factors which can be utilized to predict late radiation damage to the small intestine (Strockbine et al., 1970; Green et al., 1975; Lee et al., 1976). Time, dose, dose per fraction, and irradiated volume are the most common parameters to have been considered in this research.

The dose which can be tolerated for irradiation of the whole pelvic region is not known accurately, but there is reason to believe that the small intestine is at considerable risk at radiation doses exceeding 50 Gy (Rubin, 1974). Other predisposing factors for radiation damage to the bowel are (1) sex (the risk for women is higher than that for men), (2) age, (3) leanness, (4) hypertension, (5) previous surgery, (6) pelvic inflammatory disease, (7) vascular disease, and (8) diabetes mellitus.

Most of the patients studied had adhesions in the small intestine after surgery. In such cases, it is advisable to limit the dose to 45 Gy and to carry out accurate radiological studies before treatment (Green, 1983). The relative risk of developing late complications increases with higher doses as well as with the extent of early severe symptoms. If the latter are minimal, the frequency of later alterations will be low. It should be noted, however, that the absence of early reactions constitutes no reassurance that late complications will not occur (Bourne et al., 1983).

Preventive measures to reduce late damage have been applied to patients during the last few years, notably radiological studies of the irradiated volume, and such efforts have made it possible to limit the numbers of intestinal loops present in the treatment field, mainly in patients with adhesions (Fletcher et al., 1968). The results have been satisfactory (Green et al., 1978; Perez et al., 1978; Schraub et al., 1978; Green, 1983). In addition, there have been attempts to find probability models by analyses of radiation-related complications taking into account laparotomy, hypertension, leanness, and nutritional state (Potish et al., 1980, 1981).

Attempts to use hyperbaric oxygen (HBO) during abdominal irradiation in man (Johnson and Walton, 1974; Fowler, 1972) and in experimental animals (Johnson et al., 1972) have resulted in the demonstration that complications are markedly increased when HBO is breathed (Fletcher et al., 1977). Improved tolerance has been observed for treatment of abdominal and pelvic malignancies with two fractions per day (Loeffler, 1983).

C. Radiation Effects on the Rectum

The position of the rectosigmoid segment in the pelvis makes it less mobile and consequently more readily susceptible to radiation damage (although this segment is less radiosensitive than the small intestine). The numerous data published prior to 1955 on late damage to the rectum after treatment with radium sources are not comparable to those reported more recently because, in the earlier days, it was very difficult to determine the tissue dose to the rectum.

The rectum is believed to tolerate more than 50 Gy (Friedman, 1952; Chau et al., 1962; Rubin and Casarett, 1968b; Roswit et al., 1972a,b). Physical and biological factors which contribute to increased possibility of delayed radiation injury are the same as those for the small intestine. The presence of adhesions from previous surgery or laparotomies is very important and may lead to immobility of bowel loops.

During acute radiation injury in uterine carcinoma, more than 50% of the patients develop proctitis. Other complications are less frequent and fluctuate between 3 and 20% higher in patients who have had intraperitoneal lymphadenectomy (Anderson et al., 1955; Van Nagell et al., 1974; Joelsson, 1970; Fend et al., 1979; Green, 1983; Schellhammer and El Mahdi, 1983; Schmiedt et al., 1984).

The incidence of complications in prostatic cancer is 3–30%, with radiation damage to the urethra, bladder, and rectum (Joelsson and Raf, 1973; Schellhammer and El Mahdi, 1983). In cases of lymphomas irradiated over the abdomen, those treated with surgery and radiotherapy needed subsequent intervention much more frequently than did the cases treated with radiotherapy alone (Gennari et al., 1979). Cases requiring surgery fluctuate between 2 and 17% (Lindahl, 1970).

With radium therapy for uterine cancer, the incidence of rectal complications is less than 5% if the radiological technique is good (Fletcher et al., 1968; Rubin and Casarett, 1968b; Roswit et al., 1972a,b). As in the case of external treatment, the most severe signs of damage are ulceration, perforation, fistulae, fibrosis, and stricture. The greatest damage has been observed in the anterior wall of the rectum bordering on the posterior fornix.

D. Early and Late Intestinal Radiation Effects in Infancy

Children and infants irradiated with doses between 15 and 40 Gy to the entire abdomen for malignant diseases can show severe intestinal complications (Donaldson et al., 1975). An intestinal radiation-induced syndrome characterized by vomiting and diarrhea, sometimes presenting only in a mild form, has been observed in 70% of all cases. In 11% of the cases, symptoms of intermittent intestinal obstruction occur within 2 months after the end of radiotherapy. In 36% of the children and infants who survived for 1.6–7 years, delayed radiation enteritis occurred generally in those patients who had mild or severe early reactions during treatment.

The following factors, apart from dose, dose fractionation, and volume, modify the onset of radiation-induced enteritis: (1) age—children tolerate treatment better than do infants, (2) previous surgery, and (3) concomitant chemotherapy, particularly with actinomycin D.

In children, alterations are more severe than they are in adults. Nonetheless, conditions in enteritis may be improved if the patient is given a diet which is free from gluten, lactose, and protein from cow's milk, but which contains a limited amount of fat and low residue.

Hyperfractionated radiation to the abdomen (0.7 Gy × 5 fractions daily every 2 hr, 4 days per week, up to a total dose of 28 Gy over a period of 12 days) has also been used in children 1.5–12 years of age who had been treated with chemotherapy during the preceding months or with laparotomy before irradiation. Most patients had nephroblastoma and hepatoblastoma (Lagrange *et al.*, 1984) and a large abdominal field was irradiated. In 70% of the cases, immediate tolerance was poor. The conditions of these patients required unconventional treatments, and, because of short survival times, full evaluation of intermediate and long-term effects was not possible. Young age, recent or concomitant chemotherapy, and surgery were unfavorable factors.

E. Effects of Radiation on the Intestinal Microvasculature

A few studies have dealt specifically with the microvasculature in biopsy specimens obtained postoperatively from colectomies in patients treated previously with combined external and intracavitary radiotherapy (Carr *et al.*, 1984). Comparison with specimens from irradiated patients revealed a significant reduction in intramural vasculature in all layers of the intestinal wall. Variations in the diameter of the vessels, abnormalities in the lumen, and fibrosis in the mucosa and muscularis have been observed. Straightening of the submucosal vessels and a reduction in vascularity, particularly in the submucosa, were observed in the peripheral areas of the irradiated region. Application of a radiographic fluorescence technique revealed that the endarteritic damage was more severe than it had appeared to be on gross inspection (De Cosse *et al.*, 1969; Carr *et al.*, 1984). Late lesions involve mainly the connective tissues, and their onset was signaled by ulcerations, bowel obstructions, perforations, fistulae, lymphatic blockage in the mesentery with malabsorption, and atrophic scarring fibrosis with hyaline degeneration of smooth muscles which leads to bowel stenosis or to pathological adhesions (Green, 1983). These lesions are irreversible and may lead to intestinal obstruction by stenosis, perforation, hemorrhage, ulceration, fistulae, and malabsorption. Hypertension and diabetes can cause severe changes in small blood vessels and markedly influence the effect of radiotherapeutic treatment (Maruyama *et al*, 1974; Van Nagell *et al.*, 1974).

F. Effects of and Complications after Exposure to High LET Radiations

The effects of high LET irradiation on the small intestine, colon, and rectum have been investigated. Recovery from radiation damage occurs less readily after neutron irradiation than it does after exposure to other types of radiations, and the course of radiotherapy is less affected by environmental factors such as oxygenation of biological factors such as cell cycle phase when high LET radiations are used (Withers *et al.*, 1974). A few experimental studies have dealt with early and late effects on the colon and rectum after pelvic irradiation with 3 MeV neutrons or with γ rays from ^{137}Cs (Terry *et al.*, 1983). Changes in body weight, fecal deformity with short feces, and lethality indicate early damage due to alterations in the epithelium, whereas late damage is due to submucosal fibrosis and may lead to rectal stenosis. The relative biological effectiveness (RBE) proved to be similar to that observed in the skin and is higher for small doses per fraction of neutrons. Histological changes after neutron irradiation appear similar to, but more severe than those induced by γ rays.

Following fractionated doses, a rapid decrease in weight was observed in experimental animals between 11 and 17 days after irradiation (Terry and Denekamp, 1984). With regard to the number of deaths, there was no distinction as to the time following irradiation, but deaths were progressive. Short fecal pellets appeared within 6 months due to fibrosis constricting the rectum. The effects of recovery are practically absent after exposure to neutrons, whereas they are quite important following γ rays. High RBE and lack of repair are responsible for a much greater incidence of late damage after neutron exposure. The RBE for neutrons up to 1 Gy per fraction is 5.0 for acute and late effects. Larger doses per fraction can also lead to high RBE values for late damage.

Fast neutron irradiation has been used in the treatment of inoperable bladder and rectal cancer (Batterman and Breuer, 1981). Doses greater than 20 Gy led to ~25% severe complications in the skin and subcutaneous tissues. In the intestine, reactions requiring conservative medical or surgical care included necrosis and bowel perforation. Severe complications were caused by the neutron radiation doses to the skin and to the small and large intestines (Batterman and Breuer, 1981) using the 14 MeV D-T neutron generator.

A comparison among the effects of γ rays (^{60}Co), 8 MV X rays, and 14 MeV neutrons in patients with advanced pelvic tumors demonstrated that no deaths from intestinal complications were observed in patients treated with photons. In the neutron series, some patients died and at postmortem examination fistulae and perforation were noted in the irradiated bowel, while fibrosis, impaired mobility, and obstructions were observed in other parts of the gut (Baterman *et al.*, 1981).

Persisting complete regression of neoplasia was obtained after neutron treatment in about 50% of patients with advanced tumors that were assumed to have been incurable. Such a result is difficult to achieve with megavoltage X rays. The RBE, relative to 60% ^{60}Co γ rays, was 2.94 for serious skin damage, 3.17 for serious gut damage, and 3.19 for bladder tumor control.

G. Radiocarcinogenesis

Among the late effects which may occur after radiotherapy for carcinoma of the cervix is the appearance of rectal adenocarcinoma. A second tumor may appear to the site of radiotherapy in the damaged rectal wall from 5 to 30 years after irradiation (J. C. Smith, 1962; P. G. Smith, 1977; Castro *et al.*, 1973). In most such patients, prior radiation proctocolitis has been noted. Therefore, continuous care is necessary during the follow-up management of patients receiving irradiation to the abdomen. Also, in animals the late appearance of tumors of the colon or rectum after local irradiation was demonstrated (Hirose *et al.*, 1977; Denman *et al.*, 1978). The incidence depended on the dose administered. The tumors thus induced were adenocarcinomas like those found in man. In the stomach (Hirose, 1969; Hirose *et al.*, 1976) and small intestine (Osborne *et al.*, 1963), radiation can also cause adenocarcinomas. Some studies (Curtis *et al.*, 1984; Storm, 1983; Mettler and Moseley, 1985) also demonstrated that irradiation of the abdomen and pelvis for primary cervical or ovarian cancers or for benign gynecological disorders (Hutchinson, 1968; Dickson, 1972; Smith, 1977) could induce leukemia.

H. Radiation Effects in Experimental Animals

Results of studies designed to evaluate late radiation damage and rectal obstructions in laboratory animals have been published recently. In mice, strictures were investigated 90 days after irradiation of a small intestinal loop (Geraci *et al.*, 1977); similar phenomena were studied in rats 100 days after irradiation (Hubmann, 1981). A single dose of 22.5 Gy was needed to induce strictures in the small intestine in 50% of the animals, whereas 21.5 Gy were sufficient for the same effect in the rectum (Hubmann, 1981). After doses of 15–30 Gy in a single session, the frequency of rectal obstructions rose from 0 to 100% over a period of about 150 days. In some cases ureters were also obliterated. The effects included fibrosis of the submucosa and severe cystic fibrous proctitis. Atrophic and hyperplastic areas were observed in the mucosa. The submucosa increased in thickness due to marked fibrosis. Loss of elasticity leads to a rigid tube and transport of the formed feces appears to be difficult under such conditions. This leads to an accumulation of feces in the large intestine due to chronic obstruction coupled with a progressive enlargement of the abdomen. This dilation may cause compression of ureters and hydronephrosis.

In animals exhibiting obstruction the higher the dose, the more rapidly lethality occurs. Late-effect induction also was studied after fractionated irradiation to parts of exteriorized small intestine and alterations similar to those described already have been observed in such experiments (Black *et al.*, 1980; Hauer Jensen *et al.*, 1983).

V. Conclusions and View into the Future

Although the morphology and function of the large and small intestine have been known for a long time, research of recent years has demonstrated that much can still be learned about this vital organ. In the 1960s, the sequential processes of digestion had become well defined. Enzymatic activity was found in the brush border of the gastrointestinal tract leading to the development of the concept of "membrane digestion" which, in turn, led to the recognition that the intestinal mucosa participates actively in the hydrolysis of oligosaccharides and oligopeptides. In the 1970s, morphological investigations led to the unitarian theory of the origin of various cell types in the small intestine, leaving some kinetic aspects still to be clarified. The goal of the 1980s might well be to provide a better understanding of the mechanism of cell differentiation. These processes are instrumental in leading cells of common origin to differentiate after a limited number of divisions and to become cells with specific morphological and functional properties.

The small intestine can be used to evaluate the effects of physical and chemical agents which interfere with the process of differentiation. Thus, exposure of cell cultures or animals to ionizing radiations shows that this physical agent causes early differentiation in proliferating cells which are damaged sublethally, lose their proliferative capacity, and differentiate earlier than the nonirradiated cells.

Our studies have demonstrated this phenomenon of early differentiation during the first hours after irradiation prior to the appearance of acute damage. Furthermore, differentiation does not proceed normally during the recovery phase at a time when the epithelial cells of the small intestine have a normal morphological appearance, but are functionally deficient. The severity and consequences of this effect depend on the irradiation dose and the dose schedule. Although the phenomenon of early differentiation in irradiated cells of the small intestine appears well established, an application to radiation therapy seems far in the future. Nevertheless, the small intestine, because of its morphological and functional characteristics, may be considered a good model for investigating optimal conditions for dose–response and better tissue tolerance with less acute and late damage.

With regard to injury due to radiation therapy involving the gastrointestinal tract, the presently used dose schedules do not seem readily alterable. Although the intestinal tract is capable of a remarkable degree of recovery, the occurrence of late damage indicates that an increase of the usually employed radiation doses could be dangerous.

There are, however, some aspects of the cellular biology of the intestinal tract which could still be explored in an attempt to find new dose schedules that are better able to take into account the difference in tolerance between normal and neoplastic tissues. In addition to neutron therapy which is available in only a limited number of centers, the time–dose relationship could be exploited further. The cell cycle phenomena in the cells of the intestinal tract in relation to multiple daily fractionation could be analyzed to improve the time–dose relationship.

Only a few experimental and clinical data are available to shed light on the two above-mentioned aspects. Our studies demonstrate that acute radiation-induced injury of the small intestine of the rat is well tolerated when multiple dose fractionation is used, but these conditions are very different from those used in radiation therapy.

ACKNOWLEDGMENTS

The author's research was supported in part by the CNR special project entitled, "Control of Neoplastic Growth," grants 82.00228.96 and 83.02915.96.

The author wishes to thank Professor G. De Giuli for his encouragement, stimulation, and advice during the development of the research projects. The helpful suggestions by Professors Y. Maruyama and K.I. Altman and the revision of the English text by Dr. Ann B. Cox are also gratefully acknowledged.

A special acknowledgment belongs to my colleagues of the Radiation Biology Laboratory for their collaboration in the experimental work.

REFERENCES

Al Dewachi, H. S., Wright, N. A., Appleton, D. E., and Watson, A. J. (1976). *Cell Tissue Kinet.* **9**, 459–467.

Al Dewachi, H. S., Appleton, D. R., Watson, A. J., and Wright, N. A. (1979). *Virchows Arch.* **31**, 45–55.

Alov, I. A. (1963). *Fed. Proc., Fed. Am. Soc. Exp. Biol.* **357**, 22.

Al Nafuss, A. I., and Wright, N. A. (1982). *Virchows Arch.* **40**, 71–79.

Alpers, D. S. (1977). *Biochem. Biophys. Res. Commun.* **75**, 130–135.

Alpers, D. H., and Tedesco, F. J. (1975). *Biochim. Biophys. Acta* **401**, 28–44.

Anderson, H., Bosaeus, I., and Nystrom, C. (1978). *Acta Radiol. Oncol.* **17**, 312–318.

Anderson, R. E., Witkowski, L. J., and Pontius, G. V. (1955). *Surgery* **38**, 605–609.

Archambeau, J. O., Maetz, M., Jesseph, J. E., and Brenneis, H. J. (1967). *Arch. Surg.* **95**, 230–235.

Aschoff, J. (1979). *In* "Endocrine Rhythms" (D. T. Krieg, ed.), pp. 1–61. Raven, New York.

Battermann, J. J., and Breuer, K. (1981). *Br. J. Radiol.* **54**, 899–904.

Battermann, J. J., Hart, G. A. M., and Breuer, K. (1981). *Int. J. Radiat. Oncol. Biol. Phys.* **7**, 1039–1043.

Becciolini, A. (1980). *In* "Progress Report Radiation Protection Programme," pp. 891–895. Harwood Academic Publ., Euratom.

Becciolini, A., and Lanini, A. (1986). In preparation.

Becciolini, A., and Ravina, A. (1970). *Br. J. Radiol.* **43**, 150–151.

Becciolini, A., Ravina, A., Arganini, L., Castagnoli, P., and De Giuli, G. (1972). *Radiat. Res.* **49**, 213–225.

Becciolini, A., Castagnoli, P., Arganini, L., and De Giuli, G. (1973). *Radiat. Res.* **55**, 291–303.

Becciolini, A., Cariaggi, P., Arganini, L., Castagnoli, P., and De Giuli, G. (1974). *Acta Radiol.* **13**, 142–152.

Becciolini, A., Arganini, L., Cariaggi, P., Mazzi, S., and Franciolini, F. (1975). *Nunt. Radiol.* **39**, 13–35.

Becciolini, A., Arganini, L., Tedde, G., Vannelli, G., and Cariaggi, P. (1976). *Int. J. Radiat. Oncol. Biol. Phys.* **1**, 915–925.

Becciolini, A., Gerber, G. B., Buracchi, A., and Deroo, J. (1977a). *Strahlentherapie* **153,** 485–488.
Becciolini, A., Gerber, G. B., and Deroo, J. (1977b). *Acta Radiol. Ther.* **16,** 87–96.
Becciolini, A., Romano, S., Porciani, S., Buricchi, L., Benucci, A., and Casati, V. (1977c). *Chronobiologia* **4,** 100.
Becciolini, A., Cionini, L., Cappellini, M., and Atzeni, G. (1979). *Acta Radiol.* **18,** 145–154.
Becciolini, A., Cremonini, D., Balzi, M., Fabbrica, D., and Cinotti, S. (1982a). *Acta Radiol. Oncol.* **21,** 169–175.
Becciolini, A., Lanini, A., Giache, V., Balzi, M., and Bini, R. (1982b). *Acta Radiol. Oncol.* **21,** 273–279.
Becciolini, A., Porciani, S., Nardino, A., Lanini, A., and Giache, V. (1982c). *Strahlentherapie* *158,* 183–189.
Becciolini, A., Balzi, M., Cremonini, D., Cinotti, S., and Fabbrica, D. (1983a). *Acta Radiol. Oncol.* **22,** 201–207.
Becciolini, A., Balzi, M., Cremonini, D., and Fabbrica, D. (1983b). *Acta Radiol. Oncol.* **22,** 305–313.
Becciolini, A., Balzi, M., Cremonini, D., and Fabbrica, D. (1983c). *Acta Radiol. Oncol.* **22,** 337–344.
Becciolini, A., Cremonini, D., Fabbrica, D., and Balzi, M. (1983d). *Acta Radiol. Oncol.* **22,** 441–448.
Becciolini, A., Giache, V., Balzi, M., and Patacconi, F. (1983e). *Strahlentherapie* *159,* 508–512.
Becciolini, A., Cremonini, D., Fabbrica, D., and Balzi, M. (1986a). Submitted to *Acta Radiol. Oncol.*
Becciolini, A., Giache, V., Balzi, M., and Morrone, A. (1986b). Submitted to *Acta Radiol. Oncol.*
Bennett, L. R., Chastain, S. M., Decker, A. B., and Mead, J. J. (1951). *Proc. Soc. Exp. Biol.* **77,** 715–718.
Benucci, A., Becciolini, A., Nardino, A., Balzi, M., Cremonini, D., and Franciolini, F. (1981). *Acta Radiol. Oncol.* **20,** 193–198.
Berteloot, A., and Hugon, J. J. (1982). *Can. J. Biochem.* **60,** 434–443.
Berteloot, A., Khan, H., and Ramswamy, M. (1981). *Biochim. Biophys. Acta* **649,** 179–188.
Black, W. C., Gomez, L. S., Huhas, J. M., and Kligerman, M. M. (1980). *Cancer* **45,** 444–451.
Bloom, W. (1948). *In* "Histopathology of Irradiation from External and Internal Sources." McGraw-Hill, New York.
Bond, V. P. (1963). *Am. J. Clin. Nutr.* **12,** 194–204.
Bourne, R. G., Kersley, J. H., Grove, W. D., and Roberts, S. J. (1983). *Int. J. Radiat. Oncol. Biol. Phys.* **9,** 1445–1450.
Buchler, D. A., Kline, J. C., Peckham, B. M., Boone, L., and Carr, W. F. (1971). *Am. J. Obstet. Gynecol.* **111,** 745–751.
Cairnie, A. B., Lamerton, L. F., and Steel, G. G. (1965). *Exp. Cell Res.* **39,** 528–538.
Cameron, I. L. (1971). *In* "Cellular and Molecular Renewal in the Mammalian Body" (I. L. Cameron and J. D. Thrasher, eds.), pp. 45–85. Academic Press, New York.
Carr, N. D., Pullen, B. R., Hasleton, P. S., and Schofield, P. F. (1984). *Gut* **25,** 448–454.
Casarett, G. W. (1980). *In* "Radiation Histopathology," Vol. I, pp. 107–133. CRC Press, Boca Raton, Florida.
Casati, V., Nardino, A., Tomassi, I., Becciolini, A., Rizzi, M., and Martelli, T. (1979). *Acta Radiol.* **18,** 295–304.
Castro, E. B., Rosen, P. P., and Quan, S. H. (1973). *Cancer* **31,** 45–52.
Chau, P. M., Fletcher, G. H., Rutledge, F. N., and Dood, G. D. (1962). *Am. J. Roentgen Sol.* **87,** 22–40.
Cheng, H. (1974a). *Am. J. Anat.* **141,** 481–502.
Cheng, H. (1974b). *Am. J. Anat.* **141,** 521–536.
Cheng. H., and Leblond, C. P. (1974a). *Am. J. Anat.* **141,** 461–480.

Cheng, H., and Leblond, C. P. (1974b). *Am. J. Anat.* **141,** 503–520.

Cheng, H., and Leblond, C. P. (1974c). *Am. J. Anat.* **141,** 537–562.

Cionini, L., Becciolini, A., Dalla Palma, L., and De Giuli, G. (1971). *Acta Radiol.* **10,** 342–352.

Cionini, L., Becciolini, A., and Giannardi, G. (1976). *Strahlentherapie* **172,** 78–82.

Cochrane, J. P., Yarnold, J. R., and Slack W. W. (1981). *Br. J. Surg.* **68,** 25–28.

Cole, J. W., and McKalen, A. (1961). *Gastroenterology* **41,** 122–125.

Conard, R. A. (1953). *Proc. Soc. Exp. Biol. Med.* **82,** 333–337.

Conard, R. A. (1956). *Radiat. Res.* **5,** 167–188.

Condon, J. R., South, M., Wolverson, R. L., and Brinkley, D. (1978). *Postgrad. Med. J.* **54,** 838–839.

Congdon, C. C., Williams, F. P., Haberman, R. T., and Lorenz, E. (1955). *J. Natl. Cancer Inst.* **15,** 855–863.

Cowell, M. A. C. (1953). *Br. J. Radiol.* **26,** 652–656.

Crane, R. K. (1966). *Gastroenterology* **50,** 254–262.

Creamer, B. (1964). *Br. J. Med.* **2,** 1435–1436.

Curran, P. D., Webster, E. W., and Hovseplan, I. A. (1960). *Radiat. Res.* **13,** 369–379.

Curtis, R. E., Hankey, B. F., Myers, M. H., and Young, J. L. (1984). *J. Natl. Cancer Inst.* **72,** 531–544.

Dahlquist, A. (1962). *J. Clin. Invest.* **41,** 463–470.

Dalla Palma, L. (1968). *In* "Gastrointestinal Radiation Injury" (M. F. Sullivan, ed.), pp. 261–275. Excerpta Medica, Amsterdam.

Dalla Palma, L., and Cionini, L. (1964). *Nunt. Radiol.* **30,** 661–682.

Dalla Palma, L., Miliani, A., Cappellini, M., and Citi, S. (1962). *Radiol. Med.* **48,** 168–186.

Dalla Palma, L., Becciolini, A., Cappellini, M., Taddei, L., and Nori Bufalini, G. (1965). *Min. Radiol. Fisioter. Radiobiol.* **10,** 4–24.

Danielsen, E., Skovberg, H., Noren, D., and Sjostrom, H. (1981). *FEBS Lett.* **132,** 197–200.

De Cosse, J. J., Rhodes, R. S., Wentz, W. B., Reagan, J. W., Dworden, H. J., and Holden, W. D. (1969). *Ann. Surg.* **170,** 369–384.

De Dominicis, R., and Grechi, G. (1965). *Sperimentale* **115,** 82–103.

De Dominicis, R., Nori Bufalini, G., and Racheli, S. (1967). *Med. Nucl. Radiobiol. Lat.* **10,** 173–181.

De Dominicis, R., Aulisi, A., and Arganini, L. (1968). *Med. Nucl. Radiobiol. Lat.* **11,** 207–217.

De Dominicis, R. Arganini, L., and Aulisi, A. (1969). *Med. Nucl. Radiobiol. Lat.* **12,** 123–129.

Deeley, T. J. (1954). *J. Fac. Radiol. (London)* **5,** 289–291.

De Giuli, G. (1963). *Nuntius Radiol.* **29,** 91–93.

De Giuli, G. (1965). *Minerva Radiol. Fisioter. Radiobiol.* **10,** 231–234.

De Giuli, G. (1975). *Nunt. Radiol.* **39,** 83–94.

De Giuli, Gl., Dalla Palma, L., and De Dominicis, R. (1967). *Int. Congr. Ser.* **(105),** 1609–1616.

De Marzi, S. (1965). *Radiol. Med.* **51,** 375–401.

De Marzi, S., and Aulisi, A. (1966). *Nunt. Radiol.* **32,** 741–756.

Denman, D. L., Kirchner, F. R., and Osborne, J. W. (1978). *Cancer Res.* **38,** 1899–1905.

Dickinson, H. M. (1955). *Am. J. Physiol.* **182,** 477–478.

Dickson, R. M. (1972). *Clin. Radiol.* **23,** 528–535.

Doig, R. K., Funder, J. F., and Weiden, S. (1951). *Med. J. Aust.* **38.**

Donaldson, S. S., Jundt, S., Ricour, C., Sarrazin, D., Demerle, J., and Schweisguth, O. (1975). *Cancer* **35,** 1167–1178.

Donati, R. M., Berman, A. R., Jervis, H. R., Stromberg, L. W., and Sprinz, H. (1969). *Proc. Soc. Exp. Biol. Med.* **130,** 822–827.

Dubois, A., Jacobus, J. P., Grissom, M. P., Eng., R. R., and Conklin, J. J. (1984). *Gastroenterology* **86,** 444–448.

Dymock, I. W., (1969). *Br. J. Cancer* **20,** 236–238.

Dymock, I. W., MacKay, N., Miller, V., Thomson, T. J., Gray, B., Kennedy, E. H., and Adams, J. F. (1967). *Br. J. Cancer* **21**, 505–511.

Edwards, C. H., Gadsden, E. L., and Edwards, G. A. (1964). *Radiat. Res.* **22**, 505–511.

Fabrikant, J. I., Anlyan, W. G., and Creadick. R. N. (1959). *South. Med. J.* **52**, 1186–1191.

Fend, D. H., Gunderson, L., Krause, R., and Fisher, J. E. (1979). *Surg. Gynecol. Obstet.* **149**, 206–208.

Fletcher, G. H., Brown, T. C., and Rutledge, F. N. (1968). *Am. J. Roentgen Sol.* **79**, 421–450.

Fletcher, G. H., Lindberg, R. D., Cadero, J. B., and Wharton, J. T. (1977). *Cancer* **39**, 617–623.

Fowler, J. F. (1972). *Clin. Radiol.* **23**, 257–262.

Frick, H. C., Taylor, H. C., Guttman, R. J., Jacox, H. W., and McKelway, W. P. (1960). *Surg. Gynecol. Obstet.* **111**, 493–496.

Friedman, M. (1952). *Proc. Natl. Cancer Conf., 2nd* 390–400.

Galand, R. B., and Spencer, J. (1979). *Br. J. Surg.* **66**, 135–138.

Gennari, L., Balestrazzi, A., Zucali, R., Pizzocaro, G., Concolino, F., and Veronesi, U. (1979). *Tumori* **65**, 625–633.

Geraci, J. P., Jackson, K. L., Christensen, G. M., Thrower, P. D., and Weyer, B. J. (1977). *Int. J. Radiat. Oncol. Biol. Phys.* **2**, 693–696.

Gerber, G. B., and Altman, K. I. (1970). *In* "Radiation Biochemistry" (K. I. Altman, G. B. Gerber, and S. Okada, eds.), Vol. II, pp. 80–102. Academic Press, New York.

Gerber, G. B., Gerber, G., Altman, K. I., and Hempelman, L. H. (1963). *Int. J. Radiat. Biol.* **6**, 17–24.

Giannardi, G. F., De Marzi, S., Renzi, R., and Ciampi, G. P. (1965). *Nunt. Radiol.* **31**, 803–827.

Giannardi, G. F., De Marzi, S., Renzi, R., and Boddi, G. (1966). *Nunt. Radiol.* **8**, 723–730.

Giannardi, G. F., De Marzi, S., Renzi, R., and Atzeni, G. (1967). *Nunt. Radiol.* **33**, 2–13.

Gits, J., and Gerber, G. B. (1973). *Radiat. Res.* **55**, 18–28.

Goldberg, D. M. (1972). *Clin. Radiol.* **23**, 225–234.

Goldgraber, M. B., Rubin, C. E., Palmer, W. L., Dobson, R. L., and Massey, B. W. (1954). *Gastroenterology* **27**, 1–20.

Goodman, G. B., and Balfour, J. (1964). *J. Urol.* **92**, 30–36.

Goodner, C. J., Moore, T. E., Bowers, J. Z., and Armstrong, W. D. (1955). *Am. J. Physiol.* **183**, 475–483.

Goodrich, J. K., and Hickman, B. T. (1962). *Am. J. Roentgen Sol.* **87**, 69–75.

Green, N. (1983). *Int. J. Radiat. Oncol. Biol. Phys.* **9**, 1385–1390.

Green, N., Iba, G., and Smith, W. R. (1975). *Cancer* **35**, 1633–1640.

Green, N., Melby, E. R. W., Iba, G., and Kussin, L. (1978). *Int. J. Radiat. Oncol. Biol. Phys.* **4**, 1049–1053.

Hamilton, F. E. (1947). *Arch. Surg.* **55**, 394–399.

Hammond, C. W., Tompkins, M., and Miller, C. P. (1954). *J. Exp. Med.* **99**, 405–410.

Hauer Jensen, M., Sauer, T., Devik, F., and Nygaard, K. (1983). *Acta Radiol. Oncol.* **22**, 381–384.

Hayflick, L. (1965). *Exp. Cell Res.* **37**, 614–636.

Hendry, J. H. (1975). *Br. J. Radiol.* **48**, 312–313.

Heusinkveld, R. S., Manning, M. R., and Aristizabel, S. A. (1978). *Int. J. Radiat. Oncol. Biol. Phys.* **4**, 687–690.

Hiatt, N., and Warner, N. E. (1967). *PSEBM* **124**, 937–939.

Hirose, F. (1969). *Gann Monogr.* **8**, 75–113.

Hirose, F., Watanabed, H., Takeichi, N., Naito, Y., and Inoue, S. (1976). *Gann* **67**, 355–364.

Hirose, F., Fukazawa, K., Watanabe, H., Terada, Y., Fuji, I., and Otsuka, S. (1977). *Gann* **68**, 669–680.

Hoffman, A. F. (1967). *Gastroenterology* **52**, 752–757.

Hubmann, F. H. (1981). *Br. J. Radiol.* **54**, 250–254.

Hugon, J., and Borgers, M. (1965). *J. Microsc.* **4,** 643–656.

Hugon, J., Maisin, J. R., and Borgers, M. (1965). *Radiat. Res.* **25,** 289–502.

Hulse, E. V. (1969). *Int. J. Radiat. Biol.* **10,** 521–532.

Hutchinson, G. B. (1968). *J. Natl. Cancer Inst.* **40,** 951–982.

Jackson, B. T. (1976). *Proc. R. Soc. Med.* **69,** 683–686.

Jackson, K. L., and Entenman, C. (1959). *Radiat. Res.* **10,** 67–79.

Joelsson, I. (1970). *Acta Radiol. Suppl.* **302.**

Joelsson, I., and Raf, L. (1973). *Acta Chir. Scand.* **139,** 194–200.

Johnson, R. J., and Walton, R. J. (1974). *Am. J. Roentgen Sol.* **120,** 111–117.

Johnson, R. J., Wiseman, N., and Hogg, G. R. (1972). *Clin. Radiol.* **23,** 106–109.

Kay, R. E., and Entenman, C. (1959). *Am. J. Physiol.* **197,** 13–18.

Kaplan, A. L., Hundging, P. T., and Wall, J. A. (1965). *Am. J. Obstet. Gynecol.* **92,** 117–124.

Kaufman, M. A., Korsmo, H. A., and Olsen, W. A. (1980). *J. Clin. Invest.* **65,** 1174–1181.

Kent, T. H., Osborne, J. W., and Wende, C. M. (1968). *Radiat. Res.* **35,** 635–651.

Kiseley, P. N., and Okulov, N. M. (1962). *In* "Problems in Radiobiology" (M. V. Pobedinskii, ed.), Vol. II, pp. 166–177. Israel Program Sci. Trans., Jerusalem.

Klein, R. M. (1980). *Cell Time Kinet.* **13,** 153–162.

Kottmeier, H. L. (1953). *In* "Carcinoma of the Female Genitalia." Williams & Wilkins, Baltimore.

Kottmeier, H. L. (1964). *Acta Obstet. Gynecol. Scand. Suppl.* **2.**

Kurstin, I. T. (1963). *In* "Effect of Ionizing Radiation on the Digestive System." Elsevier, Amsterdam.

Lagrange, J. L., Roullet, B., Cosset, J. M., Demerle, J., and Sarrazin, D. (1984). *J. Eur. Radiother.* **5,** 227–286.

Laing, A. H., and Dickinson, K. M. (1965). *Clin. Radiol.* **16,** 154–164.

Lantz, B., and Einhorn, M. (1984). *Acta Radiol. Oncol.* **23,** 33–36.

Leblond, C. P., and Cheng, H. (1976). *In* "Stem Cells of Renewing Cell Populations" (A. B. Cairnie, P. K., Lala, and D. G. Osmond, eds.), pp. 7–31. Academic Press, New York.

Leblond, C. P., and Messier, P. (1958). *Anat. Rec.* **132,** 247–259.

Leblond, C. P., and Walker, B. E. (1956). *Physiol. Rev.* **36,** 255–276.

Lederman, M. V., Faria, S. L., Malzyner, A., and Vizeu, D. M. (1982). *Int. J. Radiat. Oncol. Biol. Phys.* **8,** 1447–1448.

Lee, K. H., Kagan, A. R., Nussbaum, H., Wollin, M., Winkley, J. H., and Norman, A. (1976). *Br. J. Radiol.* **49,** 430–440.

Liegner, L. M., Taylor, J. A., and Michaud, N. J. (1962). *J. Urol.* **87,** 373–380.

Lindahl, F. (1970). *Acta Chir. Scand.* **136,** 725–730.

Lipkin, M. (1965). *Gastroenterology* **48,** 616–625.

Lipkin, M. (1973). *Physiol. Rev.* **53,** 891–915.

Lipkin, M., Quastler, H., and Muggia, F. (1963). *Radiat. Res.* **19,** 277–285.

Loeffler, R. K. (1983). *Am. J. Clin. Oncol.* **6,** 619–627.

Loiudice, T., Baxter, D., and Balint, J. (1977). *Gastroenterology* **73,** 1093–1097.

Looney, W. B. (1966). *Int. J. Radiat. Biol.* **10,** 97–103.

MacDonald, W. C., Tier, J. S., and Everett, N. B. (1964). *Gastroenterology* **46,** 405–417.

Maisin, J. R. (1966). *In* "Radiation Radioprotecteur et Syndrome Gastro-Intestinal" (G. Thone, ed.). Liege.

Maisin, J. R., Dulcino, J., and Deroo, J. (1968). *In* "Gastrointestinal Radiation Injury" (M. F. Sullivan, ed.). Excerta Medica, Amsterdam.

Maryuyama, Y., Van Nagell, J. R., Utley, J., Vinder, M. L., and Parker, J. C. (1974). *Radiology* **112,** 699–703.

Mathieu, O., and Gerber G. B. (1970). *Atomkernenergie* **15,** 71–73.

Matsuzawa, T., and Wilson, R. (1965). *Radiat. Res.* **25,** 15–24.

Mead, I., Decker, A., and Bennet, L. (1951). *J. Nutr.* **43,** 485–499.

Meoz, R. T., Spanos, W. J., Dossy, L., Johnson, KR., and Wasserman, T. G. (1983). *Am. J. Clin. Oncol.* **6,** 417–422.

Mettler, F. A., and Moseley, R. D. (1985). "Medical Effects of Ionizing Radiation." Grune & Stratton, Orlando, Florida.

Miller, L. S., and Jones, J. S. (1962). *Tex. J. Med.* **58,** 716–720.

Montgomery, R. K., Sybickima, A., Forcier, A. G., and Grand, R. J. (1981). *Biochim., Biophys. Acta* **661,** 346–349.

Morgenstern, L., and Hiatt, N. (1967), *Gastroenterology* **53,** 923–929.

Morrison, R. (1975). *Clin. Radiol.* **26,** 67–75.

Nathans, D., Tapley, D., and Ross, J. R. (1960). *Biochim. Biophys. Acta* **41,** 71–282.

Newey, H., and Smith, D. H. (1959). *J. Physiol. (London)* **145,** 48–56.

Newman, A., Katsaris, J., Blendis, L. M., Charlesworth, M., and Walter, L. H. (1973). *Lancet* **2,** 1471–1473.

Nygaard, O. F., and Potter, R. L. (1959). *Radiat. Res.* **10,** 462–476.

Nygaard, O. F., and Potter, R. L. (1962). *Radiat. Res.* **16,** 243–252.

Osborne, J. W., Nicholson, D. P., and Prasad, K. M. (1963). *Radiat. Res.* **18,** 76–85.

Palmer, N. L., and Templeton, F. (1939). *J. Am. Med. Assoc.* **112,** 1429–1434.

Peckham, B. M., Kline, J. C., and Schultz, A. E. (1969). *Am. J. Obstet. Gynecol.* **104,** 485–494.

Pelu, G., Vichi, G. F., and Pistocchi, P. (1964). *Nunt. Radiol.* **30,** 816–829.

Pelu, G., Cavina, C., and Pistocchi, P. (1965). *Nunt. Radiol.* **31,** 945–956.

Perez, C. A., Korba, A., Zivnuska, F., Prasad, S., and Katzenstein, A. L. (1978). *Int. J. Radiat. Oncol. Biol. Phys.* **4,** 379–388.

Perris, A. D. (1968). *Radiat. Res.* **34,** 523–531.

Perris, A. D., Jervis, E. L., and Smyth, D. N. (1966). *Radiat. Res.* **28,** 13–29.

Pezner, R. D., Stevens, K. R., Tong, D., and Allen, C. V. (1978). *Cancer* **42,** 2563–2571.

Potish, R. A. (1980). *Radiology* **135,** 219–221.

Potish, R. A., Boen, J., Jones, T.K., and Levitt, S. H. (1981). *Radiology* **140,** 203–207.

Potten, C. S., and Hendry, J. H. (1983). *In* "Stem Cells" (C. S. Potten, ed.), pp. 155–199. Churchill Livingstone, Edinburgh.

Potten, C. S., Al Bawari, S. E., Hume, W. J., and Searle, J. (1977). *Cell Tissue Kinet.* **10,** 557–568.

Quaroni, A., Kirsch, K., Herscovics, A., and Isselbacher, K J. (1980). *Biochem. J.* **192,** 133–144.

Quastler, H. (1956). *Radiat. Res.* **4,** 303–320.

Quastler, H., and Hampton, J. C. (1962). *Radiat. Res.* **17,** 914–931.

Quastler, H., and Sherman, F. G. (1959). *Exp. Cell Res.* **17,** 420–438.

Rauch, R. F., and Stenstrom, K. W. (1952). *Gastroenterology* **20,** 595–603.

Ratzkowski, E., and Hochman, A. (1968). *Acta Radiol. Ther.* **7,** 417–432.

Reeves, R. J., Cavanaugh, P. J., Sharpe, K. W., Thorne, W. A., Winkler, C., and Sanders, A. P. (1965). *Am. J. Roentgen Sol.* **94,** 848–851.

Regaud, C., Nogier, T., and Lacassagne, A. (1912). *Arch. Electrotech. Med.* **21,** 321–334.

Ricketts, W. E., Kirsner, J. B., Humphreys, E., and Palmer, W. L. (1948). *Gastroenterology* **11,** 818–832.

Rider, J. A., Moeller, H. C., Althausen, T. L., and Sheline, G. E. (1957). *Ann. Intern. Med.* **47,** 651–665.

Rosoff, C. B. (1963). *J. Exp. Med.* **118,** 935–943.

Roswit, B., Malsky, S. Y., and Reid, C. B. (1972a). *Am. J. Roentgen Sol.* **114,** 460–475.

Roswit, B., Malsky, S. J., and Reid, C. B. (1972b). *Front. Radiat. Ther. Oncol.* **VI,** 160–181.

Rotman, M., John, M. J., Moon, S. H., Choi, K. N., Stowe, S. M., Abitbol, A., Herskovic, T., and Sall, S. (1979). *Int. J. Radiat. Oncol. Biol. Phys.* **5,** 327–332.

Rubin, P. (1974). *Semin. Roentgenol.* **9**, 51–61.

Rubin, P., and Casarett, G. W. (1968a). *Clin. Radiat. Pathol.* **I**, 153–191.

Rubin, P., and Casarett, G. W. (1968b). *Clin. Radiat. Pathol.* **I**, 193–240.

Rubin, P., and Casarett, G. W. (1972). *Front. Radiat. Ther. Oncol.* **VI**, 1–16.

Saito, M., Murakami, F., and Suda, M. (1976). *Biochim. Biophys. Acta* **421**, 177–179.

Schellhammer, P. F., and El Mahdi, A. M. (1983). *Urology* **21**, 451–457.

Schmiedt, E., Schmeller, N., and Mayer, P. (1984). *Urology* **23**, 149–152.

Schmitt, E. H., and Symmonds, R. F. (1981). *Surg. Gynecol. Obstet.* **1534**, 896–900.

Schmitz, R. L., Chao, J. H., and Juanito, B. (1974). *Surg. Gynecol. Obstet.* **138**, 29–32.

Schraub, S., Monnier, A., and Paulin, G. (1978). *J. Radiol. Electrol.* **59**, 347–352.

Schultze, B., and Kellerer, A. M. (1979). *Cell Tissue Kinet.* **12**, 347–359.

Schwartz, E. E., and Shapiro, B. (1961). *Radiology* **77**, 83–90.

Seeman, W. B., and Ackerman, L. V. (1957). *Radiology* **68**, 534–541.

Seethram, B., Kwo Yi Yeh, and Alpers, D. H. (1980). *Am. J. Physiol.* **237**, 524–531.

Semenza, G. (1968). *Handb. Physiol.* **V**, 2543–2650.

Sigdestad, C. P., and Lesher, S. (1970). *Experientia* **26**, 1321–1322.

Singh, K. (1978). *Br. J. Radiol.* **51**, 357–362.

Sleisenger, M. H., Almy T. P., and Barr, D. P. (1973). *Am. J. Med.* **15**, 666–674.

Smith, J. C. (1962). *Proc. R. Soc. Med.* **55**, 701–702.

Smith, P. G. (1977). *Cancer* **39**, 1901–1905.

Sobo, A. O., and Johnston, I. D. (1970). *Br. J. Surg.* **57**, 391.

Stevens C. E., and Leblond, C. P. (1953). *Anat. Rec.* **151**, 231–246.

Stevenson, H. R., and Fierstein, J. S. (1976). *Am. J. Physiol.* **230**, 731–735.

Storm, H. H. (1983). *Eur. J. Cancer* **19**, 1317.

Strockbine, M. F., Hancock, J. E., and Fletcher, G. H. (1970). *Am. J. Roentgenol.* **108**, 293–304.

Stryker, J. A., and Demers, L. M. (1979). *Int. J. Radiat. Oncol. Biol. Phys.* **5**, 935–939.

Stryker, J. A., Hepner, G. W., and Mortel, R. (1977). *Radiology* **124**, 213–216.

Stryker, H. A., Mortel, KR., and Hepner, G. W. (1978). *Int. J. Radiat. Oncol. Biol. Phys.* **4**, 859–863.

Sullivan, M. F. (1961). *Am. J. Physiol.* **201**, 1013–1016.

Sullivan, M. F. (1962). *Nature (London)* **195**, 1217–1218.

Sullivan, M. F. (1965). *Am. J. Physiol.* **209**, 158–164.

Sullivan, M. F. (1968). *In* "Gastrointestinal Radiation Injury" (M. F. Sullivan, ed.), pp. 216–227. Excerpta Medica, Amsterdam.

Summers, R. W., Kent, T. H., and Osborne, J. W. (1970). *Gastroenterology* **59**, 731–739.

Tarpila, S. (1971). *Scand. J. Gastroenterol.* **6**, (Suppl. 12), 9–69.

Tedde, G., Vannelli, G., Becciolini, A., and Arganini, L. (1972). *Arch. It. Anat. Embriol.* **77**, 393–410.

Tedde, G., Becciolini, A., and Vannelli, G. (1975). *Nunt. Radiol.* **39**, 71–81.

Terry, N. H., and Denekamp, J. (1984). *Br. J. Radiol.* **57**, 617–629.

Terry, N. H., Denekamp, J., and Maugham, R. L. (1983). *Br. J. Radiol.* **56**, 257–265.

Trier, J. S. (1962). *Gastroenterology* **43**, 407–424.

Trier, J. S. (1968). *Handb. Physiol.* **III**, 1125–1175.

Trier, J. S., and Browning, T. H. (1966). *J. Clin. Invest.* **45**, 194–204.

Trier, J. S., Browning, T. H., and Foroozan, F. (1968). *In* "Gastrointestinal Radiation Injury" (M. F. Sullivan, ed.), pp. 57–72. Excerpta Medica, Amsterdam.

Ugolev, A. M. (1968). *In* "Physiology and Pathology of Membrane Digestion." Plenum, New York.

Ugolev, A. M., and De Laey, P. (1973). *Biochim. Biophys. Acta* **300**, 105–128.

Van Der Werf-Messing, B. (1965). *Clin. Radiol.* **16**, 165–172.

Van Nagell, J. R., Maruyama, Y., Parker, J. C., and Dalton, W. L. (1974). *Am. J. Obstet. Gynecol.*
    118, 163–167.
Vatistas, S., and Hornsey, S. (1966). *Br. J. Radiol.* **39,** 547–550.
Webster, J. B. (1967). *Radiat. Res.* **32,** 117–124.
Wellwood, J. M., and Jackson, B. T. (1973). *Br. J. Surg.* **60,** 814–818.
Wiernik, KG. (1966). *Br. J. Radiol.* **39,** 272–279.
Wiseman, G. (1968). *Handb. Physiol.* **III,** 1277–1307.
Withers, H. R., Mason, K., Reid, B. O., Dubravsky, N., Barklery, H. T., Brown, B. W., and
    Smathers, J. B. (1974). *Cancer* **34,** 39–47.
Yeoh, E. K., Lui, D., and Lee, N. Y. (1984). *Br. J. Radiol.* **57,** 1131–1136.
Yolk, B. W., Wellmann, K. F., and Lewitan, A. (1966). *Am. J. Pathol.* **48,** 721–753.
Zsebok, Z., and Petranyi, G. (1964). *Acta Radiol.* **2,** 377–383.

Relative Radiosensitivities of the Oral Cavity, Larynx, Pharynx, and Esophagus

JOELLA F. UTLEY

DEPARTMENT OF RADIATION ONCOLOGY
SPARTANBURG GENERAL HOSPITAL
SPARTANBURG, SOUTH CAROLINA 29302

I. Introduction

Since the early days of roentgentherapy, the literature provides careful descriptions of acute and chronic radiation reactions in the tissues of the oral cavity and pharynx. Before the mid-1920s, radiation therapy varied greatly from treatment center to treatment center, but as experience grew knowledge of the most effective ways to use radiation as a therapeutic tool developed. The early investigators, using low-energy X-ray sources, made careful observations on a daily basis in order to avoid excessive acute reactions. They became aware of the need for protracted fractionation as well as of the daily and total dose that would allow tumor cures within the limits of normal tissue tolerance. The classic review by Coutard (1932), based on a series of patients treated for epitheliomas of the tonsillar region, hypopharnx, and larynx between 1920 and 1926 could be used as the description of normal tissue reactions seen in modern-day radiation clinics. It appears that time, dose, and volume may be more important factors than the energy of the radiation source to produce a specified biological endpoint.

Animal studies have played a role in developing our knowledge of tissue reactions to radiation. The mouse and rat are used most commonly and are suitable for comparison of different treatment schemes. However, the small oral cavity of these animals limits gross observations. For example, it is difficult to grade mucositis in the small animal. Larger animals with large oral cavities such as the dog have the advantage that gross observations and pathophysiologic measurements are easy (i.e., the measurement of salivary flow following irradiation).

The clinical conditions prevailing during fractionated daily radiation treatments are usually lacking in animal studies. In general, animals must be anesthetized for treatment to assure controlled settings. Daily anesthesia and radiation treatment is difficult, time-consuming, and hazardous to animals. Some of the current questions being asked clinically regard multidaily radiation treatments, and this will need to be answered at the clinical level. With the abundance of human material available for such studies, animal studies hardly seem important. However, a great deal of information has and will be derived from animal studies of microscopic, ultrastructural, and histochemical changes produced by radiation. Also, studies using radiosensitizing and protecting drugs and chemotherapeutic agents must be done first in animals to assess tolerance and effectiveness before these agents are used in humans.

Although acute radiation reactions may result in serious complications, perhaps more so in the elderly and debilitated, the late changes result in the most significant injury. There is little relation between acute mucosal membrane reactions and late damage indicating that the tissues producing the two effects are different. Cells that proliferate rapidly show early radiation death, but regenerate and repopulate quickly following injury. Mucosal epithelial cells show the earliest radiation death, but regenerate and repopulate soon following injury. Clinically, these cells show the earliest radiation reaction in the oral cavity, but the mucosa heals within a few weeks following treatment. Underlying bone and cartilage which have nonrenewing or slow proliferating rates may show no evidence of acute radiation injury, but late radiation changes occur months to years later and sometimes appear only after trauma or infection.

The individual dose per fraction is related to the severity of late effects. Large single doses deplete the normal cell population, making repair impossible. In those tissues showing late complications, a combination of parenchymal cell death, vascular injury, and fibrocyte dysfunction may operate jointly to result in complications (Kogelnik and Kärcher, 1977). Fibrocytes are abundant in subcutaneous connective tissues and fibrosis presents as a late reparative condition.

The radiation tolerance of the upper aerodigestive tract is dependent on total dose and overall time of delivery, but these two factors must be considered in conjunction with the volume of tissue irradiated. As noted by Coutard (1932), by lengthening the total duration of treatment, especially for large irradiated fields, immediate and late radiation accidents can be avoided.

Tumor destruction of the normal mucosal and underlying supportive tissues prior to irradiation may result in slower healing and secondary late changes of ulceration, infection, and necrosis. The submucosa is of variable thickness and is in close proximity to the lamina propria which lies directly over the periosteal or perichondrial tissue. Erosion of these thin epithelial surfaces may result in progressive necrosis of supporting structures.

Many technical details are important for normal tissue preservation. Careful field size planning, daily treatment of all radiation fields, shielding of sensitive organs, shrinking field techniques, and concerns for the tolerance levels of critical structures result in radiation treatment with a good therapeutic differential. In recent years, attempts to widen the therapeutic differential (the difference between radiation injury to tumors and normal tissues) have led to the development of radioprotective and radiosensitizing drugs. Brachytherapy and electron and neutron therapy provide therapeutic advantages in the proper setting.

As communication through the radiotherapeutic literature grows, treatment techniques and treatment results have become uniform. A common feeling originating during the early days of radiotherapy was expressed in an editorial (*Lancet*, 1941, **1,**140–141) that radiotherapy lacks a scientific basis and rests on an empirical footing. However, knowledge in the field of radiation therapy and radiobiology has been growing rapidly, making such statements obsolete. An increasing number of human cancers are being treated and cured with radiation therapy within the limits of good normal tissue tolerance.

II. Mucosal Epithelial Tissues

Radiomucositis is the earliest reaction seen in patients treated for malignancies of the oral cavity. The time sequence of development of lesions is dose dependent. Coutard (1932) lists, in their order of appearance, those structures showing a change during the course of irradiation. Lesions appear on the pillars and uvula on day 13, on the hypopharynx on day 14, followed by the vallecular regions, origin of the pharyngolaryngeal furrows, and floor of mouth. Next to show change is the mucosa of the cheeks, horizontal and vertical branches of the inferior maxilla on day 16, the laryngeal surface of the epiglottis and lateral parts of the intraarachnoid space and base of the tongue on day 18, and, finally, on the vocal cords as well as the dorsal and anterior surfaces of the tongue on days 20–22.

The first observable change is erythema occurring during the first week of irradiation after the standard daily fractions of 200 rad. After the second week at doses above 2000 rad, white patches appear on the mucosa, increase in size, and coalesce as the radiation dose is increased. When the entire mucosa shows this white reaction, it is referred to as a false membrane and represents dead epithelial cells in the process of sloughing. Progression of mucositis through a course of irradiation and subsequent healing is well illustrated in a series of photographs published by Rubin and Casarett (1968).

Below the layer of dead epithelial cells lies an inflamed submucosa which may become apparent as the mucosa is lost. Ulcers can develop at higher doses and a fibrinous exudate may form. Severe, acute reactions presenting as bullous edema secondary to lymphatic obstruction have been described (Rubin and Casarett, 1968). Candidiasis presenting as white plaques or nodules secondary to changes in oral flora may appear within and outside of the areas of the radiation field.

Patients complain of increasing discomfort as the reactions to radiation progress in severity. Symptoms start in the second week of fractionated radiation therapy and by the third to fifth weeks become increasingly severe, requiring supportive measures and occasional periods of rest. Painful mucositis is accompanied by changes in other organs, primarily the salivary glands, which creates discomfort in chewing. A sore throat and pain upon swallowing are accompanied by poor nutrition and weight loss (Chencharick and Mossman, 1983; Mossman and Henkin, 1978). Reducing the daily dose to 180 rads can result in improved tolerance. Heightened mucosal reactions can occur near gold fillings, which produce an electron backscatter on the adjacent mucosal surface. Patients who continue to smoke and use alcoholic beverages have exaggerated mucosal reactions. Ill-fitting dentures may enhance the mucosal reaction and increase the pain.

These acute complications usually heal rapidly. Increased cell proliferation provides repopulation of the denuded epithelium often within 2–3 weeks. After 4–6 weeks the mucosa usually has regained its normal appearance and the acute symptoms have cleared.

Late changes induced by radiation include scarring from fibrocyte infiltration and decreased vascularity within submucosal tissues. These tissues may be more susceptible to injury and less able to repair damage. Trauma, the continued insult resulting from the use of alcohol and tobacco or from ill-fitting dentures, may cause ulceration and late necrosis. Avoiding the use of dentures has been advised for as long as 1 year following therapy (Blozis and Robinson, 1968).

Careful management of ulcers overlying bone, e.g., the alveolar ridge, is especially important, since secondary infections may lead to necrosis of these underlying structures. Cautious evaluation of nonhealing ulcers is necessary because they may represent uncontrolled tumor.

Chemical radioprotection of the oral mucosa has been demonstrated using the compound WR-2721 [S-2-(3-aminopropylamino)ethylphosphorothioic acid] (Utley et al., 1978). Dogs irradiated over one-half of the oral cavity with single-dose fractions ranging from 1600 to 6400 rad received intravenously injected WR-2721 or 0.9% saline solution prior to irradiation. The reaction of the oral mucosa was reduced with a dose-modifying factor (DMF) of 1.7 at a grade 3 reaction (confluent mucositis), whereas at the highest reaction grade of 5 (necrosis) the DMF fell to 1.2. The oral radiation death syndrome (ORD) in animals causes death in 50% of the mice 10 days after irradiation of the entire head with 1600 rad (Quastler et al., 1956). The factors contributing to the muchanism of death include mucosal swelling, causing the animals to die of starvation (Goepp and Fitch, 1962). WR-2721 protects against this endpoint by a DMF of 2.1 (Utley et al., 1978; Grigsby and Maruyama, 1982). Grigsby and Maruyama

(1980, 1981) showed that WR-2721 protected against the ORD in animals carrying a buccal mucosal tumor. When they used the radiosensitizer, misonidazole, they observed radiosensitization of the tumor, but when they used misonidazole and WR-2721 in combination, tumor sensitization decreased while the protective effect of WR-2721 on the mucosa persisted.

III. Salivary Glands

There are important changes in salivary tissues following irradiation. These changes cannot be ignored because they evoke profound acute and chronic symptoms in other structures of the oral cavity. There are three major pairs of salivary glands, the parotids, submandibulars, and sublinguals as well as numerous small gland clusters in the mucosa and submucosa of the soft palate, base of the tongue, and retropharyngeal structures. The smaller glands seem to secrete continuously and furnish the liquid state of the mouth while the larger glands secrete when nerve endings are stimulated by visual, olfactory, thermal, mechanical, or chemical stimuli. The major glands contain either a pure or a mixed parenchymal cell population which secrete mucous (thick) or serous (watery) saliva. The parotid gland is composed primarily of serous cells, while the submandibular and sublingual glands contain a mixture of serous and mucinous-secreting cells. Minor salivary gland structures are purely mucinous. The gland structure includes the alveoli or acini consisting of groups of secretory cells feeding into graded-sized ducts which terminate in the gland opening in the oral cavity. The saliva provides lubrication as well as proteins, electrolytes, and enzymes used in the digestion of foods (Bloom and Faucett, 1962a–c).

In animals, radiation causes a decrease in weight and volume of salivary glands (English *et al.*, 1955) and, after large, single doses, the glands actually become difficult to locate for dissection (Cherry and Glucksmann, 1959). As the size of the gland decreases, there is a simultaneous reduction in lobes and acini and an alteration of cellular structure followed by late fibrotic ingrowth. Microscopically visualized changes include nuclear degeneration, vacuolization of the cytoplasm, and loss of secretory granules. Regeneration appears after doses of 1000 rad, but above this dose effective regeneration does not take place. After doses of 3000 rad, nonfunctioning adenomas appear as the end result of abortive regeneration (Cherry and Glucksmann, 1959). Microscopically visible radiation-induced changes in human parotid glands resemble those seen in animals, including inflammatory cell infiltrates, degeneration of acini, particularly the serous cells, and dilatation of ducts. Larger ducts withstand the damage better than the smaller ones. The parotid gland suffers greater radiation damage than the submaxillary or sublingual glands and results in the major loss of serous cells (Kashima *et al.*, 1965).

As early as 3 hr after single doses ranging from 400 to 6400 rad, cytoplasmic degeneration and the appearance of vacuoles and crystalloids can be seen at the ultrastructural level in the rat parotid gland (Sodicoff *et al.*, 1974). Macrophages increase in number and become filled with degenerative debris. By the end of 1 week, most of the damage has been removed and macrophage activity has decreased. Two days after radiation, damage to acinar cells is marked, while ductal structures appear normal. Thus, whereas the electron microscope reveals intracellular damage within a few hours, the light microscope shows damage only after several days.

Early roentgenologists using large radiation doses noted swelling of the salivary glands associated with pain within the first few hours after exposure. Doses of 1500–2750 rad produced a dry mouth 2.6 hr later and pain and tenderness within 4.6 hr, reaching a maximum between 12 and 24 hr (Kashima *et al.*, 1965). There was no associated increase in temperature or white blood count. The symptoms and swelling disappeared in 7 days.

These severe, acute symptoms are not seen with standard clinical treatment schedules, but patients often complain of a dry mouth after the first week of radiation therapy. This effect is more pronounced when larger radiation portals are necessary such as used for the treatment of malignancies of nasopharynx, base of the tongue, and tonsillar fossa. By the third week of treatment, the dry mouth is readily apparent and remains as a chronic problem for years. After a full course of radiation therapy, most patients produce too small an amount of saliva to be measured (Marks *et al.*, 1981).

Shannon reports that following 225 rad flow rates from the resting parotid gland are reduced by one-half, and following a second such dose, parotid flow is obliterated (Shannon *et al.*, 1978a). The amount of serous saliva decreases rapidly and that of mucinous saliva becomes prominent. Secretions become mechanically harder to swallow and produce gagging, cough, and even emesis. Patients learn to consume large amounts of liquids during meals and throughout the day and night to relieve the discomfort. Food collects on the teeth where it would normally be set free by the flow of saliva. The small amount of saliva that remains becomes acid (Dreizer *et al.*, 1976; Frank *et al.*, 1956; Marks *et al.*, 1981), a condition which softens the enamel of teeth (van den Brenk *et al.*, 1969). Microorganisms of a cariogenic type increase in numbers (Brown *et al.*, 1975; Llory *et al.*, 1972) and dental defects become apparent. Several investigators have studied salivary flow by sialography. The size of the gland before irradiation correlates with the salivary output, large glands having larger initial salivary flows than small ones (Cheng *et al.*, 1981). Preradiation evacuation of hypaque from the parotid glands occurs within 5 min, while incomplete evacuation lasts as long as 25 days following treatment (Kashima *et al.*, 1965). A dose of 500 rad reduces the intraductal pressure in man 80% within 24 hr and 1500 rad eliminates the pressure entirely (Jennings and Arden, 1960). Despite a profound decrease in flow, glands of irradiated patients are only slightly reduced in size at the end of radiation therapy, but become smaller over the months to a year following treatment (Eneroth *et al.*, 1971). Edema probably maintains the size of the gland immediately following treatment, whereas subsequent atrophy occurs because of stromal fibrosis and sclerosis. Shielding portions of the glands during radiation therapy can afford significant protection of the glandular secretory apparatus. Exclusion of the parotid glands from the radiation field brings about little change in stimulated salivary flow, whereas irradiation of the submandibular and sublingual glands contributes little to xerostomia (Mira *et al.*, 1981). Shielding is possible when small fields in the oral cavity are irradiated and when the contralateral gland can be spared by employing electron beam therapy to a single portal (Marks *et al.*, 1981). Split-course radiation treatment may reduce xerostomia somewhat, although this was reported only in a small, uncontrolled series (Mira *et al.*, 1981).

Amylase secretion and serum amylase levels change early in irradiated patients. Serum amylase activity shows an initial rise, but after three to four fractions most of the amylase-secreting tissue is greatly reduced, preventing the recurrence of hyperamylasemia when tissues are reexposed to irradiation (van den Brenk *et al.*, 1969). The rise of serum amylase activity corresponds to the volume of gland irradiated (Kashima *et al.*, 1965). Mossman, however, using chromatographic analysis of human parotid saliva, did not observe the fall in amylase in saliva until after a 4-week treatment period. Beyond this time salivary flow was too small to measure (Mossman *et al.*, 1981). Secretory IgA levels rose initially and then fell as patients continued treatment for parotid gland tumors (Marks *et al.*, 1981). The content of Na^+, Cl^-, Ca^{2+}, Mg^{2+} proteins is increased, while that of HCO_3^- is decreased, yielding a concentrated solution with poor buffering ability (Dreizer *et al.*, 1976).

The protection of salivary glands from radiation injury by drugs has received considerable attention in recent years. The drug WR-2721 concentrates in the major salivary glands of mice to a greater degree than in other organs (Utley *et al.*, 1976), and this led to the hope that salivary cells could be protected from radiation. Sodicoff *et al.* (1978a) measured a DMF of 2.4 for radiation-induced gland atrophy, DMF values of 3.2 for glandular amylase concentration, and of 2.0 for total gland amylase. Ultrastructural investigations show that acinar tissues are preserved when the irradiated rat parotid gland is protected by WR-2721. The same type of damage appears in the drug-treated and untreated glands, but the extent of the damage is smaller in animals receiving WR-2721 with a DMF of 2.3 (Pratt *et al.*, 1980). Secretagogues and the β-adrenergic drug isoproterenol protect the gland from weight loss and change in amylase activity (Sodicoff *et al.*, 1979, 1980). The β-adrenergic blocker propranolol counteracts the protective effect of isoproterenol (Sodicoff and Conger, 1983a). Cyclic adenosine monophosphate (cAMP), the level of which is elevated after administration of isoproterenol, mercaptoethanolamine, and cysteamine, prevents the loss of gland weight with a DMF of 1.6 (Sodicoff and Conger, 1983a). However, the desired protection from xerostomia has not as yet been demonstrated. In dogs treated with WR-2721 prior to irradiation with a single dose, no significant protection of salivary flow could be demonstrated 6 months later (Utley *et al.,* 1978). Further testing in humans is expected and needed.

These studies have created much interest and hope that there may be found a way of preventing xerostomia which at the present time remains a chronic side effect for years. Although not a threat to health, xerostomia is a constant bother in daily life.

IV. Teeth

Development of dental lesions following radiation therapy is a late change resulting from degenerative changes in surrounding tissues. Damage to teeth outside the radiation field may be similar to that to teeth directly in the path of the radiation, indicating that the changes are secondary to factors other than the radiation alone. Direct irradiation of teeth in the adult mouth produces no clinically obvious changes if salivary function is normal (Frank *et al.,* 1956). In animals, loss of saliva secondary to removal of the salivary glands, ligation of their ducts, or radiation destruction of the glands result in dental defects (Silverman and Chierici, 1965). After irradiation, teeth normally bathed by saliva of neutral pH are left in a dry, acid mouth. They retain organic material and may be invaded by bacteria (Frank *et al.,* 1956). Acid saliva softens teeth (Shannon *et al.,* 1978b) and cariogenic microorganisms increase in the irradiated mouth. The plaque that forms on human teeth in areas of carious lesions contains microorganisms which induce these lesions (Brown *et al.,* 1975), and such microbial changes can persist in patients from 1 to 4 years after irradiation (Llory *et al.,* 1972).

Del Regato (1939) describes changes in teeth of patients undergoing irradiation of the oral cavity. Acutely, patients may experience hypersensitivity to temperature and sweet foods. This hypersensitivity may recur approximately 6 months following treatment as lesions of the teeth appear. Complete loss of teeth may be seen 12–18 months after treatment, although this extreme degree of destruction is a rare complication. The appearance of such defects varies. Frequently, they present as an annular lesion around the neck of the tooth that can lead to an amputation of the crown. Other lesions appear as heavy, dark brown discoloration of the whole tooth. Lesions can also appear as diffuse, fine punctuate defects which may coalesce and cause irregular erosion of the tooth's surfaces.

Before the 1950s, it was customary to remove all teeth prior to radiation therapy of the oral cavity. This was felt necessary to avoid rampant dental destruction and to reduce the incidence of osteoradionecrosis. Wildermuth and Cantril (1953) were among the first to suggest that this regimen was not necessary in all patients. Present practice involves careful evaluation of the teeth prior to radiation therapy, removing only carious teeth, leaving healthy teeth in place, and managing these remaining teeth with great care. In the individual with a great deal of dental disease and disregard for oral hygiene, complete dental extractions may be necessary to prevent serious late bone disease. If all teeth are removed, a radical alveolectomy should be carried out to prevent sharp projecting bony spicules from penetrating the overlying mucosa. This must be followed by a sufficient time for gum healing which can last from 1 to 3 weeks. In some patients, the need to start radiation therapy is urgent, and waiting for the healing of the mucosa may jeopardize optimal curvative results. With careful management of these patients, irradiation may proceed and extractions performed later. The risk of osteonecrosis is greater with periodontal disease than caries.

The patients who are edentulous before oral cavity irradiation are less at risk of developing late sequelae than denate ones. Osteoradionecrosis developed in 24.2% of dentate patients with good dental prophylaxis, but only in 11.9% of edentulous patients (Murray et al., 1980).

Irradiation of the oral cavity of children leads to additional complications (Blozis and Robinson, 1968). The very young child has actively growing bone centers and active tooth development. The teeth that have not yet begun to develop may fail to grow. Doses of 3000 rad will stop tooth development at its prevailing level and the alveolar ridges will remain small and narrow. However, Blozis and Robinson (1968) pointed out that xerostomia after irradiation in children is usually less frequent than in adults because lower doses can often be used and salivary gland tissue in children may have a greater regenerative capacity. Children develop radiation caries less often than adults, presumably because they produce more saliva.

V. Taste

The sense of taste diminishes quickly in patients undergoing oral cavity irradiation. Loss of taste sense may be a factor in the weight loss and poor nutrition that is often seen in patients undergoing radiotherapy to the head and neck, although many patients with head and neck cancers exhibit a reduced taste sense even prior to treatment (Mossman and Henkin, 1978). Patients experience decreased salivary output and acute loss of taste during a course of radiation therapy, but these two reactions appear to be unrelated (Conger, 1973; Mossman, 1982).

The sense of taste originates in the taste buds found in the papillae of the tongue as well as on the glossopalatine arch, the soft palate, and epiglottis and pharynx to the level of the cricoid cartilage. These taste buds contain numerous taste cells, each bud recognizing one of four specific tastes: sweet, sour, bitter, or salty. There is some grouping of taste buds according to their function, with bitter sensation on the posterior part of the tongue, sour in the middle, and salty and sweet on the anterior part of the tongue. More taste buds recognize sweet than any of the other tastes (Mossman and Henkin, 1978). The taste threshold and radiation-induced loss of taste sensations have been tested in patients and showed that bitter and salty tastes were most severely impaired (Bonanni and Perazzi, 1965; Mossman and Henkin, 1978). Those patients with the most acute taste sense lose this function most rapidly, but also recover more completely following radiation treatment, while those with less acute taste sense have less to lose initially and less to regain (Conger, 1973). After irradiation taste returns, rapidly at first and more slowly thereafter. Although complete recovery of taste has been reported (Conger, 1973) by 60–120 days, some patients never feel they have returned to their preirradiation state. Radiation doses reported for taste loss vary. Conger (1973) states that a dose of 240 rad reduces the taste sensation for bitter by one-half and 400 rad reduces the taste for acid by the same amount. In his clinical series, the rate of taste loss is an exponential function when the treatment is delivered at 200 rad/day, 5 days/week, with a total dose of 3000 rad. Thereafter, the rate of taste loss decreases. Mossman (1982) fitted a sigmoid curve to the rate of taste loss. At doses up to 1000 rad little taste loss is observed in patients undergoing fractionated radiation therapy for head and neck malignancies. At doses of 2000–5000 rad, the rate of taste loss increases by a factor of four, and at doses exceeding 5000 rad, the rate of loss decreases. Taste losses after neutron irradiation of head and neck tumors revealed a relative biological effectiveness (RBE) value of 5.7, suggesting that neutron therapy offers no therapeutic gain with respect to taste preservation after radiation exposure. There is a reduction in RBE values as the neutron dose or dose per fraction increases (Mossman, 1982).

The radiation-induced destruction of taste buds and taste cells has been studied in mice following acute and fractionated irradiation (Conger and Wells, 1969). Buds began to disappear 4 days after exposure to 1000 rad, their number falling to a minimum at 9 days and showing recovery after 10 days. With fractionated radiation totaling 6000 rad, a 10% loss of taste cells and a 30% loss of taste buds were noted. This does not correlate with the marked degree of taste loss experienced by patients. However, the taste cells have extensive microvilli at their surfaces as seen with the electron microscope. Microvilli are lost from the irradiated taste cells of mice (Conger and Wells, 1969). It may be that destruction of this extensive surface area on the microvilli which contain the absorptive gustatory sites results in the profound taste loss experienced by patients.

Zinc has been suggested as useful in treating the taste impairment caused by radiation (Mossman and Henkin, 1978). However, taste, which is such an important quality of life, returns within a relatively short time.

VI. Bone, Cartilage, and Soft Tissues

Radiation damage of bone and cartilage occurs infrequently, but when this late-type injury appears, it can be quite debilitating. By far the most common injury is mandibular osteoradionecrosis. This lesion, as previously described, is almost always secondary to periodontal disease, even though devitalization of bone and cartilage occurs as a direct radiation effect. Adult dogs irradiated over the mandible either with all teeth present or after hemimandibular tooth extractions develop no osteomyelitis without ulceration of the overlying soft tissue. Necrosis of bone, when it occurs, is always associated with osteomyelitis. Removal of teeth prior to irradiation significantly decreases the incidence of osteomyelitis. In dentulous dogs, soft tissue reactions start on the gingiva around the neck of teeth. The teeth then become loose and sinus tracts develop into the body of the mandible. Pathologically, these tissues show inflammation extending into the periodontal membrane and through the lamina dura to involve the medullary cavity of the mandible and the apices of the teeth.

Many clinical reports emphasize the relationship between osteoradionecrosis and radiation damage to teeth. Silverman and Chierici (1965) summarize seven reported series showing a 3–13% incidence of osteomyelitis following irradiation of the oral cavity, 40% of which occurred in the alveolar ridge. In another series, Bedwinek et al. (1976) reported that 65% of the osteonecroses were related to dental extractions or denture irritation and only 5% were considered spontaneous. Mandibular osteoradionecrosis is far more common than maxillary necrosis owing to the more abundant blood supply of the maxilla.

The three main determinant factors associated with mandibular necrosis following radiation to the oral cavity are the site of the tumor, the tumor dose, and the dental status (Murray et al., 1980). The risk of bone necrosis is five times greater when the tumor involves or is adjacent to bone and teeth and is 2.9 times higher in patients receiving 8000 rad than in those receiving 5000 rad. The incidence of necrosis is 2.6 times greater in dentulous than in edentulous patients. Morrish et al. (1981) reported a series in which mandibular necrosis was not seen in patients receiving less than 6500 rad, but occurred in 85% of dentulous patients and 50% of edentulous patients receiving more than 7500 rad to the bone. Interstitial implants in combination with external beam therapy, especially for lesions in the anterior floor of mouth, are associated with a higher incidence of bone necrosis than treatement with external beam alone (Bedwinek et al., 1976; Fu et al., 1976a). In bone, injuries directly related to radiation appear histologically as a lack of osteoblastic and osteoclastic activity, with almost complete absence of osteocytes and areas of fatty marrow degeneration (Silverman and Chierici, 1965). This radiation-induced injury can be described an an aseptic bone necrosis which usually heals by fibrosis. Such bones are less resistant to developing infections and more prone to injury and fractures than normal bones. Osteonecrosis may appear from 3 to 24 months after treatment, although there is a continued risk of necrosis developing at later times. This is illustrated by two cases of radionecrosis of the temple bone which occurred 4 and 12 years, respectively, following radiation therapy (Schuknecht and Karmody, 1966). In both of these cases, soft tissue changes in the middle and inner ear occurred as early complications. In one case, infection proceeded through the bone into the middle cranial fossa, causing death from meningitis. Management of osteoradionecrosis should be conservative. Antibiotics should be given as well as care of ulcerated soft tissues and debridement of protruding bone spicules. With watchful waiting, most of these heal and surgical intervention is usually unnecessary. Although reports vary, ~5–10% of the necroses require surgical management (Bedwinek et al., 1976; Fu et al. 1976a; Shukowsky, 1970).

Destruction of cartilage is a clinical entity seen far less than bone necrosis. The unique structure and relatively low metabolic turnover rate of cartilage are in part responsible for the apparent resistance to radiation injury. Clusters of chondrocytes are scattered sparsely within an amorphous interstitial matrix, but contain interlacing fibrils which contain hyalin, elastin, or fibrin. Collagen fibrils receive some nutritive fluid from blood vessels in the perichondrium. Most of the cartilage about the head and neck, including the external ear, walls of the external auditory canal, eustachian tubes, epiglottis, and part of the corniculate and cuneiform cartilages, contains a matrix of elastic fibers rendering it flexible. In the adult, there is almost no proliferation of the chondrocytes (Bloom and Fawcett, 1962a–c). Radiation injury, apparent as a late event, was more common in the early days of superficial X-ray therapy. Coutard (1932), in fact, advocated resection of the thyroid and cricoid cartilage on the side of the lesion to be irradiated prior to treatment to avoid necrosis of these structures.

Tumors of the skin overlying the nose and ear are often treated by radiation in preference to surgery because of the more favorable cosmetic result. In these areas, superficial X rays are still the ideal modality. It is important to use a beam in a kilovoltage range of 100–150 kV, well filtered with 0.25–0.50 mm copper plus 1 mm of aluminum (Del Regato and Vuksanovic, 1962; Parker and Wildermuth, 1962). Fractionation is imperative and the daily dose should be adjusted according to the volume of tissue irradiated. Small tumors can be treated with daily doses of 250–300 rad, with a total dose of 4500 rad, but large volumes of skin necessitate a smaller daily dose of 150 rad given five times a week for periods of 6–7 weeks. Under these conditions, moist epidermitis develops. With proper care, late changes are infrequent, but tissues may show atrophic and fibrotic changes, and late damage may be exacerbated by trauma, extreme dryness, or frost. Megavoltage radiation sources presently used for the treatment of head and neck malignancies produce no or very infrequent necrosis of cartilage. These high-energy beams spare the skin and usually do not result in the erosion of soft tissues overlying cartilage.

Treatment of tumors of the larynx and pharynx is accompanied by acute symptoms of hoarseness, sore throat and dysphagia, beginning after the second week of treatment and healing within 2–3 weeks after completion of the therapy when doses are in the range of 6000–6600 rad. Hoarseness, however, may persist longer and may recur frequently when the voice is overused. There are certain areas of the larynx which are especially vulnerable to the development of late radiation edema such as the glossopalatine folds, the aryepiglottic folds, and the epiglottis. If these structures have been destroyed by a preexisting tumor, submucosal fibrosis may contribute to the frequency of complication. As described for other structures, the total dose and schedule of delivery of the radiation must be adjusted to the volume of tissue in the radiation field. The technique of irradiation must also take into account the geometry of the patient's anatomy. Excessive doses may be delivered to the anterior portion of the larynx if the thyroid cartilage is acutely angled, and in such patients the use of compensators and daily irradiation of two opposed fields are most beneficial. Without compensating devices, the anterior cords could receive doses as high as 7500 rad, while structures 3–6 cm posterior to the cords receive 6500–5900 rad (Kagan et $al.$, 1974). When doses to the larynx are held within a range of NSD values of 1920–2040 rets, cures of T_1N_0 laryngeal cancers are obtained in 92% of patients without necroses. Increasing the dose beyond this level increases the number of laryngeal necroses to 7% (Aristizabal and Caldwell, 1972). Shukovsky (1970) reports severe damage above NSD values of 1800 rets, when edema develops. Severe necroses occur above doses of 1900 rets. Necrosis of the larynx is an infrequent complication in ~1% of those receiving total doses exceeding 6000 rad, with a standard dose fractionation of 200–225 rad per day. Acute and chronic symptoms are exacerbated in patients who continue to smoke and drink. When these symptoms persist and laryngoscopy reveals swelling, persistence or recurrence of the tumor must be suspected. If the symptoms persist, biopsy is necessary. Tracheotomy may be necessary for severe edema and laryngectomy is necessary in some cases when deep biopsies show only necrosis, but recurrent cancer is suspected.

Severe complications are reported in patients treated for cancers of the pharyngeal wall. In a series of 52 patients, Marks reports a 31% incidence of pharyngocutaneous fistula and 14% incidence of carotid rupture. The surgical mortality was 14% and was most common in those who had received preoperative radiation. Irradiation with high doses without surgery was associated with airway obstruction in 18% (Marks et $al.$, 1978). Meoz-Mendez et $al.$ (1978) also reported a high incidence of rupture of the carotid artery, pharyngeal wall necrosis, and laryngeal edema. Tumors in this area can infiltrate and invade adjacent connective tissue, adding to the high rate of complications. When surgery and radiation are combined, surgery should precede radiation to promote good healing and diminish complications (Marks et $al.$, 1978).

VII. Esophagus

The structure of the esophagus includes a superficial squamous epithelium overlying the lamina propria which consists of loose connective tissue, collagen and elastic fibers, and lymphatic clusters. Underlying this are the smooth and striated muscles of the muscularis mucosa surrounded by loose connective tissue, the tunica adventia (Bloom and Fawcett, 1962a–c). The mucosal cells are similar to those of the oral cavity and are characterized by a rapid proliferation rate and a relatively high degree of radiosensitivity. However, animal and clinical studies show that acute injuries heal rapidly and that late ones are rare. Rats treated with single fractions of 3000 rad exhibit early submucosal congestion and infiltration by leukocytes, followed by mucosal necrosis and sloughing. This occurred after the first week of treatment. Mucosal healing progressed from the margins and from the central portions of radiation-induced ulcers (Hakansson et al., 1981) and was similar to healing seen in skin and oral mucosa. In these animals, fibrosis and scarring persisted and hyperplasia and diverticulae appeared within the muscle wall 3 months after irradiation. Similar histological changes occur in mice, but 1 year following 2500 rad, the mouse esophageal mucosa appeared completely normal (Phillips and Margolis, 1972). The dose required to produce a 5% incidence of late esophageal damage is estimated at 1850 rets (6300 rad in 30 fractions) and a 50% incidence, 200 rets (6650 rad in 30 fractions). The LD/50/20 in mice is 2500 rad (Phillips and Margolis, 1972). Similar histological observations are reported in dogs, although degeneration and regeneration were slower than in smaller animals (Engelstad, 1934).

The histological appearance of the irradiated esophagus in patients undergoing esophagectomy 3 months following a full course of radiation (5500 rad in 35–42 fractions) showed only some dilated blood vessels and slight collagenous infiltration of the muscle layers, but no perforation or mucosal atrophy (Phillips and Margolis, 1982).

Roentgenologic examination of two patients who had received doses of 7300–7600 rad (Seaman and Ackerman, 1957) showed narrowing of the esophageal lumen on barium swallow, 2 and 8 months later. Histological changes seen after esophagectomy included destruction of the mucosa and muscularis and widening of the submucosa. The muscle layers showed edema, with changes in the cell nuclei and cytoplasm. Hyaline changes or homogenization occurred with loss of nuclear detail and absence of cellular delineation. This is thought to constitute injury directly attributable to radiation and not secondarily to vascular damage. The above-cited authors feel that the level of radiation tolerated by the esophagus is limited by the tolerance of the submucosa and muscle, not the mucosa. They propose a dose of 6000 rad delivered at 1000 rad/week as the tolerance limited.

In clinical radiation therapy, mediastinal structures receive radiation most frequently during the management of carcinoma of the lung and breast. Often, the dose delivered to the esophagus is limited because of the lower radiation tolerance of adjacent structures. For example, the radiation tolerance of the spinal cord, which is in close proximity to the esophagus, is less than 5000 rad with standard fractionation. With these doses, surprisingly little has been written about damage to the esophagus. Symptoms of esophagitis are usually referred to as mild, consisting of substernal burning and dysphagia, occasionally the feeling of a "lump in the throat." Symptoms may become severe and necessitate rest periods during treatment. The pain may be associated with temporary weight loss as the patient's dietary habits change during the course of therapy. After doses of 3000 rad, these acute symptoms usually appear in the third week of treatment. Some patients experience lessening of the severity of these symptoms even during the course of therapy, indicating a rapid renewal of epithelial cells and reepithelization. Symptoms disappear in almost all patients 2–3 weeks after treatment.

Most of the clinical data on radiation reactions of the esophagus are derived from reports in the literature dealing with radiation therapy for esophageal carcinoma. Mortality from irradiation of esophageal carcinoma is extremely low, a point often raised when comparing radiation with surgical management of esophageal carcinomas. The above-mentioned reports include occurrences of radiation-induced structures requiring dilatations, radiation myelitis, pulmonary fibrosis which is usually asymptomatic, and tracheoesophageal fistulae. Fistula formation is almost always associated with tumor invasion of the esophageal wall and therefore is not directly produced by the radiation itself. The finding of a tumor-induced fistula in the esophageal wall prior to treatment is a contraindication for radiation therapy, as life-threatening mediastinitis can result. Pearson (1969) reports the occurrence of partial necrosis of the anterior portion of one or more vertebral bodies. One case of rupture of the aorta occurred in a large series reported by Marcial *et al.* (1966). However, the numbers of significant complications are small and as pointed out by Earlam and Cunha-Melo (1980a,b), in a modern department with properly planned radiation therapy, few major complications should occur.

Therefore, although acute radiation esophagitis can be quite symptomatic, it is usually self-limited and healing occurs rapidly. Long-term sequelae are rare and may be avoided with attention to total dose and daily dose fraction size. Tumor involvement of the mucosa and muscle layers of the esophagus may contribute to radiation damage and may compromise healing, but these occurrences, in fact, are infrequent.

In summary, radiation reactions in structures of the oral cavity, neck, and esophagus can be generalized and grouped according to tissue types. Cell turnover rates predict whether the reactions will appear early or late, and care of early reactions can lessen later complications. Chemical modification of these reactions is under study, but such work has yet to have great clinical significance. The time-honored parameters which determine radiobiological changes are the total radiation dose, number of radiation fractions, overall time of dose delivery, and the volume of tissue irradiated.

REFERENCES

Aristizabal, S., and Caldwell, W. L. (1972). *Radiology* **103**, 419–422.

Bedwinek, J. M., Shukovsky, L. J., Fletcher, G. H., and Daly, T. E. (1976). *Radiology* **119**, 665–667.

Blozis, G. G., and Robinson, J. E. (1968). *Dent. Clin. North Am.* **Nov.,** 643–656.

Bloom, W., and Fawcett, D. W., eds. (1962a). *In* "A textbook of Histology," pp. 144–152. Saunders, Philadelphia.

Bloom, W., and Fawcett, D. W., eds. (1962b). *In* "A Textbook of Histology," pp. 399–406. Saunders, Philadelphia.

Bloom, W., and Fawcett, D. W., eds. (1962c). *In* "A Textbook of Histology," pp. 427–431. Saunders, Philadelphia.

Bonanni, G., and Perazzi, F. (1965). *Nunt. Radiol.* **31**, 383–397.

Brown, L. R., Dreizen, S., Handler, S., and Johnston, D. A. (1975). *J. Dent. Res.* **57**, 740–750.

Chencharick, J. D., and Mossman, K. L. (1983). *Cancer* **51**, 811–815.

Cheng, V. S. T., Downs, J., Herbert, D., and Aramany, M. (1981). *Int. J. Radiat. Oncol. Biol. Phys.* **7**, 253–258.

Cherry, C. P., and Glucksmann, A. (1959). *Br. J. Radiol.* **32**, 596–608.

Conger, A. D. (1973). *Radiat. Res.* **53**, 338–347.

Conger, A. D., and Wells, M. A. (1969). *Radiat. Res.* **37**, 31–49.

Coutard, H. (1932). *Am. J. Roentgenol. Radium Ther.* **28**, 313–331.

Del Regato, J. A. (1939). *Am. J. Roentgenol. Radium Ther.* **42**, 404–410.

Del Regato, J. A., and Vuksanovic, M. (1962). *Radiology* **79**, 203–208.

Dreizer, S., Brown, L. R., Handler, S., and Levy, B. M. (1976). *Cancer* **38**, 273–278.

Earlam, R., and Cunha–Melo, J. R. (1980a). *Br. J. Surg.* **67**, 381–390.

Earlam, R. and Cunha-Melo, J. R. (1980b). *Br. J. Surg.* **67**, 457–461.

Eneroth, C. M., Henriksson, C. O., and Jakobsson, P. A. (1971). *Acta Otolaryngol.* **71**, 349–356.

Engelstad, R. B. (1934). *Acta Radiol.* **15**, 608–614.

English, J. A., Wheatcroft, M. G., Lyon, H. W., and Miller, C. (1955). *Oral Surg.* **8**, 87–99.

Frank, R. M., Herdly, J., and Philippe, E. (1956). *J. Am. Dent. Assoc.* **70**, 868–883.

Fu, K. K., Ray, J. W., Ghan, E. K., and Phillips, T. L. (1976a). *Am. J. Roentgenol. Radium Ther. Nucl. Med.* **126**, 107–115.

Fu, K. K., Lichter, A., and Galante, M. (1976b). *Int. J. Radiat. Oncol. Biol. Phys.* **1**, 821–827.

Goepp, R., and Fitch, F. (1962). *Radiat. Res.* **16**, 833–845.

Grigsby, P., and Maruyama, Y. (1980). *In* "Radiation Sensitizers" (L. W. Brady, ed.), pp. 506–509. Mason, New York.

Grigsby, P., and Maruyama, Y. (1981). *Br. J. Radiol.* **54**, 969–972.

Grigsby, P., and Maruyama, Y. (1982). *Int. J. Radiat. Oncol. Biol. Phys.* **8**, 557–559.

Jennings, F., and Arden, A. (1960). *Arch. Pathol.* **69**, 407–412.

Hakansson, C. H., Lipecki, M., and Toremalm, N. G. (1981). *Int. J. Radiat. Oncol. Biol. Phys.* **7**, 1719–1723.

Kagan, A. R., Calcaterra, T., Ward, P., and Chan, P. (1974). *Am. J. Roentgenol. Radium Ther. Nucl. Med.* **120**, 169–172.

Kashima, H. K., Kirkham, W. R., and Andrews, J. R. (1965). *Am. J. Roentgenol. Radium Ther. Nucl. Med.* **94**, 271–291.

Kogelnik, H. D., and Kärcher, K. H. (1977). *In* "Radiobiological Research and Radiotherapy," Vol. I, pp. 275–286. IAEA, Vienna.

Llory, H., Dammron, A., Gioanni, M., and Frank, R. M. (1972). *Caries Res.* **6**, 298–311.

Marcial, V. A., Tome, J. M., Ubinas, J., Bosch, A., and Correa, J. N. (1966). *Radiology* **87**, 231–239.

Marks, R. D., Jr., Scruggs, H. J., and Wallace, K. M. (1976). *Cancer* **38**, 84–89.

Marks, J. E., Freeeman, R. B., Lee, F., and Ogura, J. H. (1978). *Int. J. Radiat. Oncol. Biol. Phys.* **4**, 587–593.

Marks, J. E., Davis, C. C., Gottsman, V. L., Purdy, J. E., and Lee, F. (1981). *Int. J. Radiat. Oncol. Biol. Phys.* **7**, 1013–1019.

Meoz-Mendez, R. T., Fletcher, G. H., Guillamondegui, O. M., and Peters, L. J. (1978). *Int. J. Radiat. Oncol. Biol. Phys.* **4**, 579–585.

Millburn, L., Faber, L. P., and Hendrickson, F. R. (1968). *Am. J. Roentgenol.* **103**, 291–299.

Mira, J. G., Wescott, W. B., Starcke, E. N., and Shannon, I. L. (1981). *Int. J. Radiat. Oncol. Biol. Phys.* **7**, 535–541.

Morrish, R. B., Chan, E., Silverman, S., Meyer, J., Fu, K. K., and Greenspan, D. (1981). *Cancer* **47**, 1980–1983.

Mossman, K. L. (1982). *Radiat. Res.* **91**, 265–274.

Mossman, K. L. (1983). *Radiat. Res.* **95**, 392–398.

Mossman K. L., and Henkin, R. I. (1978). *Int. J. Radiat. Oncol. Biol. Phys.* **4**, 663–669.

Mossman, K. L., Chencharick, J. D., Scheer, A. C., Walker, W. P., Ornitz, R. D., Rogers, C. C., and Henkin, R. I. (1979). *Int. J. Radiat. Oncol. Biol. Phys.* **5**, 521–528.

Mossman, K. L., Shatzman, A. R., and Chencharick, J. D. (1981). *Radiat. Res.* **88**, 403–412.

Mossman, K. J., Shatzman, A., and Chencharick, J. (1982). *Int. J. Radiat. Oncol. Biol. Phys.* **8**, 991–997.

Murray, C. G., Herson, J., Daly, T. E., and Zimmerman, S. (1980). *Int. J. Radiat. Oncol. Biol. Phys.* **6**, 543–548.

Ng, E., Chambers, F. W., Jr., Ogden, H. S., Coggs, G. C., and Crane, J. T. (1959). *Radiology* **72**, 68–74.

Parker, R. G., and Wildermuth, O., (1962). *Cancer* **15**, 57–65.

Pearson, J. G. (1969). *Am. J. Roentgenol. Radium Ther. Nucl. Med.* **105**, 500–513.

Phillips, T. L., and Margolis, L. (1972). *In* "Radiation Effect and Tolerance Normal Tissue" J. (J. M., Vaeth, ed.), pp. 266–273. Univ. Park Press, Baltimore.

Pratt, N. E., Sodicoff, M., Liss, J., Davis, M., and Sinesi, M. (1980). *Int. J. Radiat. Oncol. Biol. Phys.* **6**, 431–435.

Quastler, J., Austin, M. K., and Miller, M. (1956). *Radiat. Res.* **5**, 338–353.

Rubin, P., and Casarett, G. W. (1968). *In* "Clinical Radiation Pathology" (P. Rubin, and G. W., Casarett, eds.), Vol.I, pp. 124–133. Saunders, Philadelphia.

Schuknecht, H. F., and Karmody, C. S. (1966). *Laryngoscope* **76**, 1416–1428.

Seaman, W. B., and Ackerman, L. V. (1957). *Radiology* **68**, 534–541.

Shannon, I. L., Trodahl, J. N., and Starcke, E. N. (1978a). *Proc. Soc. Exp. Biol. Med.* **157**, 50–53.

Shannon, I. L., Trodahl, J. N., and Starcke, E. N. (1978b). *Cancer* **41**, 1746–1750

Sholley, M. M., Pratt, N. E., and Sodicoff, M. (1981). *J. Oral Pathol.* **10**, 192–202.

Shukovsky, L. J. (1970). *Am. J. Roentgenol.* **108**, 27–29.

Silverman, S., Jr., and Chierici, G. (1965). *J. Periodontol.* **36**, 44/478–50/484.

Sodicoff, M., and Conger, A. D. (1983a). *Radiat. Res.* **94**, 97–104.

Sodicoff, M., and Conger, A. D. (1983b). *Radiat. Res.* **96**, 90–94.

Sodicoff, M., Pratt, N. E., and Sholley, M. M. (1974). *Radiat. Res.* **58,** 196–208.

Sodicoff, M., Conger, A. D., Trepper, P., and Pratt, N. E. (1978a). *Radiat. Res.* **75,** 317–326.

Sodicoff, M., Conger, A. D., Pratt, N. E., and Trepper, P. (1978b). *Radiat. Res.* **76,** 1972–1979.

Sodicoff, M., Conger, A. D., and Pratt, N. E. (1979). *Invest. Radiol.* **14,** 166–170.

Sodicoff, M., Sinesi, M. S., and Pratt, N. E. (1980). *Arch. Oral Biol.* **25,** 781–783.

Utley, J. F., Marlowe, C., and Waddell, W. J., (1976). *Radiat. Res.* **68,** 284–291.

Utley, J. F., King, R., and Giansanti, J. S. (1978). *Int. J. Radiat. Oncol. Biol. Phys.* **4,** 643–647.

Van Den Brenk, H. A. S., Hurley, R. A., Gomez, C., and Richter, W. (1969). *Br. J. Radiol.* **42,** 688–700.

Wescott, W. B., Mira, J. G., Starcke, E. N., Shannon, I. L., and Thornby, J. I. (1978). *Am. J. Roentgenol.* **130,** 145–149.

Wildermuth, O., and Cantrel, S. T. (1953). *Radiology* **61,** 771–783.

Relative Radiation Sensitivity of the Integumentary System: Dose Response of the Epidermal, Microvascular, and Dermal Populations

JOHN O. ARCHAMBEAU

DEPARTMENT OF RADIATION SCIENCE
SCHOOL OF MEDICINE,
LOMA LINDA UNIVERSITY
LOMA LINDA, CALIFORNIA 92350

> *To measure is to know.*
> (A paraphrase of Lord Kelvin)
> Jetta Hice

I. Introduction

Skin irradiation with a selected time–dose schedule produces a consistent pattern of gross and histologic changes. These are easily followed sequentially and may be quantified using noninvasive or biopsy techniques. The evolution, time course, and dose dependence of these events reflect population changes occurring concurrently in the hair, epidermis, microvessels, and dermis.

The sequence and severity of the changes are reproducible; one or more can be used to define an isoeffect that is produced by a selected dose. The doses of different time–dose schedules required to produce this parameter provide a valuable data base which may be used to (1) achieve dose equivalency between different radiations or between time–dose schedules, (2) characterize and compare different radiations, (3) estimate the associated skin population kinetics and radiobiological parameters, and (4) serve the therapist as a gauge to define the dose tolerance of tissues in a radiation field or what the response of a patient's cancer to radiation therapy will be (Archambeau 1970; Archambeau and Bennett,

ADVANCES IN RADIATION BIOLOGY, VOL. 12

1984; Archambeau *et al.*, 1968, 1969, 1971, 1984; Bewley *et al.*, 1963; Berry *et al.*, 1974a,b; Cohen, 1983; Douglas and Fowler, 1976; Ellis, 1968; Ellis *et al.*, 1969; Kirk *et al.*, 1971.

This presentation lists gross and histologic changes produced by irradiation of the skin that have been quantified. It examines available cell kinetic radiobiological and morphological variables to identify interactions that occur between component populations. The dose response data of the hair and epidermal, fibrocytic, and endothelial cell populations are examined and a rank ordering is attempted. The contribution of the radiosensitivity of these populations to defining the dose tolerance of the skin is discussed. Future clinical needs are considered. The intent is to quantify or define tissue population changes in the irradiated skin so that the data may serve as guidelines to aid the radiation therapist to select therapy schedules that preserve skin function while improving cancer control.

The emphasis is on data applicable to man that document cell kinetic and radiobiological parameters of skin changes produced by clinically accepted time–dose schedules. Data for schedules not used clinically, but which have a radiobiological or cell kinetic interest are approximated by extrapolating data from swine and mouse models. The swine is the preferred model, since the irradiation geometry of field size, irradiated organ volume, and whole body-irradiated volume ratios are similar to those of man. The evolution, time course, and dose dependence of the gross changes produced in swine approach those found for man (Archambeau *et al.*, 1968; Archambeau and Mathieu, 1969; Archambeau, 1970). The morphology, histology, and cell kinetic parameters are similar in man and swine (Montagna, 1965; Montagna and Yun, 1964; Marcarian and Calhoun, 1966; Douglas, 1972; Weinstein, 1965, 1966; Weinstein and van Scott, 1965). However, it should be noted that the swine in these studies, while of a size equivalent to man, were growing and adding 4–5 lb in weight each week (Berry *et al.*, 1974a). This results in an increase of 60% in the dimension of a field over a period of 1 year and an increase of dermal thickness by a factor of 2. As a consequence, data extrapolation from swine to adult man may not be valid.

II. Cell Kinetic Parameters and Morphological Considerations of Unirradiated Skin

A. Cell Kinetic Parameters Using Swine Skin as a Reference

The skin serves as a tough, but pliable elastic container and protective barrier (Montagna, 1965). Other functions include temperature sensation, pain, proprioceptive and touch receptor contact with the environment, body thermal regulation (hair, sweat glands, and blood vessels), and sexual attraction (hair and specialized glands). The structure of the skin accommodates and defines these multiple qualities for a variety of locations with different functional roles. The scalp, skin of the eyelids, trunk, and soles of the feet represent the functional extremes.

1. Morphology

While varying in thickness and function, the skin is essentially a shell consisting of several layers. It is covered by a dry, hard, acellular, cornified surface layer which has its own self-maintaining cell population. This epidermal layer is firmly bonded to a self-secreted, acellular basement membrane which, in turn, is firmly attached to the underlying dermis. The dermis is a 0.5- to 4 mm-thick, tough, cellular, self-repairing fibrous layer that contains the lymphatic and vascular supplies. The upper 250- to 500-μm band of the dermis, the papillary dermis, contains loose fibrous tissue and the lower portion, the reticular dermis, more dense collagen. The inner surface of the dermis consists of a variable amount of fat and loose connective tissue that anchors it to the body and contains vessels, lymphatics, and nerves. In addition to the several rather uniform layers, the skin contains hair, nerve receptors, sweat, apocrine, and sebaceous glands in isolated units. The vascular supply provides a regular arcuate distribution to the reticular and papillary dermis.

Arteries and veins from the subcutaneous plexus of vessels pass through the reticular dermis with minimal branching to terminate as a plexus of tufts or loops in the papillary dermis. This consolidation is contained in a narrow band 300 μm thick. Tufts containing arterioles, venules, and capillaries suspended in a loose connective tissue matrix extend to within 10–20 μm of the epidermal basement membrane. In microscopic sections, these vessels appear as discrete collections of circular to elongated cross sections. In swine, the tufts represent coiled and folded vessels (Archambeau et al., 1985) and in humans a vascular loop. These units are referred to as vascular islands. No island groupings are found below the 350-μm band except in the plexuses around hair shafts. The epidermal layer is avascular.

2. Epidermis

The 35- to 75-μm-thick epidermis exists as a shell. 3–6 or more cells thick covered by a cornified anuclear layer. Four morphologically distinct shells may be distinguished and are assigned to three functional compartments: (1) the proliferative (basal cell monolayer), (2) the functional prickle cell shell, 3–6 cells thick, and (3) the senescent (granular and cornified shells) (Archambeau et al., 1969; Archambeau et al., 1979, 1984; Lacassagne and Gricouroff, 1956; Bond et al., 1965).

The linear density of epidermal cells in Bouin's or formalin-fixed skin is

9402 ± 1434 cells/cm; 7245 ± 1000 cells/cm are prickle cells and 2031 ± 48 cells/cm are basal cells. Morris and Hopewell (1985) report a basal cell density of about 1400 cells/cm. These values represent an areal density of about 4×10^6 basal cell/cm^2 (Archambeau, 1969; Archambeau et al., 1979; Archambeau and Bennett, 1984; Morris and Hopewell, 1985).

The proliferative basal cell monolayer has a growth fraction of 1 for swine (Archambeau and Bennett, 1984) or estimated as 20% in adult man (Potten, 1985). The mitotic index varies from 1.8 to 3.2% for different biopsy sites, successive groups of cells, and serial sections. The distribution of mitoses among successive groups of 100 cells is random. From 12 to 18% of mitotic cells are located outside of the basal layer (Archambeau and Bennett, 1984; Morris and Hopewell, 1985), but in the autoradiographs of regenerating skin (Archambeau et al., 1979), all labeled cells are found in the basal layer of the regenerating epidermis.

The diurnal variation of the mitotic index shows four peaks with a 5- to 7-hr period. About 0.5% of the basal cells enter DNA synthesis and mitosis each hour (Archambeau and Bennett, 1984; Morris and Hopewell, 1985). The tritiated thymidine labeling index (LI) varies from 3.9 to 7.9%. There is a diurnal variation of the LI, with a peak at 1600 hr and a nadir at 0900 hr.

The average cell generation time in swine ranges from 10 to 16 days in the unirradiated epidermis, but falls to 13–15 hr during regeneration. This 30-fold variation is achieved by shortening the duration of the postmitotic or T_{G1} phase. The duration of the phases of the mitotic cycle and the average generation times (T_C) are shown in Table I for swine, man, and mouse.

Using the grain count-halving technique, in immature Yorkshire swine the length of the phase of DNA synthesis (T_S) is 9 hr, and the sum of the duration of the DNA synthesis phase (T_S), the premitotic rest phase (T_{G2}), and one-half of the mitotic phase ($T_{1/2\ M}$) is 19 hr ($T_S + T_{G2} + T_{1/2\ M} = 19$ hr). These values agree well with those obtained using intravenous labeling techniques, but are about twice the value of those estimated for another strain using intradermal double labeling techniques (Morris and Hopewell, 1985).

The minimum transit time of cells through the prickle layer is 13 days, with an upper limit of about 25 days using basal cell loss from irradiated epidermis as a criterion. The average lifetime is estimated at 17 days. The upper limit of the turnover time is estimated at 25 days and the lower limit as 7 days for reepithelialization following irradiation.

3. Microvasculature

There are about 172 ± 7 vascular islands in the 350 × 10,000-μm band and an average of 17 ± 5 vascular lumens per island. The average vascular diameter is 5.1 ± 0.04 μm in a Bouin's-fixed biopsy specimen. The tritiated thymidine

TABLE I

CELL KINETIC PARAMETERS[a]

	MI (%)	LI (%)	T_S (HRS)	T_C (HR)	TT (HR)	Cell flux into mitosis/HR (number/100/hr)	GF
Epidermal							
Man[b]	0.06–1.7	3.5–7	10.3 ± 0.6 (9.6–16)	18.4 ± 26 (140–232)	288–336		1
Swine	1.7–3.2 0.1–1.0	3.9 ± 0.1	9.4–15.2	235 ± 53	312	0.005	
			(9 ± 1.1)	295.2 ±58 (235–389)			
		7.9 ± 1.5	8.8 ± 1.4	127 ± 26			
		7.0 ± 2.3	9.0 ± 0.5	155 ± 26			
Mouse	~1.2	~4.6	12.6	138	216	0.002–0.009	0.12–0.54
Endothelial							
Man	~0						
Swine	~0	1.5					
Mouse		0.56–0.82					

[a]Abbreviations: MI, mitotic index; LI, labeling index; S, DNA synthesis phase of the cell mitotic cycle; T_S, duration of DNA synthesis phase; T_C duration of the average cell cycle time; TT, average transit time of cells from the basal layer through the prickle layer; GF, growth fraction.

[b]References: Man: Camplejohn et al. (1984), Heenen and Galand, (1971), Heenen et al. (1973), Izquierdo and Gibbs, (1974), Jurnovoy et al. (1975), and Weinstein, (1965); Swine: Archambeau and Bennett, (1984), Archambeau et al. (1984), and Morris and Hopewell (1985); Mouse: Denekamp et al. (1971), Hirst et al. (1980), Potten, (1985), and Fowler and Stern, (1963).

labeling index in these growing swine averages 1.5% (Archambeau *et al.*, 1984; Hopewell, 1983; Young and Hopewell, 1980).

Histologic descriptions, cell kinetic data, and morphological measurements characterize epidermal, microvascular, and dermal population parameters. These values suggest the type of population changes to expect after irradiation. The proliferative status and kinetic parameters anticipate the time course and evolution of cell population changes whereas the radiobiological parameters indicate the shape of the dose survival curve. The degree of the histologic changes, in turn, indicates the type and severity of the gross changes that are produced (see Section III). It is anticipated that having sufficient comparative data available, gross changes can be used to estimate cell kinetic, cell population, and radiobiological parameters (Cohen, 1966, 1978, 1983; Cohen and Creditor, 1983; Cohen and Redpath, 1977).

B. Anatomical and Kinetic Considerations of the Dose Response

Several histologic and cell kinetic considerations influence the interpretation of the dose response of the skin. Grossly, the skin is viewed as a covering. Histologically, cell layers are seen in two dimensions, whereas the third dimension which recognizes these layers as shells is not apparent. Vessels are seen in cross section and the coiling and folding are overlooked. Dermal collagen appears as flat chips rather than as bundles (Archambeau and Bennett, 1984; Archambeau *et al.*, 1979, 1985).

1. Epidermal Structural Models

In one model, the basal cell layer of the epidermis may be considered similar to the *in vitro* plateau phase, confluent monolayer of proliferative (''stem'') cells (Archambeau and Bennett, 1984). The immature swine basal cell monolayer has a growth factor of 1 and cell kinetic and radiobiological parameters as listed in Table I. As the total dose is increased, the number of surviving epidermal cells is reduced. Surviving cells are recognized morphologically as a discrete group of proliferating cells (clone) among the degenerating epidermal remnant. Three distinct island groups (possibly subpopulations) are recognized by their individual cell kinetic and radiobiological parameters. These suggest that there are different proliferative epidermal cell populations that respond differently to irradiation.

In a second model, an ordered vertical hexagonal columnar arrangement of the cells is described morphologically in the rodent epidermis (MacKenzie, 1975; MacKenzie *et al.*, 1981; Potten, 1975; Potten and Hendry, 1983, 1985). This is not recognized in immature swine (Archambeau and Bennett, 1984;

Potten and Hendry, 1985; Shymko and Archambeau, 1985) or in actively turning over human skin. This structural arrangement identifies a basal layer parquet consisting of a self-sustaining unit with a central stem cell surrounded by 4–5 non-self-sustaining proliferative cells that give rise to the 4–6 nonproliferative cells found in the prickle and cornified layers. This grouping is referred to as the epidermal proliferative unit (EPU) and comprises at least two populations of proliferative cells with different cell kinetic and radiobiological parameters. In this model, the growth fraction is 12–20%.

It is not known whether the swine confluent monolayer and the rodent EPU models are equivalent (Shymko and Archambeau, 1985). However, both models indicate that there are subpopulations of cells whose cell kinetic and radiobiological characteristics determine the cell population response and the subsequent evolution, time course, and apparent dose dependence of the gross changes that are observed.

2. Dose Response

The interpretation of gross and histologic changes requires a strict definition of the dose response to irradiation. Consistent with dose response *in vitro*, cells are not killed outright by irradiation, but are rendered nonproliferative, still capable of a prolonged period of metabolic activity and apparent viability morphologically. Cell loss in this context occurs only when the cell reaches the end of its life span or attempts mitosis (Archambeau and Shymko, 1984).

Proliferative status and the cell kinetics of the target population dictate the population response to irradiation. In an active cell renewal system in a steady state, cell loss equals cell replacement. If the number of proliferative cells capable of sustained proliferation is reduced to a low number (fraction of cells surviving), the remaining nonproliferative cells continue to live their normal life span and are lost. If the rate of loss exceeds replacement in the epidermis, serum leakage occurs and a crust is formed. As the surviving proliferative cells continue to proliferate at an increased rate, the cells lost are replaced and the moist reaction heals. This population sequence is documented by the evolution and time course of the gross and histologic changes of the moist reaction. The time course of these gross changes correlates with the principal histologic cell density changes (Archambeau *et al.*, 1965, 1968, 1969, 1979; Etoh *et al.*, 1977; Yamaguchi and Tabachnick, 1972).

If the population is a nonproliferative one with little cell loss, irradiation will produce no immediate changes. If the population is a slowly proliferating one with a turnover time of months to years, cell loss and replacement if any will be prolonged and occur much later than the epidermal response. If the population is a potentially proliferative one, cell density changes will not be seen until cells respond to the stimulus of injury and become proliferative (hepatic, connective tissue, and endothelial cells). These kinetic considerations are used in interpreting the population changes occurring months to years following the evolution of erythema and a moist desquamation (Michalowski, 1981; Lajtha and Oliver, 1962; Bond *et al.*, 1965).

3. Anatomical Relationships

The microscopic morphology of the skin demonstrates the close spatial relationship between the epidermal, the microvascular, and the dermal cell populations. Since the range of the radiations used is large compared with the distance between the different tissues that make up the skin, it is impossible to irradiate one population selectively. Thus, the anatomical target is composed of three cell populations rather than one, and radiation-induced change reflects alterations in cell density and function in each population. Consequently, changes in a radiosensitive population can produce changes in an associated population that is intact and more radioresistant.

4. Significance of the Population Spatial Configuration

The tissue anatomical and spatial configuration establishes population limits. For example, the dermis represents a thick, relatively homogenous volume (shell). A proliferative cell at one location can replace a cell lost at another. Similarly, in the proliferative basal monolayer, continued proliferation and migration of a cell at one location replaces a cell lost at another. Data on epidermal survival indicate that if on average one cell per square centimeter survives, epidermal reepitheliatization will occur.

In contrast, the hair bulbs and the dermal microvascular tufts have an isolated unit configuration, as do the proliferating populations in the intestinal crypts, pulmonary alveolei, and renal tubules. In addition, the dermal tuft configuration (vessels of less than 10 μm in diameter) differs from the other configurations in that the endothelial cells are present in a linear arrangement rather than a multicellular sheet or surface (see diagrammatic representation of papillary tuft in Fig. 1.) Endothelial proliferation of surviving cells in one vascular loop will not repopulate an adjacent loop. Similarly, repopulation of the cells in the hair bulb or in one crypt or tubule does not replace cell populations lost in adjacent hair bulbs, crypts, or tubules. The linear array of endothelial cells alters the apparent dose response in that the loss of a single cell from the line can break the functional continuity of a capillary, resulting in the loss of a vessel (Stearner and Christian, 1969, 1971; Dimitrievich *et al.*, 1984, 1985; Van Den Brenk, 1959).

These anatomical considerations are expected to affect the recovery of tissues following irradiation. After irradiation, isolated small blood vessels are found

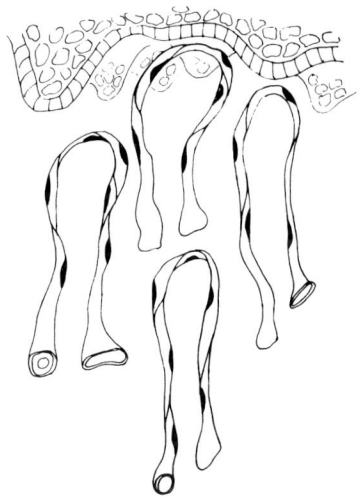

FIG. 1. A diagrammatic representation of the spatial configuration of the subpapillary micro-vasculature, epidermis, and dermis. The microvascular loops (tufts) exist as isolated anatomical units (see text).

repopulated with a hyperplastic endothelium (Hopewell, 1980, 1983), whereas adjacent small blood vessels contain no functional endothelium. As a consequence, even though a large fraction of irradiated cells survive, the original endothelial population number is not restored as is that of the epidermis where there is no barrier to cell migration.

These relationships are well illustrated in the skin by an example. If the fraction of cells surviving irradiation is 0.01, the perturbation caused by irradiation will be temporary in the basal cell monolayer where the initial population density is 10^6 cell/cm^2. The number surviving is 10^2–10^4 cells/cm^2, but re-epithelialization occurs. However, in the dermal microvasculature where there are of the order of 10 endothelial cells in each tuft, the survival fraction may still be 0.01, which represents the survival of 1 cell in 10 loops. As a consequence, cells would not survive in about 9 loops and repopulation would not occur uniformly. Furthermore, if the loss of a single cell in a linear array (capillary endothelium) causes disruption of lumen continuity, vascular loss can occur before cells can be replaced from adjacent surviving endothelium (Dimitrievich *et al.*, 1984; Van Den Brenk, 1959). Consequently, the apparent survival fraction is less than 0.01, since the loss of a single cell among 10 is sufficient to cause the loss of the 10.

III. Gross and Histologic Skin Changes Produced by Irradiation

A. General

Irradiation of the skin with increasing total absorbed doses from various types of radiation or different time–dose fraction schedules produces a regular sequence of change (Becquerel and Curie, 1901; Fajardo, 1982; Lacassagne and Gricouroff, 1956; Rubin and Casarett, 1968; Warren, 1943). The full range of changes begins with erythema and progresses, as the dose is increased, through pigmentation and epilation to a dry or moist desquamation, which may or may not heal by 50–70 days after the start of irradiation. Weeks to months later dryness, scaling, atrophy, telangiectasia, fibrosis, or necrosis may form regardless of the occurrence of the earlier changes (Liegner and Michaud, 1961). The evolution and time course of these changes are consistent, and only the absorbed dose at which the change occurs varies for different radiations and time–dose schedules.

The radiation-induced sequential changes occur in two time periods. The early reactions of erythema, dry desquamation, and moist reaction occur prior to 70–120 days. The late reactions of atrophy, telangiectasia, fibrosis, and necrosis occur from 4 to 6 months to years later (Rubin and Casarett, 1968). The changes taking place at increasing total doses are itemized in Table II.

The development of the histologic changes in man during and following a course of external irradiation has not been well documented. Histologic, morphological, and cell density parameters in man have not been quantified. Also, the changes in the population density in mice are not well documented. Alterations in cell kinetics subsequent to strontium plaque irradiation of guinea pig skin have been measured (Song and Tabachnick, 1969; Yamaguchi and Tabachnick, 1972; Etoh et al., 1977). The epidermal and endothelial cell population changes have been quantified in immature swine following single dose fractions and confirm the gross changes (Archambeau and Bennett, 1984; Archambeau et al., 1984).

The gross changes brought about by increasing absorbed doses are described in the following sections. The emphasis is on the sequence of these changes. The evolution in time of gross, functional, and cell population changes resulting from two different single dose fractions is compared for swine in Fig. 2. Similar data for multidose fraction schedules are not available.

The erythema produced by irradiation has served as an important parameter to characterize the skin response to irradiation. Changes in the intensity of the erythema are used as indicators of epidermal changes. The time course and evolution of erythema in swine after large dose fractions are shown in the first charts of Fig. 2. These changes may be compared with alterations in the epider-

TABLE II

CHANGES PRODUCED BY INCREASING TOTAL DOSE[a]

Schedule dose range		Gross change	Onset of change	Functional change	Histologic change
Dose Fraction					
Single (cGy)	Multiple (200 cGy/day)				
500–700	2000	Epilation	~18 days		Empty follicle
1000–2000	2000–4000	Erythema	12–17 days	Hyperemia	None noted
2000–3000			2–6 days		
1000–2000	~4500	Pigmentation	30–70 days	None	Increased melanin
1000–2000	~4500	Dry desquamation			Hyperplasia
2000–2400	4500–5000	Moist desquamation that heals	30–50 days	Serum leakage; healing regenerates functional barrier	Linear decrease in cell density; exponential cell replacement
>2400	>5000	Moist reaction does not heal >50%	30–50 days	Loss of protective barrier	Linear decrease in cell density
	>6000				
1700–2400	4500–5000	Telangiectasia	6 months–years	None	Cell and vessel loss, lumen dilatation
>2700	>6000	Necrosis nonhealing	Months, years	Loss of protective barrier	Necrosis

[a] All doses are given in units of centigrays.

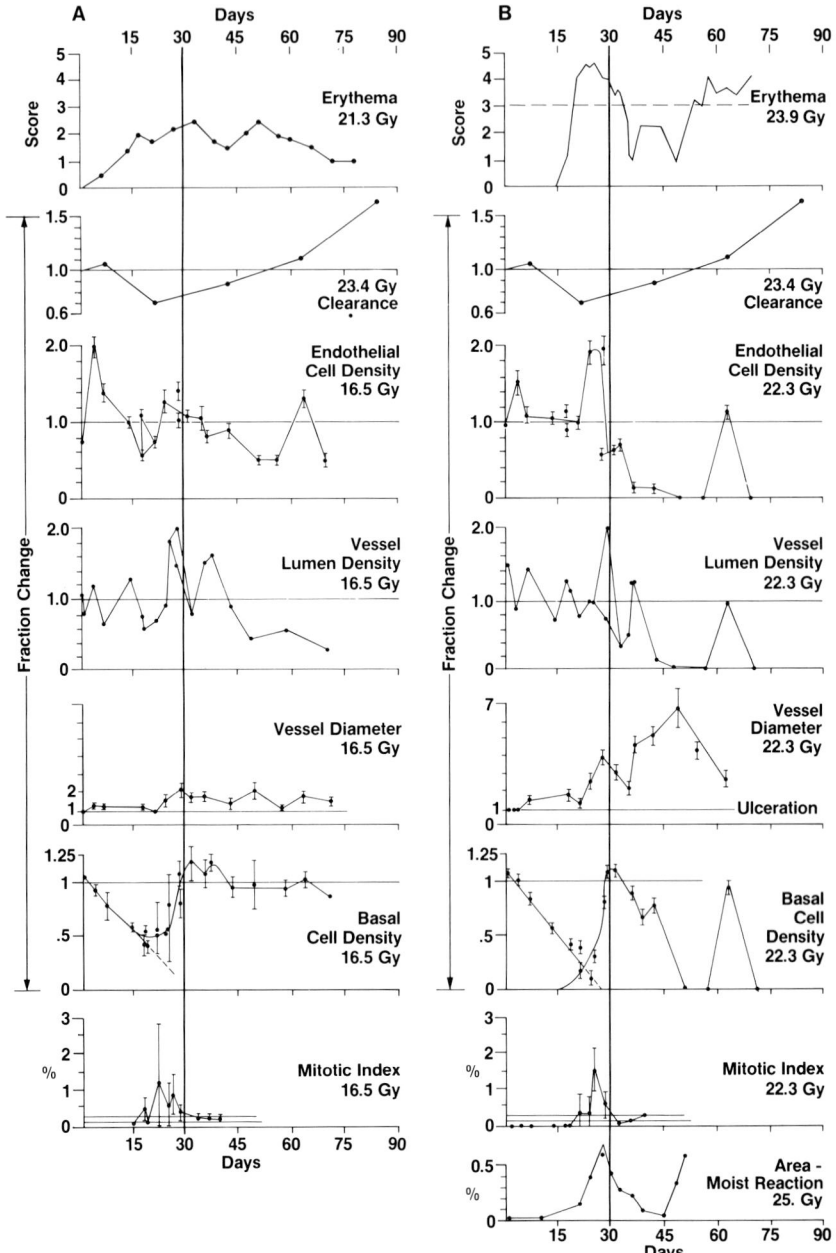

mal and microvascular cell population resulting from the same dose in the succeeding charts. Such a comparison places the erythema into perspective and shows that it is not as sensitive an indicator of change as other parameters.

B. Early Reactions

1. Erythema

In man, characteristic reddening of the skin appears after a single dose fraction of 500 cGy or more and after multiple dose fractions when the total dose exceeds 1200–2000 cGy. The reddening occurs in two phases, but the first phase is recognized only following single doses larger than 1500 cGy. This first and more variable phase appears at 1–2 days, reaches a maximum, and declines at 5–6 days postirradiation. The second more consistent phase begins at about 10–12 days and reaches a maximum at about 20 days. This reaction may have faded at 30–40 days to a uniform discoloration that will persist for a prolonged period or may merge with changes of dry or moist desquamation. When 5000 cGy are delivered in 25 fractions over 35 days, erythema appears at about 12–20 days, reaching a maximum at 36–42 days. The erythematous dose response varies among individuals and is masked by pigment content, skin thickness, ambient temperature, and individual activity (Charyulu et al., 1966; Chu et al., 1960, 1970; Nias, 1963; Turresson and Notter, 1975, 1976a,b, 1979, 1984a–c). In swine, erythema cannot be used to determine dose response by comparing change in a large 10 cm in diameter (78 cm^2) field on different animals because the formation of erythema is inconsistent and varies widely among animals (Archambeau et al., 1968). When multiple small (20 cm^2) fields on the same animal are irradiated, each with a different dose, the fields respond differently so that a dose difference of 200 rads can be resolved (Bewley et al., 1963; Fowler, 1963; Fowler et al., 1965).

The time course of evolution of the erythema in man is given in Table III. The intensity of redness can be graded from 1 to 3 or more by an established arbitrary scoring system for mild, moderate, and severe reactions (Bewley et al., 1963; Fowler et al., 1965). Change in this parameter can be verified consistently

---

FIG. 2. In A, the changes are those produced by single dose fractions of 1600–2340 cGy, whereas those in B are the result of single dose fractions of 2200–2500 cGy. The principal distinction between these two columns is that at the lower dose fraction a moist reaction and necrosis are not formed, whereas at the larger dose fraction a moist reaction and subsequent necrosis are formed. The critical time reference is the day 30, marked by a vertical line running through all graphs. The data should be examined first to show the relative response of the reaction in the skin as a whole with those of the epidermal or endothelial cell populations. Later in the text, the population changes and specific charts will be referred to. These charts are redrawn from published values. (From Bewely et al., 1963; Fowler, et al., 1965; Moustafa and Hopewell, 1979; and Archambeau et al., 1979, 1984.)

TABLE III

The Evolution and Time Course of Epilation, Formation of Erythema, and a Moist Reaction

Reaction	Man, gross (day)	Swine, gross (day)	Mouse, gross (day)
Epilation			
Single dose fraction	400–1400 cGy[b]		
Day of first appearance	14.5 ± 0.7	8.7 ± 0.3	
Day of maximal involvement	28.0 ± 2.2	—	
Multiple dose fraction	200–300 cGy/day		
Day of first appearance	15–20		
Erythema[a]			
Single dose fraction	400–1400 cGy[b]		
Day of first appearance	~1		
Day of maximal involvement	33 ± 0.4		
Moist reaction			
Single dose fraction	400–1400 cGy[b]	1645–2619 cGy	
Day of first appearance	~15 / 17.4 ± 1.2	17.5–21 / 17.5 ± 0.6	8–12
Day of maximal involvement	22–28 / 29.6 ± 1.5	25–28.5 / 24.9 ± 0.5	15–17
Day of maximal healing	36–50 / 40.3 ± 2.5	36–38 / 36 ± 1.0	26–30
Multiple dose fraction	5 × 900 cGy (in days)		
Day of first appearance	?	17–22	
Day of maximal involvement	?	22–27	
Day of maximal healing	?	28–33	
Multiple dose fraction	30 × 180 cGy (in 45 days)	30 × 200 cGy (in 42 days)	
Day of first appearance	~37–45	Moist reaction did not form	
Day of maximal involvement	~56		
Day of maximal healing	~66		

[a]Coutard (1934, 1935) and Baclesse (1958) first noted and Friedman (1939) confirmed cycles of erythema and dry desquamation formed and healed during course of irradiation lasting 8–12 weeks.
[b]High LET $^{10}B(n,\alpha)^7Li$ radiation.

by different observers. The use of this relative scale has proved to be effective and reliable, giving reproducible results in evaluating erythema in animal models and in man.

In man, the arbitrary scale has been replaced. Since skin color is determined by the presence of melanin and oxyhemoglobin, both of which absorb light at characteristic wavelengths, the intensity of the erythema can be quantified by measuring the degree of light absorption with a reflectance spectrophotometer (Chu *et al.,* 1960; Turresson and Notter, 1975).

Using the light absorption technique, a 5% difference in dose level can be determined (Turesson and Notter, 1975). The amount of change and the day when a maximal change occurs are well defined, and the shape of the response curve is easily separated from those produced utilizing other schedules (Turesson and Notter, 1975, 1976a,b, 1984a–c). The resolution into dose and time response is such that these parameters or the average response occurring within a specified time period are used as isoeffects to specify the response produced by different time–dose schedules (Turesson and Notter, 1984a–c).

These clinical studies quantify the evolution, time course, and dose dependence of the erythema produced by irradiation of 5×12 cm parasternal fields in patients with carcinoma of the breast. The values of these parameters are listed in Table III. The dose response parameters derived using erythema as an isoeffect are listed later in Table VII and compared with values obtained in other clinical studies and animal models.

The evolution of the erythematous reaction is similar for all schedules using regularly spaced dose fractions. The time of the maximum reaction is constant for all schedules occurring 8–14 days following the end of irradiation or 37–43 days following the start. Subsequent to the maximal reaction, the rate of decrease in intensity is similar to that of the increase. The full width-half maximum of the peak is about 10–14 days. However, the reflectance level does not return to control levels in the upper portion of the field until about 1 year after irradiation, but is at control level in the lower portion of the field at 4 months.

In those patients receiving a split-course schedule with 2 weeks of daily irradiation followed by a 3-week rest and 2 additional weeks of daily irradiation, the rate of formation of erythema is similar to that found for a regular uninterrupted schedule, remains constant during the 3-week gap, and then increases to a second maximum 3 weeks later at about 57 days after the start. This is about 7 days after the end of irradiation. The rate of decrease was similar to that of the regular schedule.

The bright erythema of the irradiated skin blanches under pressure, suggesting an increased dermal subpapillary blood concentration. This may occur either because of increased blood flow or vascular dilatation.

Blood flow in skin of swine is increased during the first week following large single exposures and decreased from the second until the seventh week (Moustafa and Hopewell, 1979). Erythema is associated with an increased vascular permeability as demonstrated by the extravasation of intravenously injected, radioactively labeled proteins or of vital dyes (see Law, 1981 and Hopewell, 1983 for excellent reviews). An increased interstitial blood–lymphatic circulation can be demonstrated by clearance studies of intradermally injected radioactive isotopes (Fig. 2, the second chart). The clearance of intradermally injected (^{99m}Tc) pertechnetate in the skin of swine following single dose fractions varying from 800 to 2340 rad is increased during the first 6 weeks, whereas at 6–12 weeks, the clearance decreased. Clearance returns to control levels between 12 weeks and 1 year (Moustafa and Hopewell, 1979). The clearance following multiple dose fractions follows a similar course. After total doses exceeding 3600 cGy, the clearance of ^{22}Na increases in one-third of the patients or remains unchanged up to 10 years after exposure (Roswit *et al.*, 1953).

No morphological epidermal, endothelial, or fibrocytic cellular changes are noted. Subpapillary dermal edema has been reported. Dilation of the dermal microvasculature is not well documented. (See microvascular changes in Section III,B,5, and refer to Fig. 2.

2. Pigmentation

Skin pigmentation in man appearing during and following irradiation represents an increased melanin content of cells in the basal layer. The degree of pigmentation may be quantified using the light reflection technique. The evolution of the time course and dose dependence of the increased pigmentation mirrors the change in the intensity of erythema. However, differences between radiation schedules are less pronounced.

The fading of erythema is not complete, but merges with an increased pigmentation. This pigmentation may be permanent or continue to fade. The loss of pigmentation may be gradual, lasting for a period of weeks, or may be relatively rapid over a period of days as a dry desquamation proceeds, leaving an intact surface that is again erythematous.

Pigmentation has not been used as a separate parameter to measure skin response. It is used clinically to indicate that the field has been irradiated and where the borders are located. Histologically, an increased number of cells containing melanin are found.

3. Epilation

Epilation in man occurs following single dose fractions of 500–600 cGy and multiple fractions greater than 1000–1200 cGy. It first appears about 14–18 days after the start of irradiation and is complete at 4 weeks (Archambeau, 1970). The evolution and time course are not dose dependent. Regrowth of scalp hair after

4000–4400 cGy in patients with Hodgkin's disease is complete, but is sparse following 5000 cGy and minimal after 6000–6500 cGy. The density and kinetics of cells of the hair bulb have not been quantified in man. The time course of the evolution of epilation in man is listed in Table III.

In swine, epilation becomes manifest 7–9 days postirradiation using 250 kV X rays and is complete at about 28 days later (with a range of 20–44 days) (Archambeau et al., 1968, 1969). The most complete quantitative data on the dose response of hair have been reported for mice (Potten, 1985; Malkinson and Griem, 1966). The D_0 for hair dysplasia is 260 cGy, with a dose recovery (D_Q) of 173 cGy and an extrapolation number (n) of about 2. See Malkinson and Keane (1981, 1983) for excellent reviews of original work.

4. Dry Desquamation

Dry desquamation represents a loss of epidermal cells accompanied by cell replacement which takes place before the loss of cells becomes so great that serum exudation and crust formation may result. The changes in the epidermal cell population are preceded by decreasing erythema and an increasing pigmentation. In man, the skin becomes roughened and dry, with epidermal scaling on the ventral and flexor surfaces. A thicker, larger scale may be found on the dorsal and extensor surfaces and sometimes on face and neck. In other instances, the entire epidermis may peel off as a sheet, leaving an erythematous, but intact epidermal surface.

The evolution, severity, and time course of a dry desquamation are variable. The degree of change depends on location and skin type, but can be used effectively in man as a parameter to estimate dose response (Dutreix et al., 1973).

In man, the epidermal surface becomes hyperplastic, with a thickened cornified layer during the 2- to 3-week period following 4000–5000 cGy given as daily dose fractions. In swine, these changes occur in the period of 28–42 days after single dose fractions of 1500–2700 cGy. (Refer to Fig. 2 for charts displaying the fraction change in basal cell density of swine.)

5. Moist Desquamation

A moist reaction represents the loss of sufficient epidermis to allow serum leakage and crust formation (Lacassagne and Gricouroff, 1956) and is histologically characterized by depletion of the epidermal cell population. The moist reaction begins as erythema and is replaced by epidermal peeling that leaves a bright red base. This lasts for a day or two and is followed by serum leakage and crust formation. [See Table III for time course of these changes (Archambeau et al., 1968, 1969; Archambeau, 1970) and refer to Fig. 2, sixth chart.]

The onset of the moist reaction may vary depending on the time–dose schedule. Following acute single exposures in swine, an area of erosion or a vesicle appears at about day 17 that progresses to involve an increasing area of the field.

In man irradiated with 200 cGy daily dose fractions to a total dose of between 4500 to 5500 cGy, the erythema present during the fourth and fifth week changes hue and turns mauve. Dermal edema is present, followed by epidermal peeling, erosion, and leakage of serum.

If the total dose delivered exceeds the skin tolerance, the skin becomes pale, dermal edema increases, and desquamation is rapid, producing an ulcer which over the next days to a week extends into the dermis. Little healing occurs and necrosis results.

In clinical radiation therapy, a moist reaction may be produced following a course of skin irradiation for basal and squamous carcinomas of head and neck or carcinoma of the breast. However, this turn of events is not the aim of therapy, and treatment is usually terminated as soon as these reactions occur. As a consequence, the evolution and time course of the moist reaction in man has not been used to define dose response. Some information pertaining to the evolution, time course, and dose dependence of the formation of the moist reaction in man is derived from casual observations. The values are similar to those reported for swine. Therefore, the quantification of the time course and dose dependence of the reaction obtained from studies utilizing the skin of swine provides insight into the response in man.

As the total dose, delivered as a single dose fraction, to the skin of swine is increased, a level is reached between 1500 and 2700 cGy, where a moist reaction may or may not form between 17 and 21 days. If a moist reaction does form, the area involved increases progressively to involve 5–50% of the irradiated field between 25 and 28 days. Healing occurs during the fifth week from the edges of the field, hair follicles, and isolated skin islands found within the area of moist reaction. Healing is either complete or partial during this period. Maximal healing occurs at about 36 days following irradiation, and no further healing is noted in fields 10 cm in diameter except from the edges of the field (Archambeau *et al.*, 1968, 1969, 1979; Bewley *et al.*, 1963; Fowler *et al.*, 1963, 1964, 1965). The evolution of the moist reaction is complete at 49 days. A second moist reaction that progresses to ulceration and necrosis may occur during the interval from 43 to 120 days (end of period of observation).

The time course of the evolution of the moist reaction for man and swine is summarized in Table III. As the total dose is increased, the area of field involvement is increased, and the dose at which 50% involvement occurs is 2444 cGy at the time of maximal involvement or 2774 cGy at the time of maximal healing (Archambeau *et al.*, 1968). Operationally, the area of the field involved is a useful parameter, but it has no histologic (radiobiological) explanation or justification and is not used as an isoeffect parameter (Archambeau *et al.*, 1968). As the total dose is increased, the fraction of the fields not healed by 49 days increases; the dose at which 50% of the irradiated fields are not healed is 2437 cGy. This is the preferred isoeffect parameter, since all irradiated fields are included in the evaluation and the gross changes correlate with the histologic changes of cell loss and reepithelialization.

When the total dose is given in two fractions separated by 6 or 24 hr or in daily dose fractions 5 times a week for several weeks, the evolution, time course, and dose dependence of the formation of the moist reaction change. As the fraction size is decreased, a larger total dose is required to produce a moist reaction. As the time period between dose fractions is increased or the time period over which multiple dose fractions are used increases, the total dose required to produce the reaction increases. When clinically equivalent dose schedules of 200 cGy/day, 5 days/week for 5–6 weeks (5000–6000 cGy) are used in swine, a moist reaction does not form. When such a schedule is used for patients, a moist reaction does result. This difference in response is explained on the basis of basal cell proliferation which occurs in the immature swine during the period of the irradiation, thus maintaining the epidermal cell population, whereas in man, cell proliferation during irradiation is not as active. However, after the irradiation is finished, the surviving cells continue to proliferate and the surface is reepithelialized in about 7–10 days (Archambeau *et al.*, 1968, 1969, 1979; Archambeau and Mathieu, 1969).

When the total dose is given to swine in two dose fractions separated by an interval of 6 hr, the evolution and time course of the moist reaction is similar to that found for a single dose fraction. When the total dose is given in two dose fractions separated by an interval of 24 hr, the evolution of the moist reaction changes (no data are found evaluating this change in man) (Archambeau *et al.*, 1969). Following single dose fractions, only a small area of the irradiated field is involved in a moist reaction. If more than about 20% of the field is involved, healing is never complete. Following two dose fractions separated by 24 hr, about 100% of the irradiated field is involved in a moist reaction. Complete healing occurs in 80% of these fields. However, when a moist reaction appears after five daily dose fractions, complete healing does not occur (Archambeau *et al.*, 1969). These data indicate that the dose response of basal cells varies nonuniformly across the field and that the radiosensitivity changes during the 24-hr interval.

When the total dose is given in equal increments at an interval of 6 hr, the dose at which 50% of the fields are not healed (ED_{50}) is 2794 $\pm$ 92 cGy. When equal increments are given at an interval of 24 hr, the ED_{50} is 3360 cGy, and following five daily increments, the ED_{50} is 4761 $\pm$ 366 cGy. The average recovery dose is 581 cGy for each dose fraction. Following 30 daily fractions in

43 days, the ED_{50} is ~10,670 cGy and the average dose recovery is 284 cGy. These values are larger than those reported for man (see Section IV,A,3).

The time course of the evolution of the moist reaction following all dose schedules is similar. Values of these parameters for man, swine, and mouse are presented for comparison in Table III.

The histologic data on the moist reaction in man are fragmented and incomplete. In swine, after large single dose fractions, the epidermal histologic changes of cell morphology, linear density, and mitotic activity can be separated distinctly into three phases: the degenerative (cell loss), the regenerative (cell replacement), and the postregenerative (hyperplasia followed by cell loss). These phases are characterized by changes in cell population density (see Fig. 2, sixth chart). Cell degeneration and cell loss predominate during the first 2–3 weeks. Regeneration begins during the third, predominates during the fourth, and is complete in the fifth week. The epidermis is then maintained intact or a second cycle of cell loss may occur during weeks 6–10 postirradiation (Archambeau *et al.*, 1979; Etoh *et al.*, 1977; Potten, 1985).

During the degenerative phase, the morphological changes are minimal during the first 2 weeks following irradiation. The mitotic index is low, and abnormal mitoses are observed. At 14 days, the regular architectural pattern becomes distorted. The basal and prickle cell populations decrease progressively to a low value or are lost. Splitting and separation of the epidermis from the basement membrane and dermis become manifest. In some areas, an epidermal remnant persists while in others the surface is denuded and the basement membrane lost. At 17–21 days, a crust overlies the ulcerated areas and spreads to cover also the adjacent epidermis. If cell replacement does not occur, the epidermal remnant and crust are lost as ulceration progresses.

The loss of prickle and basal cells is linear at about 4.6% of the population per day. This loss is independent of the dose schedule. The constant for daily cell loss indicates that these cells have a fixed lifetime, with a compartment transit time of about 27 days (Archambeau *et al.*, 1979; Etoh *et al.*, 1977; Potten, 1985).

The regenerative phase appears between 17 and 28 days after a single dose fraction, at which time groups of varying numbers of normal-looking cells are found distributed randomly in the epidermal remnant, over the ulcerated surface beneath the crust, or along hair follicles and gland ducts. These groups, called "islands," begin and end without relation to a specific structure and may or may not be covered by an epidermal remnant and/or crust. When first noted, they are between 14 and 20 cells long, but later the length and height of the islands increase, as do the mitotic and labeling indices. The islands vary in diameter or chord length, but may be separated into three groups: small, intermediate, or giant, based on the linear density of the chord length. If healing is to be complete, adjoining islands must fuse at between 28 and 32 days. The crust over the reepitheliating areas is lost at between 32 and 45 days. If healing is incomplete, the islands increase in diameter, but are still separated by areas of ulceration (Archambeau *et al.*, 1979).

Regeneration proceeds to completion at 28–32 days and the time required for regeneration is dose related: Healing is complete after 1645 cGy by day 28 and after 2619 cGy by day 32. The basal cell density in giant islands contributes from 36 to 70% of the epidermal cell density as the dose is increased from 1645 to 2619 cGy. The cell density of small islands contributes an increasingly smaller fraction to the epidermal cell density as the dose level increases over the same range (Archambeau *et al.*, 1979). This observation indicates the rate of proliferation is lower in the small islands than in the giant.

The fields utilized in these studies were 10 cm in diameter, 78 cm^2. While some healing occurred at the margin of the irradiated field, it was not responsible for healing taking place in the center of the field. The effect of any marginal healing was eliminated by evaluating only those microscopic sections obtained from the central area 5 cm in diameter (Archambeau *et al.*, 1979).

The cell density increases exponentially during the regenerative phase, with an average doubling time ($1/e$) of 2.55 days. When this is evaluated to account for the increase in nonproliferating prickle cells and the basal cells, the estimated cell generation time is 13–15 hr. The average doubling time of cells in the small islands is 9 days. Based on the relative contribution of basal cells to the total number of cells at any time, it is surmised that cell proliferation in the giant islands is responsible for reepithelialization (Archambeau *et al.*, 1979).

The linear density of giant, intermediate, and small regenerating islands along the basement membrane is dose dependent. An average dose response (D_0) of 337 cGy (but later corrected to 250 cGy) has been reported using the combined density for the three island groups (Shymko *et al.*, 1984). The D_0 for small islands is 272 cGy (corrected to 241 cGy), 568 cGy for intermediate islands (corrected to 654 cGy), and 1602 cGy for the giant islands (corrected to 1490 cGy).

Reepithelialization is completed between 28 and 32 days after exposure, and cell density exceeds control levels in the immediate postregenerative period from 28 to 36 days. The hyperplastic epidermis returns to control levels at 1649 cGy and to less than control levels by 35 days, producing a second ulceration between 35 and 57 days at 2231 or 2619 cGy. During this period, the mitotic and labeling indices are at control levels or above (see Fig. 2, sixth and seventh charts, for basal cell density and mitotic index).

The basement cell density is correlated inversely with the area of the field involved in a moist reaction (see Fig. 2B; compare the sixth and eighth charts). The times of onset, maximal reaction, and complete healing (loss of crust) do not agree well with the time of the cell density nadir and the time of return to control levels, since the persistence of the crust masks the amount of epidermal regeneration.

Characteristic dermal and vascular histologic changes accompany the epidermal degenerative, regenerative, and postregenerative phases. During the degenerative phase, extending from day 0 to 25, no morphological changes in the microvasculature are noted until the epidermal remnant is lost. At that time, an inflammatory infiltrate and focal edema appear in the upper dermis beneath the eroded area (Archambeau et al., 1984, 1985). During the regenerative phase, lasting from day 17 to 28, there are few morphological changes. Where there is ulceration, an inflammatory infiltrate is seen which, however, disappears as reepithelialization proceeds. During the postregenerative phase, i.e., on day 28 and beyond, the endothelial cell density decreases abruptly in a dose-dependent manner between days 28 to 43, the endothelial nuclei becoming larger and paler. Occasional nuclei are pyknotic. Endothelial cell loss is more marked in the subpapillary than in the reticular plexus. The decrease in the cell density is accompanied by a parallel loss of lumen cross sections and a progressive dilatation, particularly of the capillaries and the venules. These changes are dose dependent and more marked at 2619 cGy than at 1645 cGy (Archambeau et al., 1984, 1985) (refer to Fig. 2 for charts of fractional endothelial cell density; third chart).

After irradiation with 1645 cGy, endothelial cell density does not fall below 50% of the control value (Archambeau et al., 1984). Vascular dilatation is minimal and epidermal cell loss does not occur. Focal areas of edema are found in the upper dermis and there is no fibrosis (Archambeau et al., 1985).

Following doses of 2231 or 2619 cGy, the endothelial cell density decreases below 50% of the control value (Archambeau et al., 1984); epidermal cell loss becomes apparent (Archambeau et al., 1979) and progresses to ulceration by day 49. The diameters of vascular lumina increase, reaching a maximum of 7 times those of the control levels by day 49, after a dose of 2231 cGy, and somewhat less of an increase at day 43 after 2619 cGy (refer to Fig. 2 for the time course of the change in lumen diameters, fifth chart). The upper dermis becomes edematous and the lower reticular dermis shows an increase in fibrosis. In areas of ulceration, the inflammatory infiltrate extends into the reticular dermis. Vessels containing microthrombi are present and are more numerous at the largest dose (Archambeau et al., 1985).

The principal morphological consequence of this sequence of events is the fragmentation and alteration of the vascular islands. There is cell loss followed by a decrease in the number of vessel lumen cross sections and a change in the density of the surrounding connective tissue. At day 43, after exposure to 2619 cGy, and day 49, after 2231 cGy, the vascular units appear as isolated, dilated single vessels set in an edematous matrix. The predominant feature is that lumina of the few remaining vessels are more elongated, suggesting that they have been cut longitudinally. Regeneration of vessels, endothelial proliferation, and intimal hyperplasia do not occur in this period. Similar morphological findings can be traced sequentially in the rabbit ear window preparation (Dimitrievich *et al.*, 1984; Griem *et al.*, 1985). These changes occur earlier and at a lower dose in the small vessels (<10 μm) than reported for the swine.

The survival dose of capillary loops in skin grafts 3 weeks after irradiation and 48 hr after grafting has a D_0 of 300 cGy in one study and 1030 cGy in another (Hopewell, 1983; Hopewell and Patterson. 1972). In other species, the D_0 for the capillary response to proliferative stimulation following irradiation varies over a range from 168 to 237 cGy (Fike *et al.*, 1979; Hopewell, 1983; Law, 1981; Reinhold, 1972; Reinhold and Buisman, 1973).

C. Late Reactions

The late skin changes of telangiectasia, dermal fibrosis, epidermal and dermal atrophy, and skin necrosis may occur without an earlier acute reaction (Archambeau *et al.*, 1968, 1969; Liegner and Michaud, 1961). Alternatively, these changes may follow in sequence weeks to months or even years later. The time of onset is, in part, dose dependent, and once present, a reaction may progress or remain static.

1. Telangiectasia

In the months following irradiation, a time is reached when superficial, elongated, and dilated vessels appear. During the ensuing months there is a gradual increase in the number of these vessels and the area covered by them, attaining a maximum involvement in 5–6 years after irradiation. The skin may be soft and pliable, depending on whether dermal or subcutaneous fibrosis is present. The epidermis is atrophic and dry, and epilation as well as loss of sweat and sebaceous glands is observed. Depending on the severity of the epidermal response, areas of focal keratosis and dysplasia may be present (Fajardo, 1982).

The occurrence, development, and severity of telangiectasia in man has been quantified (Turesson and Notter, 1984a–c). The formation of telangiectatic vessels is dose dependent, but the severity depends to a greater extent on the size of the daily fraction used to deliver the total radiation dose than on the total dose or time over which this dose was given. A few large fractions produce a more severe reaction at a lower dose than multiple small fractions (Withers *et al.*, 1978a,b).

The extent of telangiectasia has been utilized to quantify the skin dose response in man following clinical time–dose schedules. The absolute degree of severity is scored on an arbitrary scale of 0–3. Since the severity varies along the axis of the 5 × 12 cm fields and the area of greatest involvement is in the upper portion of the field, it is necessary to compare similar regions on adjacent fields (Turesson and Notter, 1984a–c).

The difference in the degree of telangiectasia in adjacent fields serves successfully as an isoeffect to characterize or compare delayed effects. Dose schedules using once weekly or twice weekly fractions have been compared with a standard of five daily fractions and a schedule of hyperfractionation using three daily fractions with 4–6 hr separating each of the fractions. As the size of the fraction is reduced, the relative severity of the involvement decreases (Turesson and Notter, 1976a).

These results are based on clinical studies which demonstrate the dose equivalency of the time–dose schedules used. Each time–dose schedule produced an equivalent erythemalous response, but the late telangiectactic changes were not the same. This suggests that a continuity between erythema and telangiectasia does not exist and that the population response producing erythema is different from that responsible for telangiectasia. These findings confirm the investigations which show that the mathematical expressions anticipating a dose response are not reliable in the documentation of a dose equivalency between various time–dose schedules (see later).

Biopsies obtained at various times after irradiation show a loss of microvasculature and an increase in the inside diameter of the remaining vessels (Archambeau *et al.*, 1979, 1984, 1985). The most prominent change is in the microvasculature of the subpapillary and reticular dermis. There is a decrease in endothelial cell density associated with a progressive increase in the lumen diameter and a loss of vessel lumina cross sections. After the initial rapid onset, these changes are slowly progressive. Changes in vessels larger than 10 μm may be minimal to marked depending on the dose delivered and time observed.

The microvascular remodeling which produces telangiectasia has been evaluated morphologically in swine following 2231–2619 cGy single dose fractions (Archambeau *et al.*, 1985). The suggested sequence in the subpapillary dermis is that as endothelial cells are lost, the microvessels shorten, uncoil, and dilate. The apparent decrease in lumen density in the vascular tufts may be explained on the basis of the configuration of the vessels in the tuft. If the tuft consists of multiple parallel vessels, as in a manifold (glomerulus), the decrease represents a loss of vessels. If the tufts consist of a single coiled and folded vessel, the apparent loss of lumina would reflect the uncoiling and shortening of the vessel as cells are lost. However, the available "two-dimensional" histologic data do not distinguish between these two configurations; three-dimensional reconstructions are needed (Archambeau et al., 1985).

Changes of endothelial cell density in larger vessels appear to be different from those in smaller ones, since focal endothelial hyperplasia is reported in arterioles (Hopewell, 1980; Young and Hopewell, 1980), but not in capillaries (Archambeau, et al., 1984, 1985). The endothelial cell density in vessels of the mouse mesentary shows a decrease in density at 3 months, with a return to control levels by 1 year (Hirst et al., 1979, 1980), and cell loss in the dermal microvasculature is followed by vessel loss (Archambeau, et al., 1984, 1985).

2. Fibrosis

This late radiation-induced skin change is characterized by increased induration, stiffening, and thickening of the dermis in the irradiated fields. It occurs during the months and years following the radiation exposure. Fibrosis may appear following the acute responses of erythema, dry and moist desquamation, or in the absence of acute reactions. It is most severe in those areas where there was an earlier moist reaction with ulceration that healed. Onset and formation of fibrosis are dose dependent and, once started, are slowly progressive. As the degree of fibrosis increases, the probability that necrosis will be the end result also increases. Fibrosis of the irradiated field is more extensive in areas with a reduced vascularity overlying bone, e.g., sacrum, or in fields with abundant subcutaneous fat.

The formation of fibrosis indicates that the field has received a near-tolerance dose. While accepted as a characteristic of late radiation change, fibrosis has not been used as an isoeffect parameter to evaluate the dose response.

The morphology of the fibrotic, irradiated field is characteristic. The epidermis may be thickened, hyperplastic, and distorted with blunt elongated papillae extending into a thickened dermis or may be atrophic with few small shortened papillae. The dermis consists of a layer of multiple distorted collagenous bundles that are relatively acellular in the upper dermis. In other areas, there is extensive, cellular fibrosis that lacks a consistent pattern.

The degree of cutaneous atrophy and contraction in fields 8 × 8 cm on young swine is attributed to cell population density changes (Withers et al., 1978b, 1980). These changes are suggested by comparing the field dimensions of the irradiated fields with those of unirradiated control fields. However, viewed from the perspective of animal growth, these changes may also represent the failure of continued tissue growth rather than cell loss. Dimensions of the irradiated fields do not increase as much as those of the control fields in these immature, rapidly growing swine. At the highest dose, the field dimensions do not decrease unless there has been subcutaneous necrosis or field ulceration and healing.

In patients, the fields irradiated at a high dose are indurated and depressed in the center compared with the surrounding skin. Field contraction or decrease in dimensions and loss of fibrous tissue have not been reported, and reduced dimensions are seen only after healing (late) of previously ulcerated fields.

There are various models used to account for the late change manifesting as fibrosis. One proposal suggests that the increased fibrosis reflects continued proliferation of the fibrocytic cell renewal system (Withers *et al.*, 1980). A second view considers fibrosis as reflecting the repair and consolidation of inflammatory or necrotic foci, e.g., fat necrosis (Archambeau and Vora, unpublished). The distorted morphology of the fibrotic dermis found in healed scars, abscesses, and inflammatory foci can also be found in the irradiated dermis. Fibrosis represents a specific response of dermal fibrocytes to damage to the dermis and associated cell populations.

3. Necrosis

The end stage of the dose response is necrosis. This may be focal or include the entire irradiated field (Archambeau and Mathieu, 1969; Archambeau *et al.*, 1968, 1969). The time of onset and occurrence is dose and fraction size dependent.

The late onset of necrosis follows as an end stage to progressive fibrosis and loss of vascularity, but the time of onset and evolution of necrosis vary. Following single dose fractions in swine, the lowest dose at which necrosis develops is 1843 cGy. At 3000 cGy and above, all fields become necrotic within a period of 36–70 days. After a total dose exceeding 2522 cGy and given in two dose fractions at an interval of 6 hr or one greater than 4365 cGy and given in five daily exposures resulting in a moist reaction, the field becomes necrotic in 70 days. Yet, when a dose of 3360 ± 86 cGy is given in two fractions separated by an interval of 24 hr, the moist reaction heals in 80% of the fields without progressing to necrosis. The dose recovery in this instance is 923 cGy compared to about one-half of this following other schedules (see discussion later and data in Table VI). Assuming that necrosis results from vascular damage, this observation suggests that the radiosensitivity of the microvasculature changes markedly during the 24-hr interval.

Necrosis or nonhealing of the field is a useful isoeffect parameter defining relative skin tolerance. In swine, the 50% effective dose, i.e., the dose at which 50% of the irradiated fields do not heal, has been defined for several time–dose schedules. These values are listed in Table VI and plotted in Figs. 3 and 4, with data defining dose tolerance in man.

It is a well-documented clinical observation that a moist reaction may evolve and heal without being followed by a progressive necrosis reported for swine. As

TABLE IV

FACTORS ALTERING DOSE RESPONSE

1. Type and energy of radiation used
2. The time–dose schedule used, dose rate
3. The responding cell population (basic tissue unit)
4. The proliferative activity of the target cell population, position in the mitotic cycle, and distribution in the cell cycle
5. The capacity (biochemical) to repair DNA and other injury
6. The volume of tissue irradiated
7. The physiologic states of temperature oxygen tension, as well as the presence of chemical sensitizers or protectors

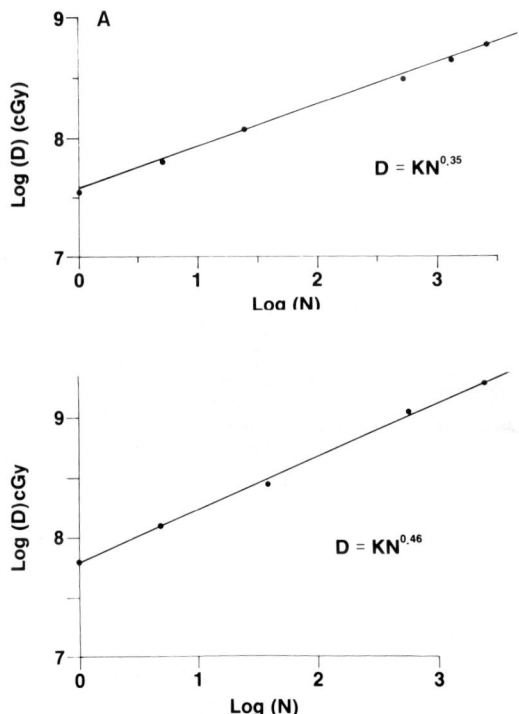

FIG. 3. (A) Log–log dose number of dose fractions isoeffect curve for patient skin "tolerance" treated with daily irradiation (derived from patient data published by Patterson, 1963). The best fitting linear approximation has a slope of 0.35. (B) Log–log dose number of dose fractions isoeffect curve for swine skin 50% effect dose using daily irradiation (Archambeau *et al.*, 1968, 1969, and unpublished data). The best fitting linear approximation has a slope of 0.46.

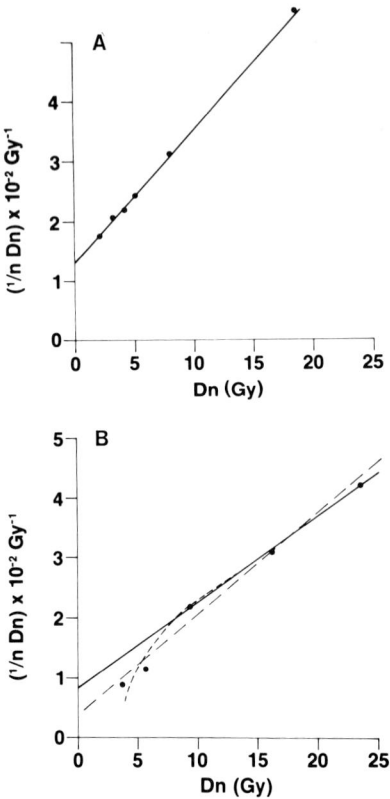

F ig. 4. (A) A reciprocal dose–dose fraction isoeffect curve for patient skin ''tolerance'' treated with daily irradiation (derived from patient data published by Patterson, 1963); the α/β ratio is estimated to be 599 cGy for the best fit line. (B) Reciprocal dose–dose fraction isoeffect curve for swine skin 50% effect dose using daily irradiation (Archambeau *et al.*, 1968, 1969, and unpublished data). The α/β ratio is 229 cGy for the best fit line for all data (dashed line) or 569 cGy for first three data points (solid line).

a consequence, the extrapolation of dose tolerance data from swine to man may not be justified. This difference between man and swine in the evolution of the response may be referable to the use of swine which are immature and still growing as contrasted with adult man. Assuming that necrosis reflects the loss of the dermal microvasculature, the difference in late response between man and swine may also reflect a difference in vascular endothelial dose tolerance or kinetics.

IV. Clinical and Radiobiological Considerations: Skin Dose Response

The skin dose response characteristic of the radiation or time–dose schedule used is specified as a dose or a range of doses that will produce a selected isoeffect. If the clinical or investigative data were suitable, a better approach would be to state the probability that a given dose will elicit a selected isoeffect (Cohen, 1983; Moulder *et al.*, 1975). Any time–dose schedule or type of radiation of sufficiently high total dose will result in the skin change selected for evaluation. When gross parameters are defined, histologic parameters are implied.

The dose tolerance of skin cannot be defined by a single value for all time–dose combinations, radiations, or conditions. Instead, there is a range of values which depend on multiple physical and biological factors. These factors are listed in Table IV. Before skin tolerance for a selected time–dose schedule can be applied to treatment planning, the effect of the time, dose, and dose fraction variables must be taken into account.

A. Response to Various Radiation Qualities and Time–Dose Schedules

1. Response to Various Radiation Qualities

The evolution and time course of the early and late skin changes induced by different types of radiations and various time–dose schedules are similar. However, the dose required for the induction of these changes is characteristic of the type of radiation employed rather than the tissue irradiated.

The dose response for different types of radiations as determined by using skin isoeffects of erythema, area of field involved in moist reaction, fraction of fields nonhealed, and degree of "field contraction" (lack of growth?) are listed in Table V. The dose response to various radiation qualities reflects the linear energy transfer (LET) or ionization density along the track of the radiation as it traverses the tissues. Densely ionizing radiations such as fast neutrons and α particles produce the same response at a lower dose than that required by photons, as listed in Table V(A). This decreased dose requirement is inferred from the ratio of the dose required to produce an isoeffect by X rays and by the high LET radiation [Table V(B), column labeled Ratio].

2. Response to Different Time–Dose Schedules

The skin response resulting from single dose fractions is constant for a specified proliferative and functional state of the tissue's cell population. How-

TABLE V

THE DOSE RESPONSE OF THE SKIN OF SWINE AND MAN PRODUCED BY VARIOUS RADIATIONS

A:	Radiation	Study	Schedule	Isoeffect	Dose (cGy)
I.	250 kv X ray[a]	Swine	Single	ED^{50} (nonheal)	2364 ± 86
II.	Fast neutrons (fn). 50 MeV, d-Be	Swine	2–4 times/week for 6½ weeks	Contraction	2200
III.	Degraded fission neutrons (fn) fn 67%, γ 32%	Swine	Single dose fraction	ED^{50} (nonheal)	944 ± 109
IV.	Thermal neutrons, γ 68%, fn 14%, $^{14}N(n,p)^{14}C$ 25%	Swine	Single dose fraction	ED^{50} (nonheal)	1469–107
V.	$^{10}B(n,a)^7Li$, γ 29%	Man	Single dose fraction	ED^{50} (nonheal)	700–1100
VI.		Swine	Single dose fraction	ED^{50} (nonheal)	1100 ± 112
			5 dose fractions 3 times/week for 12 days		~1100–1200
VII.	Fast neutrons + γ	Swine	Single dose fraction	Necrosis	>1725
VIII.	Fast neutrons + γ	Swine	2–4 times/week for 6½ weeks	25% contraction	2200
				50% contraction	2340

176

B: Radiation	Study	Schedule	Standard	Isoeffect	Ratio[b]	Reference
Fast neutrons, 42 MeV, d-Be	Swine, 16 × 4 cm, 6 × 4 cm	Single	X ray	1.5–2.5 average score, 16 weeks	1.4–1.6	Field et al. (1976); Hopewell et al. (1982)
				2 average score, 3–9 weeks	~1.2	
				Necrosis	1.35	
				Contraction	1.5	
Fast neutrons, 50 MeV d-Be	Swine, 8 × 8 cm	2 times/week for 6½ weeks	Cobalt	Average score (8 weeks)	<2	Hussey et al. (1982)
		4–5 times/week for 6½ weeks		Contraction (8 months)	2.2–2.6	Withers et al. (1978a,b)
				Score (8 weeks)	<2.5	
				Contraction	3.1–3.4	
Pions	Swine	5 times/week for 2 weeks	280 kV P X rays	Score 2–6 weeks	~1.4	Goodman et al. (1982); Skarsgard et al. (1982)
				1.5–7 months?	1.4– >1.6	
	Man	5 times/week for 2 weeks	280 kV P X rays	Moist reaction	1.4–1.7	Bush et al. (1982); Raju and Tokita (1982)
	Clinical Rx	10–14 day 3 fx		Erythema	1.3	
	Clinical Rx	13–15 fx			1.4	
	Clinical Rx	10 fx			1.41–1.48	
		3 fx		?	1.34	

[a] References: I, Archambeau et al. (1968); II and VIII, Withers et al. (1978a,b); III–VI, Archambeau et al. (1971); V, Archambeau (1970); VII, Berry et al. (1974a,b).

[b] Dose ratio is the ratio of the dose of standard photon radiation (column 4) to the dose of the test radiation to produce the isoeffect. It is not an RBE as such because the beams contained other radiations, e.g., γ rays or neutrons.

ever, the response resulting from X-ray dose schedules using multiple dose fractions changes uniformly when the time or number of dose fractions are varied over a range of values. When the number of equal-sized dose fractions or the time over which the irradiation is given is increased, the dose required to produce a selected effect increases. This requirement to increase the total dose in order to produce the same effect indicates that the dose response is not cumulative and that during the interval between dose fractions, some biological or biochemical recovery from radiation damage caused by the previous dose fraction has taken place.

When the total dose is given in multiple dose fractions using high LET radiations, the total dose required to produce the isoeffect does not have to be increased. This fact suggests that little or no dose recovery occurs in the interval between dose fractions. Thus, the dose effect produced by high LET radiation is cumulative unless the interval between dose fractions is prolonged to several days and cell proliferation occurs. This response is interpreted to mean that the dose survival response is exponential and dose/recovery or repair of injury does not occur (see next section).

3. Dose Recovery

The total dose of X rays required to produce a selected isoeffect is larger when it is delivered in multiple dose fractions than as a single dose fraction. The dose difference reflects a difference in the extent of radiation-induced tissue changes or a degree of injury from which the tissue has recovered. This value is referred to as dose recovery (D_Q) and is expressed as the average dose increase required for each dose fraction to produce the isoeffect:

$$D_Q = \frac{D_n - D_1}{n - 1} \tag{1}$$

where D_Q is the dose recovery, D_n the total dose given in n dose fractions to produce the isoeffect, D_1 the total dose given in a single dose fraction required to produce the isoeffect, and n the number of dose fractions.

The dose recovery for skin using different isoeffects and different time–dose schedules are listed in Table VI (Archambeau and Mathieu, 1969; Archambeau et al., 1968, 1969; Bewley et al., 1963; Dutreix et al., 1973; Fowler et al., 1963, 1965). Note that the recovery dose determined by graphic analysis of data using multiple schedules varies as a power function of time or number of dose fractions with the power having a range of values (see Table VII, Dose recovery factor, and Section IV,A,4).

Using split-dose techniques in which two equal dose fractions are given at increasing intervals, the average dose recovery increases as the interval between

TABLE VI

Dose Recovery in Swine[a]

Number dose fraction/time	Interval	Isoeffect	Total dose (cGy)	Dose Recovery (cGy)	Reference
Single dose		50% nonhealing, 49 days	2437 ± 89		Archambeau et al. (1968, 1969); Archambeau and Shymko (1984)
2/6 hr	6 hr		2794 ± 92	357	
2/24 hr	24 hr		3360 ± 86	923	
5/4 day	Daily		4761 ± 366	581	
30/40 day	Daily; 5 times/week		10.670	284	
2/24 hr	24 hr	Score ~2, 30–70 day, Erythema		580	Fowler et al. (1965)
2/5 hr	2 day			730	
3/2 hr	Daily			640	
5/4 hr	Daily			465, 403	
5/28 day	Weekly			520	
9/28 day	2 times/week			378	
21/28	Daily			228	
Single dose		Dry desquamation	2900		Berry et al. (1974a)
6/18 day	2 times/week		4250	270	
30/39 day	Daily		5700	96	

[a] Dose recovery is defined as $D_n - D_{1/n-1}$, where D_n is total dose given in n dose fractions required to produce the isoefect, D_1 is total dose given as a single dose fraction required to produce the isoeffect, and n is the number of dose fractions.

dose fractions increases. The early increase in recovery dose which occurs as the interval increases from 1 to 6 hr is attributed to the recovery from sublethal injury (Elkind and Sutton, 1960), whereas a subsequent increase in dose recovery seen from 6–12 to 24 hr, concurrent with a change in the pattern of recovery, suggests the movement of basal cells into phases of the cell cycle having different dose responses. Increases in the recovery dose at longer intervals may derive from basal cell proliferation (Denekamp et al., 1966, 1969, 1971; Dutriex et al., 1973; Withers, 1967a,b).

An interpretation of the data on the time–dose isoeffect response implies that the erythema and moist reaction isoeffects represent changes primarily in the epidermal basal cell population. When the nonhealing isoeffect is used, the available histologic data indicate that changes in the microvascular endothelium contribute to the production of the isoeffect (see histologic description of the microvasculature). As a consequence, using the nonhealing isoeffect, the increased recovery found when the two dose fractions are given at in interval of 24 hr in swine may reflect recovery in the endothelial population (see discussion of the variation in the slope of dose response curves for different tissues).

4. Time–Dose Isoeffect Plots

When different dose schedules are used in radiation therapy, it is necessary to compensate for dose recovery by altering the schedule in order to achieve a constant therapeutic effect. The variation in dose response, i.e., the dose recovery associated with different time–dose schedules, is smooth and predictable (within limits). It is easily demonstrated by plotting the change in total dose required to bring about a selected isoeffect as a function of time or number of dose fractions. The graphic plot presents a clear and convenient accounting of the complex time–dose fraction–recovery dose relationship and allows a definition of the dose response of the skin. At the present time, this determination of the dose response dominates the clinical approach to selecting equivalent therapy schedules.

When the log of the total dose is plotted against (1) the log of time over which the irradiation is given, or (2) the log of the number of dose fractions given at some uniform repetitive rate [daily, 5×/week, 3×/week, 1×/week [Eq. (2)], the response is approximately linear with a positive slope (see the power function log–log plot in Fig. 3).

A linear relationship with a positive slope is also obtained when the reciprocal of the total dose ($1/D_n N$) is plotted against the daily dose fraction (D_n) [Eq. (3)] (see the linear quadratic function and plot in Fig. 4). However, this relationship has not been observed in the rat (Moulder et al., 1975).

The power (log-log) and linear quadratic (Lin-Q) time–dose fraction isoeffect plots are represented by different mathematical expressions (see Goitein,

1976 and Barendsen, 1982 for a comparison of power constructs, and Douglas and Fowler, 1976 and Withers *et al.*, 1982a for a discussion of the linear quadratic equation).

Power Function:

$$D = kT^a = kN^b \quad \text{or} \quad kT^cN^d \quad \text{and} \quad a = b = c + d \qquad (2)$$

where D is the total dose (cGy) required to produce the isoeffect; T the length of time during which the total dose D was given using multiple dose fractions, i.e., elapsed time; N the number of dose fractions used to deliver the total dose (N is accepted as being of equal size and given in a repetitive pattern); a, b, c, and d are the fractional exponents of the T (a) or N (b) or the product $T \times N$ ($c + d$) variables and represent the slope of the log–log plot; and k is the proportionality constant (see NSD later).

Linear quadratic equation: $1/D_nN = N(\alpha + \beta D_n)$ (3)

where D_n is the size of the dose fraction in cGy; N the number of dose fractions; ND_n the total dose (cGy) required to produce the isoeffect and equal to D as defined above; α the constant coefficient of the D variable with units of cGy$_{-1}$ and represents nonrecoverable (single particle) injury, i.e., the initial slope of the dose survival curve, or as the intercept of the ordinate of the reciprocal of the total dose–dose fraction isoeffect curve; and β the constant coefficient of the D^2 term, with units of cGy^{-2}, and represents recoverable (multiparticle) injury, i.e., the slope of the reciprocal of the total dose fraction isoeffect curve.

Since the gross isoeffect is brought about by fluctuations and interactions of various cell populations, the parameters of the above equations are presumed to represent, but not identify, changes in cell population density.

a. Power Function.   The slope (rate of dose recovery) of the log–log time–dose isoeffect plot generated by clinically used time–dose schedules varies, as shown in Table VII. The best (?) clinical value for the slope is 0.35. The plots for the control of squamous skin cancer have a corrected slope of 0.24. The difference of 0.11 (some feel it should be 0.07) (Herbert, 1981) between the slopes of these two plots represents the recovery dose occurring in time for skin, but not for cancer (Ellis, 1968). This differential in recovery time shows the effect of time on recovery dose, whereas the difference, 0.24, shows the effect of the number of fractions on the recovery dose as time or number of dose fractions are varied. The resulting expression is the Ellis (Ellis, 1986; Ellis *et al.*, 1969) or NSD equation which duplicates the dose response for early skin reactions:

$$D = \text{NSD} \times N^{0.24} \times T^{0.11} \qquad (4)$$

NSD refers to the nominal standard dose measured in rad equivalent therapy (ret) and is a proportionality constant dependent on the degree of radiation

TABLE VII

Test of the Values for the Dose Recovery Factor b and the α/β Ratio Obtained from Number of Dose Fractions, Total Dose, and Reciprocal Total Dose – Dose Fraction Isoeffect Data

Source/dose schedules used	Dose recovery factor b[a]	Dose survival (α/β)[b] (cGy)	Insoeffect	Reference
Human				
16 fx in 22 days	0.24	833		Turesson and
12 fx in 40 days	0.12–0.22	625, 1670		Notter (1984a,c)
30 fx in 40 days		Depends on overall time	Erythema	
12 fx in 40 days	0.26–0.35	270–454	Telangiectasia	
30 fx in 40 days		Depends on level of effect		
16 fx in 22 days, 4 fx in 22 days	0.34	322		
Multiple clinical schedules	0.35	599	Tolerance?	Patterson (1963)
		365		Cohen (1983)
Swine				
Single	0.21	3020(?)	Dry desqua- mation	Berry et al. (1974a,b)
6 fx in 18 days	0.16	5035(?)	Moist reaction	
30 fx in 39 days	0.33	1390(?)	Dose that did not produce necrosis	
Single				
2 fx in 2 days	0.46	560[c]	50% nonheal- ing at 49 days	Archambeau et al. (1968, 1969)
5 fx in 5 days, 16 fx in 19 days		229[d]		
30 fx in 39 days		613		
13 fx in 44 days, 32 fx in 44 days	0.45–0.47	305	Contracture	Withers et al. (1978a,b)
6 fx in 18 days, 30 fx in 39 days	0.37–0.39	220–360	Necrosis	Berry et al. (1974a,b)

[a]See Eq. (1) under power function.
[b]See Eq. (2) under linear quadratics.
[c]For three schedules.
[d]Best curve fit for five schedules.

response considered. T, N, and D are as defined above. A variable to account for the quality (RBE) of the radiation (Q) and the volume (V) of tissue irradiated [$(\text{vol}/1000 \text{ cm}^3)^{0.16}$] may also be included in the equation.

b. Linear Quadratic Function. The time–dose isoeffect relationships were reexamined emphasizing cell kinetic and radiobiological considerations. When the reciprocal of the total dose ($1/ND_n$) that produced the isoeffect is plotted against the individual dose fraction, a linear relationship is obtained (Fig. 4), the Lin-Q concept.

The linear quadratic equation is based on the assumption that the survival dose (for single dose fractions) of the cell population is expressed by

$$E = \alpha\, D + \beta\, D^2 \tag{5}$$

and for multiple dose fractions by

$$E = n[\alpha\, D_n + \beta(D_n)^2] \tag{6}$$

$$E = -\ln(N/N_0/\alpha D) = (1 + D_n/\alpha/\beta) \tag{7}$$

where E represents the effect produced or the isoeffect parameter, and N/N_0 the fraction of cells surviving;

$$E = -\ln(S/\alpha\, D) = (1 + D_n/\alpha/\beta \tag{8}$$

where S is the fraction of surviving cells and is equal to N/N_0.

The skin change associated with the isoeffect parameter is assumed to represent the survival of a fixed fraction of the population (Barendsen, 1982; Douglas and Fowler, 1976). If it is assumed that cell recovery between fractions is complete, that there is no proliferation, and that the α/β ratio is known, Eq. (8) can be used to determine the dose response and cell survival following different time–dose schedules (see Section IV,B,2).

The Lin-Q expression (see above) implies that the size of the dose fraction is more important in eliciting the dose response in skin than the time during which the dose fractions are given or the number of fractions used (Fowler *et al.*, 1963, 1965). Although different sets of parameters are used in this construct, the results obtained do not deviate markedly from those obtained using the NSD construct, since the total dose is the product of the size of the dose fractions and the number of fractions used. Therefore, while variables T and N are not used directly, their effect is retained implicitly, since dose fractions of equal size are given at a uniform rate, i.e., fractions per day or week over a known period of time. The α/β ratio that defines the survival dose response now becomes a parameter that can be readily derived from isoeffect plots of reciprocal total dose versus dose fraction. The intercept of the ordinate (reciprocal dose) when D_n is zero is α, and

β is the slope of the line. The significance of the α and β parameters is taken to be the same as those used in the dose survival expression.

The numerical value of the α/β ratio is characteristic for different tissues (Barendsen, 1982; Withers *et al.*, 1980, 1982a–c, 1983). This value is greater than 500 cGy for rapidly proliferating tissues, e.g., about 1000 cGy for mouse epidermis, 500 cGy for swine epidermis using the dose at which 50% of the irradiated fields were not healed (ED_{50}) as the isoeffect, and about 620 cGy for the production of erythema in man. The ratio is less than 500 cGy for slowly proliferating tissues. The value for late isoeffects such as skin contraction and telangiectasia, which are considered to result from changes in the micro-vasculature and dermis, is 270–450 cGy for telangiectasia in man, 406 to 734 cGy for skin contracture in the legs of mice, and 270 and 360 cGy for skin contracture in swine (Berry *et al.*, 1974a,b; Withers *et al.*, 1978a,b). The α/β ratio for acute changes in endothelial cells of the microvasculature in the ear window chamber of the rabbit is 550 cGy (Griem, 1985), whereas the value for dermal fibrous tissue is not known.

c. Computer-Derived Model. Cohen (1983) has shown that tolerance dose limits for skin and other organs irradiated with different time–dose schedules can be defined with a known degree of confidence. Using cell population kinetics and radiobiological parameters derived from published time–dose isoeffect data, the dose tolerance for different tissues has been defined following several clinically appropriate time–dose schedules. A cell lethality and repopulation model was designed to simulate the effects of fractionated radiation therapy. The model employs a single hit, multitarget-single hit, or linear quadratic survival dose expression, whichever is appropriate. Repopulation during the interval between exposures and cell survival following the irradiation is described by a recursive form of the logarithmic growth function. The value of the parameter (b) in the power function construct equation (2) used for skin (see above) obtained with this model is 0.45. The α/β ratio is 365 cGy. These values are similar to those obtained using the other constructs (see Table VII).

B. Swine Skin Model for Radiation Response

Knowledge of cell kinetics and radiobiological parameters will permit the clinician to anticipate gross and cell population changes in the skin produced by various radiations and different time–dose schedules. These data can then be used to select a time–dose schedule that will produce a desired response. The parameters that quantify the evolution, time course, and dose dependence of skin changes discussed in Section II are listed in Tables II and III and displayed for swine in Fig. 2. These results depict the formation and evolution of erythema, a moist reaction that heals, and the formation of changes in the microvasculature.

These data have been used to derive estimates of the cell kinetics and radiobiological parameters of the different skin populations. The epidermal and basal cell kinetic parameters of unirradiated and irradiated skin are presented in Table I. Comparison of the derived and measured population parameters indicates that the models (expressions) representing the skin response to irradiation are incomplete. If specified time–dose schedules are to be used to modify the dose response of the skin or are to be used to produce a particular response, the population changes and dose response must be quantified.

1. Prediction of Radiation Response Based on Kinetic Parameters

The gross and histologic changes produced by irradiation that are anticipated by a simple model are compared with the observed findings in Table VIII. This model assumes that radiation tolerance of cells is randomly distributed across the field, that cell loss from the basal layer is the result of cells maturing and passing into the prickle cell layer or of cell death at mitosis, and that the rate of cell loss is not changed by irradiation. It is anticipated that the size of the fraction of cells surviving irradiation will be dose dependent. At high dose fractions, only one or a few cells per square centimeter need to survive to assure repopulation; the time required for healing reflects the rate of cell proliferation. Subpopulations with different survival characteristics may be present which will alter the dose response (Archambeau and Mathieu, 1969). All basal cells are proliferative in growing swine (Archambeau and Bennett, 1984; Shymko and Archambeau, 1985), and all basal cells participate in reepithelialization.

The agreement between observed and measured parameters ranges from poor to good. Analysis of these deviations suggests new interpretations for the observations. If the value of the time course for gross changes were estimated from the histologic parameters using an average generation time of 12.4 days and a minimum transit time through the prickle cell layer of 13 days, the loss of epidermis would be progressive and complete at about 25 days. If 100 cells survived in 4×10^6 cell/cm^2 ($D_0 \sim 250$ cGy), a total of about 10 cell doublings would be required to replace the cells lost. If the rate of cell proliferation remained unchanged, replacement of cell loss would require about 120 days provided that all cells are proliferating. The epidermal cell population returns to control levels by 28–32 days, indicating a shortening of the average regeneration time to 3 days or less. Estimates of the average generation time (T_C) that accounts for prickle cell replacement suggest a T_C of about 13–15 hr (Archambeau et al., 1979). A T_C of 22 or 36 hr has been estimated for the regenerating basal cells in the mouse (Withers, 1967a,b; Emery et al., 1970). These estimates are confirmed by kinetic data from guinea pig and human skin (Song and Tabachnick, 1969; Tabachnick and Weiss, 1959; Yamaguchi and Tabachnick, 1972; Etoh et al., 1977).

TABLE VIII

COMPARISON OF PARAMETERS REPRESENTING THE EVOLUTION AND TIME COURSE OF THE MOIST
REACTION OBSERVED FOLLOWING IRRADIATION WITH THOSE PREDICTED BY A SIMPLE MODEL

Parameter	Model value	Observed			Agreement
		Gross	Histologic		
Time of onset of reaction (time of transit through prickle layer)[a]	>13 days	17.5 days	>13 days		Good
Time of maximal involvement	Average generation time of basal cells + transit time of prickle layer, 12 days + 13 days; Not dose dependent	25–28.5 days Not dose dependent	22–24		Good
Time of maximal healing	Variable, 25–45 days	Fixed, 36–38 days	Variable, 28 days at 1645 cGy; 28–30 days at 2231 cGy; 32 days at 2619 cGy		Fair to good; curst masks amount of healing

Area of involvement	Dose dependent 100% involved	Not dose dependent Variable, always <80% area	Dose dependent Cell loss noted in all areas	Poor
Time to evolve	Not dose dependent Variable, 1–20 days	Dose dependent! 10–11 days, all doses	Dose dependent! 6–8 days	Fair Fair–good
Completeness of healing	Dose dependent Complete or not, no partial healing	Not dose dependent Variable area healed	Dose dependent Complete! With highest dose used	Fair healing complete under the crust; partial healing may represent early onset of second breakdown
Dose response		Fraction not healed increases with dose $ED_{50/49}$ days	$D_0 = 250 \pm$ cGy	Good
Microvasculature	Sequence not specified, but with loss, ulceration and necrosis occurs	Necrosis seen after healing 36–49 days and later	Confirmed	

[a] The average basal cell generation time is ~12 days. The transit time through the prickle layer is ~13 days.

A mitotic index near zero and a labeling index of 0.01 in the endothelial population of the microvasculature indicate that this is a slowly proliferating population with a prolonged average generation time (Tannock, 1968; Tannock and Hayashi, 1972). Based on the propositions set forth above, cell loss would be slow to develop and cell replacement would be prolonged beyond that required of cells in the epidermis. The loss of the endothelial cell population at 32–43 days postirradiation was neither anticipated nor satisfactorily explained (Archambeau *et al.*, 1984, 1985). Endothelial cell proliferation was observed neither in the dermal microvasculature (Archambeau *et al.*, 1984, 1985) nor in the the rabbit ear window chamber (Van Den Brenk, 1959; Dimitrievich *et al.*, 1984) nor in the yolk vitelline vessels (Stearner and Christian, 1969, 1971). The endothelial cell repopulation of larger vessels takes place slowly during the year following irradiation (Hopewell, 1983; Dimitrievich *et al.*, 1984, 1985; El-Naggar *et al.*, 1978; Fischer-Dzoga *et al.*, 1984, 1985; Griem *et al.*, 1985; Hirst *et al.*, 1979, 1980). Arteriolar and venous endothelial hyperplasia has been reported (Hopewell, 1980, 1983). The continuing dilatation (telangiectasia) and loss of microvessels with time after irradiation suggest that endothelial cells in vessels of less than 10 μm have a fixed life span and are not replaced by adjacent cells. Thus, with continued cell loss, continuity of the lumen is broken and the vessels contract (Archambeau *et al.*, 1984; Dimitrievich *et al.*, 1984; Van Den Brenk, 1959). However, wound healing with revascularization occurs in irradiated fields.

Studies on endothelial cell survival using indirect capillary sprouting techniques reported a D_0 of 168–1030 cGy. Direct counting of cells in the rabbit ear window reveals the presence of radioresistant and radiosensitive cell populations with different α/β ratios (Dimitrievich *et al.*, 1984; Griem *et al.*, 1985), and these data suggest that the dose sensitivity of the sensitive component of the endothelium is similar to or greater than that of the epidermis. The apparent radiosensitivity of the endothelial cells may be interpreted as arising from (1) the apparent lack of repopulation to replace cells lost (Stearner and Christian, 1969, 1971; Dimitrievich *et al.*, 1984; Van Den Brenk, 1959), (2) cell loss that causes an interruption of the linear arrangement of endothelial cells, retraction, and loss of the vessel (Dimitrievich *et al.*, 1984), and (3) a low total population number of $N_0 \sim 10$ such that a small fractional cell loss represents a large actual loss of function per cubic centimeter which cannot be compensated for by cell proliferation (see considerations in Section II). Clinical data that show that the formation of telangiectatic vessels is reduced when the size of the dose fraction is decreased and given daily as compared to once or twice weekly suggest that the endothelial cell is capable of considerable dose recovery, i.e., sublethal injury repair.

Few data are available characterizing the dermal fibrocytic population. The population parameters obtained by histologic observation suggest that this is a potentially proliferative cell renewal system which responds to injury or inflammation by proliferating. The average cell generation time is not known and survival dose parameters cannot be estimated. If the microvasculature is intact, dermal fibrocytes are present. After doses greater than 6000–8000 cGy, intact fibrocytes are found in the dermis adjacent to tissue volumes treated with interstitial radiation therapy using implants of radium or radioactive iridium. These fibrocytes are proliferative and form new fibrous tissue in areas of inflammation, suggesting a high dose tolerance for these cells (Archambeau and Vora, unpublished data; Fajardo, 1982; Lacassagne and Gricouroff, 1956; Warren, 1943; Rubin and Casarett, 1968).

2. Comparison of Derived and Measured Population Parameters

The range of values derived for the population survival parameters using the time–dose equations is similar in magnitude to that measured in cell cultures. It is not known whether these values reflect the response of a specific target cell population or that of a composite tissue unit.

The dose response with respect to gross skin changes elicited by different time–dose schedules may be represented accurately by a power function. However, when the mathematical formulations are applied to changes in cell population, the results are inconsistent with radiobiological and cell kinetic findings, as discussed at length by Barendsen (1982) and Herbert (1981). For example, in the NSD formula expressing equivalence among different time–exposure schedules [Eq. (4)], the time factor is typically taken to represent the effects of proliferative recovery during the course of irradiation. In the form given, the implication of the fractional exponent for T is that the relative recovery is greater for smaller values of T and therefore that the effects of proliferation are greatest early in the exposure schedule. Direct histologic observation reveals, however, that the greatest increase in cell numbers occurs at 2–3 weeks following irradiation. As a consequence, while this equation accounts for the dose response of the skin as the number of dose fractions is increased or as the time is increased, it does not reflect the population change.

When the NSD concept is used in clinical practice to select time–dose schedules that produce equivalent acute tissue responses, the late tissue changes that are produced are unacceptable. This difference in response increases as the number of dose fractions decreases and the dose per fraction increases (Singh, 1978; Turresson and Notter, 1976b, 1984b; Withers et al., 1978a,b). The acute reaction is thought to be localized in the rapidly proliferating epidermal cell population while the late reactions occur in slowly proliferating tissues such as the microvasculature and connective tissue. This indicates that the dose response of these cell populations is different.

Gross skin changes brought about by different time–dose schedules may also be represented accurately by the linear quadratic model in which cell survival is given by Eq. (7). This model which is based on the cell population response to irradiation gives more meaning to and insight into the skin response. However, this model-based approach has limitations because it does not account for cell proliferation during the delivery of the time–dose schedule. This represents a real restriction, since it is known that proliferation does occur in the epidermis of swine and mice and becomes important beginning at 2–3 weeks following the start of irradiation when the average generation time decreases (Archambeau, unpublished; Cohen, 1983; Denekamp, 1973; Archambeau *et al.*, 1979; Denenkamp *et al.*, 1966, 1969, 1971).

The reciprocal total dose–dose fraction isoeffect plot for swine skin deviates from linearity, which suggests that proliferation may be occurring (Fig. 4). The five data points can be fitted statistically to a straight line, giving an α/β of 269 cGy. However, the deviation of the last two points from a straight line fitted to the first three points strongly suggests proliferative activity. Proliferation has also been demonstrated histologically in swine receiving daily irradiation for 6 weeks or more (Fig. 5) (Archambeau, unpublished data).

The occurrence of cell proliferation in man can be inferred from skin changes observed during a schedule of irradiation that lasts 10–12 weeks. Two cycles are noted in which erythema and dry desquamation are formed and are followed by clearing. This indicates that cell loss followed by repopulation occurs in a 6- to 7-week cycle during the period of irradiation (Baclesse, 1958; Coutard, 1935; Friedman, 1939). A similar, but shorter cyclic response is seen in the oral mucosa in the course of radiation therapy of head and neck cancer.

The length of the preferred treatment schedules for patients exceeds 3 weeks. Therefore, if proliferation occurs after 3 weeks in one population and not in another, the cell kinetic and radiobiological parameters derived from the analysis of the bulk tissue response to these schedules are probably in error unless the critical target cell populations are clearly identified and proliferation is taken into account appropriately. This argument is well illustrated in the following discussion in which basal cell survival parameters obtained by measurement and derived from different models are compared (Table IX).

The basal cell survival data listed in Table IX, column 2, were obtained from histologic measurement of the linear "regenerating island" density in microscopic sections following selected single doses (Archambeau *et al.*, 1979). The dose required to produce nonhealing in 50% of the irradiated fields (ED_{50}) following single dose fractions is 2364 ± 86 cGy (Archambeau *et al.*, 1968, 1969). It was assumed in these calculations that the observed nonhealing isoeffect indicates a fixed level of cell density. It is not specified whether this density reflects the number of cells surviving irradiation or the number of cells that survived irradiation and proliferated to produce the specified level. Using the measured survival data and the known N_0 population density, the cell density would be 4.6×10^{-6} basal cells/cm^2 after 2364 cGy.

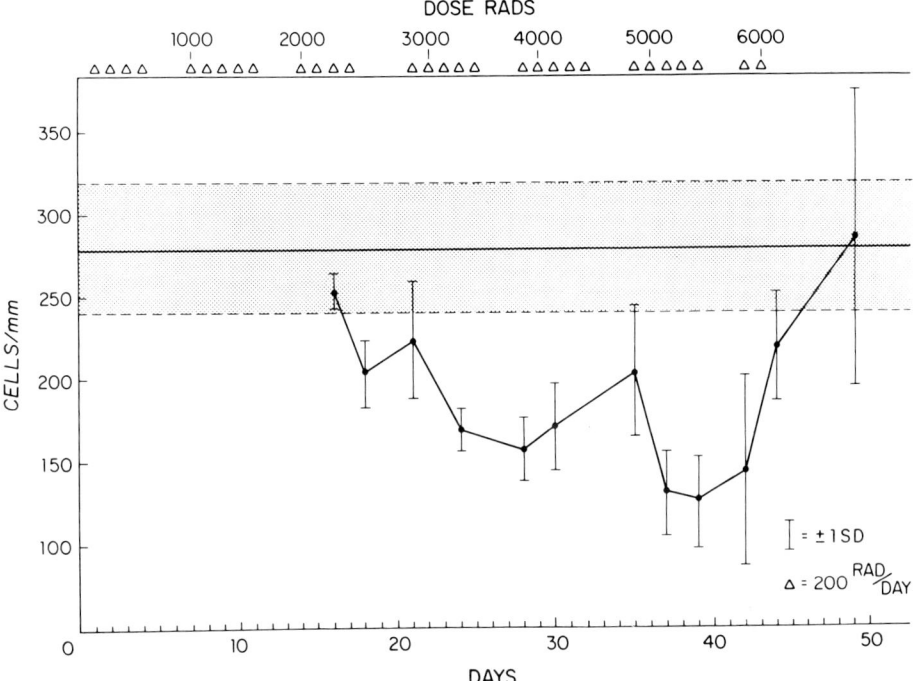

FIG. 5. The change in basal cell density during the period in which swine received 6000 cGy in 30 200-cGy dose fractions 5 days a week for 43 days. The open triangles indicate the days on which irradiation was given.

The increase in cell density occurring before the end of the therapy schedule indicates cell proliferation. (From Archambeau, unpublished data.)

A relationship between epidermal cell population and dose response has been derived from the data on the time–dose isoeffect response of swine skin (Shymko *et al.*, 1984; Archambeau and Bennett, 1984). Isoeffect analysis of these data using a multitarget, single hit model yields an extrapolation number, $n = 78$, a 37% survival dose, $D_0 = 145 \pm 2$ cGy, and a dose recovery factor, $D_Q = 632$ cGy (Table IX, column 3). However, the directly measured basal cell dose survival parameters obtained after single dose fractions yielded different cell survival values and a D_0 of 250 cGy (Table IX, column 2). Other values for these parameters are obtained by using Cohen's Rad 2 model (Table IX, column 4) and the linear quadratic approach (Table IX, column 5). These findings reveal a discrepancy between the observed and derived survival data. This lack of correlation suggests that the ED_{50} time–dose isoeffect parameter is not dependent on the fraction of basal cells surviving the direct effects of radiation. This conclusion is confirmed by the histologic finding that the epidermal repopulation

TABLE IX

CELL KINETIC AND DOSE SURVIVAL PARAMETERS DERIVED FROM MODELS SIMULATING TIME–DOSE ISOEFFECT RESPONSE COMPARED WITH PARAMETERS MEASURED HISTOLOGICALLY

Parameter[b]	Derived using[a]			
	Measured (1)	Fixed survival model (2)	Rad 2 model f (3)	Lin-Q model (4)
N_0	4.6×10^6 cell/cm²			
T_C unirradiated	12.4 ± 2.4 days			
Growth fraction	1			
T_C irradiated	13–15 hr	22 hr	1.75 hr	
D_Q	c	632 cGy	633 cGy	
D_0	250 cGy	145 ± cGy	255 cGy	
Slope at isoeffect	+32	+32		183 cGy
Extrapolation number	c	78–22	12	
Cell survival dose (cGy)	$-\ln(N/N_0)$ (N/N_0)	$-\ln(N/N_0)$ (N/N_0)	$-\ln(N/N_0)$ (N/N_0)	$-\ln(N/N_0)$ (N/N_0)
970		613 cGy	578 cGy	
1649	−9.01 (1.22×10^{-4})	−7.03 (8.85×10^{-4})	−4.19 (1.51×10^{-2})	−3.85 (2.13×10^{-2})
2231	−9.43 (8.03×10^{-5})	−11.01 (1.65×10^{-5})	−6.70 (1.20×10^{-3})	−6.70 (1.23×10^{-3})
2619	−11.93 (6.59×10^{-6})	−13.66 (1.17×10^{-6})	−8.50 (2.03×10^{-4})	−8.87 (1.41×10^{-4})
	−13.28 (1.71×10^{-6})			

[a] Subpopulations were determined by measuring the dimension of regenerating islands and determining D_0 for these islands; (1) Archambeau et al. (1979); Archambeau and Bennett (1984); (2) value obtained from the isoeffect data of (1) by Shymko et al. (1984); (3) value obtained from the isoeffect data of (1) using the Rad 2 model of Cohen (1983); (4) value obtained from the isoeffect data of (1) using the linear quadratic (Lin-Q) model of Douglas (1982).

[b] N/N_0 fraction of the population surviving irradiation; N_0 cell density of the unirradiated population; T_C average generation time of the basal cells; D_Q dose recovery; D_0, average survival dose derived from the exponential portion of the dose–survival curve; Lin-Q, linear quadratic model of Douglas (1982); and Rad 2, a model from Cohen (1983).

[c] Unable to measure these values because the data extrapolated at zero dose to an initial survival fraction of between 0.0001 and 0.1.

of cells to control levels occurs by days 28–32 for all single doses tested and is followed by a second loss of cells prior to 49 days in an increasing proportion of fields. Therefore, the probability of healing is not related to the number of initially surviving basal cells.

A comparison of changes in the endothelial cell population with those of the epidermal cell population reveals that the second breakdown of the epidermal population is related to the loss in endothelial and vascular density (refer to Fig. 2; compare population changes in charts 3 and 6) (Archambeau et al., 1984, 1985). These population changes suggest that the isoeffect as measured and used to define the skin response represents survival parameters in the combined endothelial–epidermal populations. Consequently, the skin isoeffects specified using gross parameters are not to be considered representative of changes in a single cell population.

C. Skin Tolerance Dose

The human skin tolerance dose based on skin changes following a single dose fraction lies between 2000 and 2400 cGy. This dose increases to 6000 cGy when the dose is given in 30 daily fractions of 200 cGy, given 5 days a week over a period of 40–42 days (Ellis, 1968; Patterson, 1963; Douglas, 1982). The best estimate of the human skin tolerance dose has been set at 5500 cGy for the production of a 5% incidence of complications at 5 years or at 7000 cGy for the production of a 50% incidence of complications at 5 years after radiation therapy with time–dose schedules employed clinically (Rubin and Casarett, 1968; Rubin et al., 1975, 1983).

The NSD for skin and connective tissue is 1758 ret when the tolerance dose of 6000 cGy is given in 30 fractions over a period of 40–42 days, although the standard deviation or confidence limits are not specified (Ellis, 1968; Ellis et al., 1969). The equivalent dose required is 2700 cGy when the total dose is given in three 900-cGy daily fractions, 4000 cGy when given in 10 weekly 400-cGy fractions (elapsed time 13 weeks), or 4824 cGy in 12 twice weekly 400-cGy fractions (6 weeks elapsed time). [See Fig. 3 for approximate values for the selected example which plots dose vs. number of fractions (Ellis, 1968).]

The value of the α/β ratio for the skin tolerance isoeffect in man is estimated to be 500 cGy (Barendsen, 1982; Withers et al., 1980, 1982a–c). When this value is used to determine dose equivalency with respect to various time–dose schedules, the values for fraction size, number of fractions, and total dose are similar to those obtained by using the NSD approach.

Cohen's extensive analysis of available time–dose data gives a similar equivalent dose response for the same schedules (Cohen, 1983). The survival parameters for dry desquamation in human skin have a D_0 of 108 cGy and n of 21 or an α/β ratio of 365 cGy using the linear quadratic format (Cohen, 1983). Dry desquamation occurs after a total dose of 4650 cGy given in 30 daily doses over a period of 39 days or a dose of 5050 cGy given in 30 weekly dose fractions over a period of 204 days. The survival dose parameters for necrosis of human sub-cutaneous tissue are $D_0 = 149$ cGy, $n = 5$, and $D_Q = 225$ cGy. The tolerance dose for a 16% incidence of necrosis is 2000 cGy after a single dose fraction, 8461 cGy after 30 daily dose fractions spread over 39 days, or 9161 cGy after 30 weekly dose fractions given over 204 days (Cohen, 1983).

Qualitatively, the formation of a moist reaction that heals is attributed to changes in the epidermal cell population. The late changes of telangiectasia and necrosis are attributed to changes in the vasculature, whereas fibrosis and indura-tion are brought about by changes in the dermal cell population. Epilation repre-sents a loss of proliferative cells of the hair bulb and is nearly complete when the moist reaction becomes manifest. The moist reaction occurs and heals at total doses at which the late changes of fibrosis and necrosis occur.

Based on the preceding observations, the cells of the hair bulb are considered the most radiosensitive and the epidermal population as the most resistant, fol-lowed by the dermal fibrocytic and the endothelial cell populations. The rank ordering of the fibrocytic and the endothelial cell populations is not clearly defined. Withers attributes the loss of the endothelial cell population to a con-tinuing fibrocytic change which, if true, would reverse the rank ordering (With-ers *et al.*, 1980).

The sequence of cell density changes in the epidermal and endothelial popu-lations in swine skin after single dose fractions indicates that there is a difference in relative radiosensitivity between the epidermal and endothelial cells (refer to Fig. 2, compare the third and sixth charts). The second loss of epidermal cells tracks the decrease and loss of the endothelial cell population at 43 days or later following 2231 or 2619 cGy. This sequence indicates that the endothelial cells are relatively more radiosensitive than the epidermal cells. The manifestation of this difference is masked unless the period of observation is sufficiently long to show the endothelial loss. During this period of varying endothelial and epider-mal population densities, the fibrocytic cell population remains unchanged.

As the endothelial cell density of the dermal microvasculature decreases, the number of vessels is reduced accompanied by distortion and dilatation of the remaining vessels. However, the endothelial population in large arteries and veins of the dermis remains intact. Whether a difference in dose response exists between the endothelium of the large vessels and that of the capillaries is not clear.

In the section on anatomical and kinetic considerations of the dose response, the endothelial cell population density per vessel of the isolated microvascular unit is defined as about 10 cells distributed in a linear configuration, thus differ-ing from the monolayer configuration in the larger vessels. Endothelial prolifera-tion is demonstrable in larger vessels by repopulation of the endothelial monolayer, intimal thickening, and arteriolar endothelial hyperplasia. These endothelial proliferative changes are not found in vessels less than 10 μm in diameter.

The apparent difference between large vessel and microvascular response is a consequence of cell loss in that the loss of endothelial cells from vessels less than 10 μm in diameter may result in the loss of lumen continuity, vessel retraction, and a net decrease in the number of vessels. When an endothelial cell is lost from the endothelial monolayer of the larger vessels, the muscular and medial components of the vessel wall serve as splints, while repopulation occurs from adjacent cells. In this instance, the proliferative capacity of the cells and the vessel size can have a greater effect than dose tolerance in determining the tissue response.

Irradiation of the skin of swine indicates that the epidermis can tolerate a dose in excess of 2619 cGy (the epidermis survives following single dose fractions of 4500 cGy), while the endothelium can tolerate a single dose fraction larger than 1649 cGy, but smaller than 2231 cGy.

The D_0 value for the microvasculature of the rabbit ear window is similar to that for the microvasculature of swine and man. The derived values of the α/β ratio for late changes range from 220 to 560 cGy (Table VII) compared to 556 cGy for the rabbit ear window (Griem et al., 1985; Dimitrievich et al., 1984, 1985).

No cell survival data are available for dermal fibrocytes. Attempts at establishing a rank ordering of radiosensitivities of skin tissue populations are not appropriate, since the overall tolerance dose is dictated by the most radiosensitive tissue. The epidermal, endothelial, and fibrocytic cell populations receive the same dose and are in close proximity. Therefore, changes in one component population will affect the other ones.

Based on the above analysis, the endothelial population is the most sensitive and a loss in the microvasculature represents a loss in nutritional support which results in epidermal and dermal necrosis. Therefore, it is not possible to use the intact skin as a test system for the purpose of distinguishing between values for the dose response of the epidermal, endothelial, and fibrocytic cell populations.

D. Clinical Need

Clinically a need exists for a data base of quantified cell kinetic and radiobiological parameters of skin and other tissues. Such data would define the skin response and tolerance dose limits for a selected time–dose schedule and would permit the therapist to select a treatment plan which would preserve tissue function. If similar data were available for cancers to be treated, a dose schedule could be selected that would optimize the differential dose response between the skin and the cancer. At present, the clinician is better able to define dose limits than to alter them. It is to be hoped that the definition and understanding of the dose response of skin populations will indicate approaches that can be used to modify this response.

The NSD, Lin-Q, and Cohen equations which determine dose equivalency between dose schedules and estimate tissue tolerance mix empirical observation with radiobiological and cell kinetic insights. Each accounts for the result by specifying the form in which recovery between repetitively administered dose fractions takes place. The type and degree of recovery varies for each of the three aforementioned approaches, but the cell population response is not defined.

The acceptance of these equations by clinicians varies. When clinical problems that require a change in the time–dose schedule are approached using one of these equations, the change in the total dose, dose fraction size, or fraction number predicted to establish dose equivalency between schedules has about the same value using any of the three equations. This indicates that the models represent composite population responses and do not distinguish between the component populations. In addition, the equations do not distinguish between the different types of dose recovery. Recovery from sublethal injury to skin is complete when an interval of 24 hr or longer intervenes between dose fractions. The D_Q or width of the shoulder on the survival dose equation for cells (composite population) is of the order of 180–600 cGy or larger and thus, nonexponential variations in cell survival that are not large may not be recognized. The gross isoeffects used do not distinguish among the responding cell populations. For example, the intensity of the erythematous dose response increases linearly as the dose increases. However, changes in the basal cell population density (responsible for moist reaction) are nonlinear (see section on dose response). Thus, erythema should not be expected to correspond to a uniform basal cell response. These results demonstrate that changes in one tissue do not necessarily directly correlate with those in another. By the same argument, the equations used to define skin response are not suitable to define dose response for other organs.

In spite of the above comments, the continued use of the time–dose equations is valuable. The users, when reporting treatment results, are aware of the necessity of specifying (1) the total dose, (2) the number of exposures, (3) the size of the individual dose fractions, (4) the time interval between dose fractions, (5) length and number of rest periods in the schedule, (6) the size of the irradiated volume, and (7) the quality of radiation employed. These efforts compel the clinician to focus on details that are needed to improve the efficiency of the clinical program. It is hoped that the necessary values for the cell parameters will become available to permit the use of an appropriate construct.

In order to quantify the time–dose relationship, the cell recovery constants of normal tissue and cancer must be measured. Until now, the investigative approach has relied on measuring the evolution of gross skin isoeffects (which are inadequate, as stated above). Cell kinetic or population parameters in man have not been measured. Population parameters cannot be derived accurately from gross changes in swine skin, as shown in the last section. Few data are available for human malignancies. Such quantified data are needed in order to specify a therapeutic ratio.

One apparent solution for defining the time–dose relationship is to use histologic techniques to measure the cell population parameters. However, efforts to obtain such data come immediately into conflict with the moral and ethical guidelines for the conduct of a clinical practice. It is difficult to justify sequential biopsies in a patient in order to measure parameters that have little bearing at this point in time on disease management. Even if this hurdle were circumvented, there are no available techniques except the very inefficient and work-intensive technique of direct cell counting to give an accurate determination of these parameters.

Another ethical impasse for the clinician is that he cannot justify the use of time–dose schedules that differ from those generally accepted and proved by careful and prudent radiation therapy practice. Yet, in order to sample a full range of skin (or other tissue) responses, schedules would have to be varied and would have to include some schedules which would provide less than optimal cancer control. This, of course, would be unacceptable.

The use of animal models and the correlation of results obtained with these models with those observed in man remains the most suitable investigative approach to create a useful data base. The continued use of the well-designed prospective controlled clinical trial under the supervision of a national organization of clinical oncologists remains the appropriate method for gathering data on patients that can be used to establish an accurate model.

The clinical evaluation determining the incidence of cancer control and tissue response of selected time–dose schedules has been productive. Prospective studies evaluating schedules employing single, daily, or weekly dose fractions do not have to be repeated as such. There is sufficient experience with large single dose fractions to know that few cancers are controlled locally and that severe late skin changes result when such schedules are employed. Schedules using one or two dose fractions per week produce acceptable early reactions, but unacceptable late reactions for the same incidence of cancer control. Schedules using daily dose fractions extending to 10–12 weeks and split-course schedules having a 2- to 3-week rest period between two cycles of daily irradiations allow delivery of large total absorbed doses with normal tissue recovery, although the incidence of disease control is not increased. Schedules using equal daily dose fractions of 180–200 cGy 5 days per week with varying total doses have become the standard. While these schedules may not be considered optimal, they do produce a uniform and predictable incidence of control of the cancer with an acceptable normal tissue change.

Currently, clinical trials that utilize multiple daily dose fractions are being evaluated to see if there is an increased incidence in cancer control. These carefully structured trials are also designed to yield information concerning dose recovery resulting from repair of sublethal injury and cell proliferation. These new investigations fall into two categories. In the first approach, the customary total daily dose of about 200 cGy is divided among multiple daily dose fractions. This approach is referred to as hyperfractionation and delivers the same or higher total doses using more daily fractions over the same period of time. In the second approach, the customary dose fraction at about 200 cGy is maintained and multiple dose fractions per day are given. As a consequence, the total daily dose is increased and the time required to deliver the total dose is reduced by about 50%. This approach is referred to as "accelerated fractionation." The total dose delivered may be increased or decreased, depending on the tissue response (Thames *et al.*, 1982, 1983; Withers *et al.*, 1982a,b, 1983; Arcangeli *et al.*, 1979).

A third approach (single pilot study) involves irradiation 7 days per week, eliminating the weekend rest (Archambeau and Desai, 1981; Archambeau *et al.*, 1984; Lipsett *et al.*, 1984). It should be noted that the difference in time between the first treatment 0800 Monday and the last at 0800 Friday is 96 hr. The weekend lasts from the treatment at 0800 hr Friday to the treatment at 0800 hr Monday and is 72 hr and not 48 hr long. This 72-hr rest period represents 43% of the week. It allows the patient a period for recovering his energies and for resting (Archambeau *et al.*, 1984). It also permits recovery of losses of normal tissue cell populations, but for the same reason tumor cell repopulation may also occur. The rationale supporting the criticality of this rest period is based on the average cell generation time. If this time is of the order of 1 month, very little proliferation occurs. If the generation time is as short as 15–20 hr, as has been reported, 3–4 population doublings may occur (Archambeau and Desai, 1981).

The intent of the clinical trials used to evaluate these new schedules is to search for differences in the dose response of the normal tissues that improve the therapeutic ratio. The rationale of these new schedules was suggested by the dose recovery measured using the reciprocal total dose–dose fraction isoeffect data for cell populations with different rates of proliferation. If the results of these trials show an improved incidence of cancer control, they will justify performing studies designed to quantify population cell kinetic and radiobiological parameters of cancers and normal tissues. Data on dose recovery would be needed for schedules which employ equal dose fractions at different intervals and unequal large dose fractions at selected intervals.

The use of hyperfractionation (two or more smaller doses per day) anticipates that the recovery of sublethal injury is slower and redistribution of cells into a more sensitive phase in the cell cycle is faster for the cancer than for the normal tissues. As a result, by changing from daily to several times daily irradiation, cell survival in the cancer is relatively lower than for normal tissues. Hyperfractionation schedules also capitalize on the differences between the dose survival response for early- and late-responding tissues, since early-responding or rapidly proliferating tissues exhibit less capacity for recovery than do the late-responding or slowly proliferating ones (Thames *et al.*, 1982, 1983; Withers *et al.*, 1983). Preliminary clinical data reveal an improved patient tolerance and decreased reactions of normal tissue when hyperfractionated schedules are used.

The use of accelerated schedules is based on the presumption that less cell proliferation occurs in cancer than in normal tissues in the 4- to 6- and the 18- to 20-hr intervals between dose fractions for twice-a-day irradiations. However, these accelerated schedules include a weekend rest, during which proliferation can occur. The 7-day-a-week irradiation schedule, of course, eliminates this rest period as well as the inconvenience of spending nearly a day in the radiation therapy department.

Clinical and animal examples of rapid proliferation with a short average generation time are available in rapidly proliferating normal and abnormal tissues, e.g., gastrointestinal tract, skin, and tumors. The classic clinical example is that of Burkitt's lymphoma (Norin and Onyango, 1977) as well as other examples (Thames *et al.*, 1982, 1983).

Skin reactions become manifest at the expected time following twice-a-day accelerated schedules. If the total dose delivered remains unchanged, the skin response will remain unchanged. Skin changes following hyperfractionation and 7-day-a-week schedules using daily dose fractions of 180–220 cGy are not different from those found for the 5-day schedules (Choi and Suit, 1977; Lipsett *et al.*, 1984). The tolerance dose for the formation of the moist reactions is not increased and remains unchanged or increased for late reactions. Erythematous and mucosal reactions are more severe following three dose fractions per day than after a single daily irradiation (Turesson and Notter, 1976a; Thames *et al.*, 1982, 1983), but the time course remains the same regardless of the number of dose fractions.

If there is an improved therapeutic ratio, the above-mentioned results suggest that this improvement is derived from an increased recovery dose in the slowly proliferating cell populations of the microvasculature and dermis. Extrapolation of the dose response curve for skin cell populations having α/β ratios of 500 cGy or less suggests that these cell populations will have an increased tolerance dose. The exact level of this tolerance dose remains to be determined.

ACKNOWLEDGMENTS

I appreciate the painstaking help, suggestions, and interactions of Dr. K. Altman in editing this effort. I want to thank Dr. R. Shymko for his many discussions with me and his help in achieving perspective. I am particularly grateful to Jetta Hice, Marie Cool, and Lisa Lehmann for their devoted, patient, and persistent completion of the typing and editing.

REFERENCES

Archambeau, J. O. (1969). *J. Invest. Dermatol.* **52**, 399.

Archambeau, J. O. (1970). *Radiology* **94**, 179–187.

Archambeau, J. O., and Bennett, G. W. (1984). *Radiat. Res.* **98**, 254–273.

Archambeau, J. O., and Desai, K. (1981). Protocol at City of Hope National Medical Center.

Archambeau, J. O., and Mathieu, G. R. (1969). *Radiat. Res.* **40**, 285–287.

Archambeau, J. O., and Shymko, R. M. (1984). *Curr. Probl. Cancer* **8**, 6–56.

Archambeau, J. O., Fairchild, R. G., and Commerford, S. L. (1965). *Proc. Int. Symp., Richland, Washington, July 19–21,* 463–489.

Archambeau, J. O., Mathieu, G. R., Brenneis, H. J., Thompson, K. H., and Fairchild, R. G. (1968). *Radiat. Res.* **36**, 299–326.

Archambeau, J. O., Mathieu, G. R., Brenneis, H. J., and Thompson, K. R. (1969). *Radiat. Res.* **37**, 141–160.

Archambeau, J. O., Fairchild, R. G., and Brenneis, H. J. (1971). *Radiat. Res.* **45**, 145–165.

Archambeau, J. O., Bennett, G. W., Abata, J. J., and Brenneis, H. J. (1979). *Radiat. Res.* **79**, 298–337.

Archambeau, J. O., Ines, A., and Fajardo, L. F. (1984). *Radiat. Res.* **98**, 37–51.

Archambeau, J. O., Ines, A., and Fajardo, L. F. (1985). *Int. J. Radiat. Oncol. Biol. Phys.* (in press).

Archangeli, G., Mauro, F., Morelli, D., and Nervi, C. (1979). *Eur. J. Cancer* **15**, 1077–1083.

Baclesse, F. (1958). *In* "Progress in Radiation Therapy" (F. Buschke, ed.), pp. 128–143. Grune & Stratton, New York.

Barendsen, G. W. (1982). *Int. J. Radiat. Oncol. Biol. Phys.* **8**, 1981–1997.

Becquerel, H., and Curie, P. (1901). *C.R. Acad. Sci.* **132**, 1289–1291.

Berry, R. J., Wiernik, G., and Patterson, T. J. S. (1974a). *Br. J. Radiol.* **47**, 185–190.

Berry, R. J., Wiernik, G., Patterson, T. J. S., and Hopewell, J. W. (1974b). *Br. J. Radiol.* **47**, 277–281.

Bewley, D. K., Fowler, J. F., Morgan, R. L., Silvester, J. A., Turner, B. A., and Thomlinson, R. H. (1963). *Br. J. Radiol.* **36**, 107–115.

Bond, V. P., Fliedner, T., and Archambeau, J. O. (1965). "Mammalian Radiation Lethality: A Disturbance in Cell Kinetics." Academic Press, New York.

Bush, S. E., Smitle, A. R., and Link, S. (1982). *Int. J. Radiat. Oncol. Biol. Phys.* **8**, 2181–2186.

Camplejohn, R. S., Gelfant, S., Chalker, D., and Sittampalam, Y. (1984). *Cell Tissue Kinet.* **17**, 315–322.

Charyulu, K. K. N., Tyree, E. B., Glicksman, A. S., and Nickson, J. J. (1966). *Radiology* **87**, 916–921.

Choi, C., and Suit, H. (1977). *Radiology* **116**, 703–707.

Chu, F. C. H., Conrad, J. T., Bane, H. N., Glicksman, A. S., and Nickson, J. J. (1960). *Radiology* **75**, 406–415.

Chu, F. C. H., Glicksman, A. S., and Nickson, J. J. (1970). *Radiology* **94**, 669–672.

Cohen, L. (1966). *In* "The Biological Basis of Radiation Therapy" (E. E. Schwartz, ed.), pp. 208–348. Lippincott, Philadelphia.

Cohen, L. (1978). *Int. J. Radiat. Oncol. Biol. Phys.* **4**, 267–271.

Cohen, L. (1983). "Biophysical Models in Radiation Oncology." CRC Press, Boca Raton, Florida.

Cohen, L., and Creditor, M. (1983). *Int. J. Radiat. Oncol. Biol. Phys.* **9**, 1065–1071.

Cohen, L., and Redpath, J. L. (1977). *Radiat. Res.* **69**, 387–401.

Coutard, H. (1934). *Lancet* **227**, 1–8.

Coutard, H. (1935). *Proc. Inst. Med. Chicago* **10**, 310–323.

Denekamp, J., Fowler, J. F., and Kragt, K. (1966). *Radiat. Res.* **29**, 71–84.

Denekamp, J., Ball, M. M., and Fowler, J. F. (1969). *Radiat. Res.* **37**, 361–370.

Denekamp, J. (1973). *Br. J. Radiol.* **46,** 381–387.

Denekamp, J. (1982). *Br. J. Cancer* **45,** 136.

Denekamp, J., Emery, E. W., and Field, S. B. (1971). *Radiat. Res.* **45,** 80–84.

Dimitrievich, G. S., Fischer-Dzoga, K., and Griem, M. L. (1984). *Radiat. Res.* **99,** 511–535.

Dimitrievich, G. S., Fischer-Dzoga, K., Griem, M. L., and Krowitz, A. L. (1985). *Proc. Annu. Sci. Meet. Radiat. Res. Soc., 33rd, Los Angeles, May 8.*

Douglas, B. G. (1982). *Int. J. Radiat. Oncol. Biol. Phys.* **8,** 1135–1142.

Douglas, B. G., and Fowler, J. F. (1976). *Radiat. Res.* **66,** 401–426.

Douglas, W. R. (1972). *Space Life Sci.* **3,** 226–234.

Dutreix, J., Wambersie, A., and Bounik, C. (1973). *Eur. J. Cancer* **9,** 159–167.

Elkind, M. M., and Sutton, H. (1960). *Radiat. Res.* **13,** 556–593.

Ellis, F. (1968). *Curr. Top. Radiat. Res. Q.* **4,** 357–397.

Ellis, F., Winston, B. M., Fowler, J. F., and DeGinder, W. L. (1969). *Br. J. Radiol.* **42,** 716–717.

El-Naggar, A. M., El-Baz, L. M., Carsten, A. L., Chanana, A. D., and Cronkite, E. P. (1978). *Int. J. Radiat. Biol.* **34,** 359–366.

Emery, E. W., Denekamp, J., and Ball, M. M. (1970). *Radiat. Res.* **41,** 450–466.

Etoh, H., Taguchi, Y. H., and Tabachnick, J. (1977). *Radiat. Res.* **71,** 109–118.

Fajardo, L. F. (1982). "Pathology of Radiation Injury." Masson, New York.

Field, S. B. (1976). *Curr. Top. Radiat. Res.* **11,** 1–86.

Fike, J. R., Gillette, E. L., and Clow, D. J. (1979). *Int. J. Radiat. Oncol. Biol. Phys.* **5,** 339–342.

Fischer-Dzoga, K., Dimitrievich, G. S., and Griem, M. L. (1984). *Radiat. Res.* **99,** 536–546.

Fischer-Dzoga, K., Dimitrievich, G. S., and Griem, M. L. (1985). *Proc. Annu. Sci. Meet. Radiat. Res. Soc., 33rd, Los Angeles, May 8.*

Fowler, J. F. (1963). *In* "Biological Effects of Neutron and Proton Irradiations," Vol. 2, p. 185.

Fowler, J. F., and Stern, B. E. (1963). *Br. J. Radiol.* **36,** 163–171.

Fowler, J. F., Morgan, R. L., Silvester, J. A., Bewley, D. K., and Turner, B. A. (1963). *Br. J. Radiol.* **36,** 188–196.

Fowler, J. F., Kragt, K., Ellis, R. E., Lindop, P. J., and Berry, R. J. (1964). *Int. J. Radiat. Biol.* **9,** 241–252.

Fowler, J. F., Bewley, D. K., Morgan, R. L., and Silvester, J. A. (1965). *Br. J. Radiol.* **38,** 278–284.

Friedman, M. (1939). *Radiology* **33,** 633–643.

Glicksman, A. S., Chu, F. C. H., Bane, H. N., and Nickson, J. J. (1960). *Radiology* **75,** 411–414.

Goitein, M. (1976). *Clin. Radiol.* **27,** 389–404.

Goodman, G. B., Douglas, B. G., Jackson, S. M., Kornelsen, R. O., Lam, G. K. Y., Ludgate, C.M., and Skarsgard, L. D. (1982). *Int. J. Radiat. Oncol. Biol. Phys.* **8,** 2187–2190.

Griem, M. L., Dimitrievich, G. S., and Krowitz, A. (1985). *Proc. Annu. Sci. Meet. Radiat. Res. Soc., 33rd, Los Angeles, May 8.*

Hall, E. J. (1978). "Radiobiology for the Radiologist," 2nd Ed. Harper, Hagerstown, Maryland.

Heenen, M., and Galand, P. (1971). *J. Invest. Dermatol.* **56,** 425–429.

Heenen, M., Achten, G., and Galand, P. (1973). *Cancer Res.* **33,** 123–127.

Herbert, D. E. (1981). *Med. Phys.* **8,** 813–847.

Hirst, D. G., Denekamp, J., and Travis, E. L. (1979). *Radiat. Res.* **77,** 259–275.

Hirst, D. G., Denekamp, J., and Hobson, B. (1980). *Cell Tissue Kinet.* **13,** 91–104.

Hopewell, J. W., and Patterson, T. J. S. (1972). *Biorheology (Oxford)* **9,** 46–47.

Hopewell, J. W. (1980). *In* "Radiation Biology in Cancer Research" (R. E. Meyn and H. R. Withers, eds.), pp. 449–459. Raven, New York.

Hopewell, J. W., Barnes, D. W. H., Goodhead, D. T., Knowles, J. F., Wiernik, G., and Young, C. M. A. (1982). *Int. J. Radiat. Oncol. Biol. Phys.* **8,** 2077–2082.

Hopewell, J. W. (1983). *In* "Cytotoxic Insult to Tissue: Effects on Cell Lineages" (C. S. Potten and J. H. Hendry, eds.), pp. 228–257. Churchill Livingstone, Edinburgh.

Hussey, D. H., Jardine, J. H., Raulston, G. L., Stephens, L. C., Gray, K. N., Maor, M. H., and Withers, H. R. (1982). *Int. J. Radiat. Oncol. Biol. Phys.* **8,** 2083–2088.

Izquierdo, J. N., and Gibbs, S. J. (1974). *Cell Tissue Kinet.* **7,** 99–111.

Jurnovoy, J. B., Forbes, P. D., and Johnson, W. C. (1975). *Dermatologica* **150,** 2–15.

Kirk, J., Gray, W. M., and Watson, E. R. (1971). *Clin. Radiol.* **22,** 145–155.

Lacassagne, A., and Gricouroff, G. (1956). "Action des radiations ionisantes sur l'organisme," pp. 21–36. Masson, Paris.

Lajtha, L. G., and Oliver, R. (1962). *Br. J. Radiol.* **35,** 131–140.

Law, M. P. (1981). *Adv. Radiat. Biol.* **9,** 37–73.

Liebner, E. J., Moos, W. S., Hochhauser, M., and Harvey, R. A. (1962). *Proc. Annu. Meet. Am. Roentgen Ray Soc., 62nd, Miami Beach, 1961* **88,** 976–987.

Liegner, L. M., and Michaud, N. J. (1961). *Am J Roentgenol.* **85,** 533–549.

Lipsett, J. A., Desai, K., Pezner, R., Vora, N., Chong, L. M., and Archambeau, J. O. (1984). *Int. J. Radiat. Oncol. Biol. Phys.* **10,** 1049–1052.

Maciejewski, B., Preuss-Bayer, G., and Trott, K. (1983). *Int. J. Radiat. Oncol. Biol. Phys.* **9,** 321–328.

MacKenzie, I. C. (1975). *J. Invest. Dermatol.* **65,** 45–51.

MacKenzie, I. C., Zimmerman, K., and Peterson, L. (1981). *J. Invest. Dermatol.* **76,** 459–461.

Malkinson, F., and Griem, M. (1966). *Arch. Dermatol.* **94,** 491–498.

Malkinson, F. D., and Keane, J. T. (1981). *J. Invest. Dermatol.* **77,** 133–138.

Malkinson, F. D., and Keane, J. T. (1983). *In* "Biochemistry and Physiology of the Skin" (L. A. Goldsmith, ed.), pp. 769–814. Oxford Univ. Press, New York.

Marcarian, H. Q., and Calhoun, M. L. (1966). *Am. J. Vet. Res.* **27,** 765–772.

Michalowski, A. (1981). *Radiat. Environ. Biophys.* **19,** 157–172.

Montagna, W. (1965). *Sci. Am.* **212,** 56–66.

Montagna, W., and Yun, J. S. (1964). *J. Invest. Dermatol.* **43,** 11–21.

Morris, G. M., and Hopewell, J. W. (1985). *Cell Tissue Kinet.* **18,** 407–415.

Moulder, J. E., Fischer, J. J., and Casey, A. (1975). *Radiology* **115,** 465–470.

Moustafa, H. F., and Hopewell, J. W. (1979). *Br. J. Radiol.* **52,** 138–144.

Nias, A. H. W. (1963). *Br. J. Radiol.* **36,** 183–187.

Norin, T., and Onyango, J. (1977). *Int. J. Radiat. Oncol. Biol. Phys.* **2,** 399–406.

Paterson, R. (1963). "The Treatment of Malignant Disease by Radiotherapy," 2nd Ed. Williams & Wilkins, Baltimore.

Potten, C. S. (1975). *J. Invest. Dermatol.* **65,** 488–500.

Potten, C. S. (1985). "Radiation and Skin." Taylor & Francis, London.

Potten, C. S., and Hendry, J. H. (1983) "Cytotoxic Insult to Tissue—Effects on Cell Lineages." Churchill Livingstone, Edinburgh.

Potten, C. S., and Hendry, J. H. (1985). *Int. J. Radiat. Oncol. Biol. Phys.* **11.**

Raju, M. R., and Tokita, N. (1982). *Int. J. Radiat. Oncol. Biol. Phys.* **8,** 2133–2136.

Reinhold, H. S. (1972). *Proc. L. H. Gray Conf., Inst. Phys., 4th, London.*

Reinhold, H. S., and Buisman, G. H. (1973). *Br. J. Radiol.* **46,** 53–57.

Roswit, W., Wisham, L. H., and Sorrentino, J. (1953). *An. J. Roentgenol. Radium Ther. Nucl. Med.* **69,** 980–990.

Rubin, P., and Casarett, G. W. (1968). "Clinical Radiation Pathology," Chap. 2. Saunders, Philadelphia.

Rubin, P., Cooper, R. A., Jr., and Phillips, T. L., Eds. (1975). "Radiation Biology and Radiation Pathology Syllabus." American College of Radiology, Chicago.

Rubin, P., Keys, H., and Poulter, C. (1983). *In* "Biological Bases and Clinical Implications of

Tumor Resistance'' (G. Fletcher, C. Nervi, and N. R. Withers, eds.), pp. 175–195. Masson, New York.

Shymko, R. M., and Archambeau, J. O. (1985). *Int. J. Radiat. Oncol. Biol. Phys.* **11** (in press).

Shymko, R. M., Hauser, D. L., and Archambeau, J. O. (1984). *Int. J. Radiat. Oncol. Biol. Phys.* **10**, 1079–1085.

Singh, K. (1978). *Br. J. Radiol.* **51**, 357–362.

Skarsgard, L. D., Palcic, B., Douglas, B. G., and Lam, G. K. Y. (1982). *Int. J. Radiat. Oncol. Biol. Phys.* **8**, 2127–2132.

Song, C. W., and Tabachnick, J. (1969). *Int. J. Radiat. Biol.* **15**, 171–174.

Stearner, S. P., and Christian, E. J. B. (1969). *Radiat. Res.* **38**, 153–160.

Stearner, S. P., and Christian, E. J. B. (1971). *Radiat. Res.* **47**, 741–755.

Suit, H. D., Silver, G., Sedlacek, R. S., and Walker, A. (1983). *Radiat. Res.* **95**, 427–433.

Tabachnick, J., and Weiss, C. (1959). *Radiat. Res.* **11**, 684–699.

Tannock, I. F. (1968). *Br. J. Cancer* **22**, 258–273.

Tannock, I. F., and Hayashi, S. (1972). *Cancer Res.* **32**, 77–82.

Thames, H. D., and Withers, H. R. (1980). *Br. J. Radiol.* **53**, 1071–1077.

Thames, H. D., Withers, H. R., Peters, L. J., and Fletcher, G. H. (1982). *Int. J. Radiat. Oncol. Biol. Phys.* **8**, 219–226.

Thames, H. D., Peters, L. J., Withers, H. R., and Fletcher, G. H. (1983). *Int. J. Radiat. Oncol. Biol. Phys.* **9**, 127–138.

Turesson, I., and Notter, G. (1975). *Acta Radiol. Ther. Phys. Biol.* **14**, 475–483.

Turesson, I., and Notter, G. (1976a). *Acta Radiol. Ther. Phys. Biol.* **15**, 162–176.

Turesson, I., and Notter, G. (1976b). *Radiology* **120**, 399–404.

Turesson, I., and Notter, G. (1979). *Int. J. Radiat. Oncol. Biol. Phys.* **5**, 1773–1779.

Turesson, I., and Notter, G. (1984a). *Int. J. Radiat. Oncol. Biol. Phys.* **10**, 593–598.

Turesson, I., and Notter, G. (1984b). *Int. J. Radiat. Oncol. Biol. Phys.* **10**, 599–606.

Turesson, I., and Notter, G. (1984c). *Int. J. Radiat. Oncol. Biol. Phys.* **10**, 607–618.

Van Den Brenk, H. A. S. (1959). *Am J Roentgenol.* **81**, 859–884.

Warren, S. (1943). *Arch. Pathol.* **35**, 340–347.

Weinstein, G. D. (1965). *J. Invest. Dermatol.* **44**, 413–419.

Weinstein, G. D. (1966). *In* ''Swine in Biomedical Research'' (L. K. Bustad and R. O. McClellan, eds.), pp. 287–297. Battelle Memorial Institute, Ruchland, Washington.

Weinstein, G. D., and Van Scott, E. J. (1965). *J. Invest. Dermatol.* **45**, 257–262.

Withers, H. R. (1967a). *Br. J. Radiol.* **40**, 187–194.

Withers, H. R. (1967b). *Radiat. Res.* **32**, 227–239.

Withers, H. R., Thames, H. D., Flow, B. L., Mason, K. A., and Hussey, D. H. (1978a). *Int. J. Radiat. Oncol. Biol. Phys.* **4**, 595–601.

Withers, H. R., Thames, H. D., Hussey, D. H., Flow, B. L., and Mason, K. A. (1978b). *Int. J. Radiat. Oncol. Biol. Phys.* **4**, 603–608.

Withers, H. R., Peters, L. J., and Kogelnik, H. (1980). *In* ''Radiation Biology in Cancer Research'' (R. E. Meyn and H. R. Withers, eds.), pp. 439–448. Raven, New York.

Withers, H. R., Peters, L. J., Thames, H. D., and Fletcher, G. H. (1982a). *Int. J. Radiat. Oncol. Biol. Phys.* **8**, 1807–1809.

Withers, H. R., Thames, H. D., and Peters, L. J. (1982b). *Int. J. Radiat. Oncol. Biol. Phys.* **8**, 2071–2076.

Withers, H. R., Thames, H. D., and Peters, L. J. (1982c). *In* ''Progress in Radio-Oncology'' (K. H. Kärcher, H. D. Kogelnik, and G. Reinartz, eds.), pp. 287–296. Raven, New York.

Withers, H. R., Thames, H. D., and Peters, L. J. (1983). *Radiother. Oncol.* **1**, 187–191.

Yamaguchi, T., and Tabachnick, J. (1972). *Radiat. Res.* **50**, 158–180.

Young, C. M. A., and Hopewell, J. W. (1980). *Midcrovasc. Res.* **20**, 182–194.

Relative Radiosensitivity of the Human Lung

ELIZABETH L. TRAVIS

DEPARTMENT OF EXPERIMENTAL RADIOTHERAPY
THE UNIVERSITY OF TEXAS M. D. ANDERSON HOSPITAL AND TUMOR INSTITUTE
HOUSTON, TEXAS 77030

I. The Problem

The lung is a major dose-limiting organ in the delivery of sufficiently high doses of radiation to eradicate a variety of malignant diseases involving the thorax, including primary lung tumors, Hodgkin's disease, esophageal cancer, breast cancer, and occult metastases from distant primary tumors. In addition, new techniques using upper half body and total body irradiation have overcome the bone marrow toxicity initially associated with this treatment so that the lung has become the dose-limiting organ for this treatment.

The problem is not insignificant. Lesions in the lung are common after radiation of even part of the thorax (Gross, 1977; Carmel and Kaplan, 1976; Jennings and Arden, 1962; Warren and Spencer, 1940). The clinical expression of this injury is relatively uncommon, at least in a nonstressed situation, because of the considerable anatomical and functional reserve of this tissue. One estimate is that 10% of these patients exhibit pulmonary function changes even when they are grossly asymptomatic (Gross, 1977). Thus, the lung is an organ sensitive to radiation and, despite the absence of symptomatic changes, almost all patients receiving radiotherapy involving even a small portion of the thorax will receive some degree of pulmonary damage.

Obviously, "sensitivity" can be a somewhat ambiguous term, particularly when it is used to describe the response of an organized tissue to irradiation. It may have different meanings, depending on the context in which it is used; i.e., the clinical radiotherapist's definition of radiosensitivity may differ from that of the "mouse radiotherapist" or cellular radiobiologist. Thus, it is important to establish criteria for defining the term radiosensitivity and to identify ways of

ADVANCES IN RADIATION BIOLOGY, VOL. 12

measuring sensitivity in organized tissues like the lung for which no *in vivo* clonogenic assays of target cell survival are available.

Although the objective of this article is to discuss the radiosensitivity of the human lung, research in the past decade on the lungs of experimental animals, particularly rodents, has greatly increased our knowledge of the response of this tissue to radiation. Thus, this article will draw from these studies, comparing and contrasting the response of small animals' lungs to the available human data, with the goal of obtaining some understanding of radiosensitivity in relation to the radiation response of the lung.

II. Pulmonary Structure and Function

A. Lung Anatomy

The lung constitutes only 1% of body weight and is divided in half at the bifurcation of the trachea into the right and left lung, each consisting of independent units, the pulmonary lobes. Each lobe is further divided into two to six polyhedral-shaped segments which, although they vary in size and shape according to position within the lobe, have at least one of their phases exposed upon the surface of the lung. The main tissue components of the pulmonary segments are bronchi, pulmonary arteries, and veins, which together form the structural framework of the pulmonary parenchyma. The whole lung contains an average of 18 of these segments. The surface of the lung is covered by the pleura, connective tissue of mesothelial origin, which tends to be thicker in large than in small animals.

The structural unit of the lung is the pulmonary lobule and in man consists of 5–10 acini[1] and 3–5 terminal bronchioles. Individual lobules vary in size from 0.5 to 2 cm in diameter and extend from the central portion of the lung (deep lobules) to 3–4 cm beneath the lung surface (peripheral lobules). Contiguous lobules are limited either partially or totally by connective tissue, the interlobular septa. Many studies have shown that physiological differences exist between these two sets of lobules; e.g., after pneumothorax, the most profound collapse occurs in the peripheral lobules. Pulmonary metatases tend to localize mainly in the subpleural regions (Nagaishi, 1972) and radiation pneumonitis and fibrosis occur most often in subpleural areas or in areas adjacent to the lung surface, suggesting perhaps that the peripheral lobules are more susceptible to tissue injury.

---

[1]An acinus is the gas-exchange unit of the lung consisting of all structures distal to the terminal bronchiole.

B. Lung Structure

The structure of the lung is divided into two distinct anatomical and physiological units, each of which has a specific function in respiration. The first unit, the respiratory tract, extends from the bronchus to the terminal bronchioles and is a system of branching tubes (airways) of decreasing diameter; their only function is to move air in and out of the lungs (ventilation), and they do not participate in gas exchange. Bronchial branching occurs up to 25 times, starting from the main bronchus at the hilum of the lung and terminating in the periphery of the lung with the terminal bronchioles. The bronchial mucosa is lined by pseudostratified ciliated columnar epithelium with some goblet cells and undifferentiated basal cells, the stem cells for the bronchi. Beneath the mucosa are seromucinous glands, which become mucous glands as the bronchi branch. The glands disappear altogether at the level of the terminal bronchioles. Bronchial muscle surrounds the mucosa and extends up to the level of and into the alveolar ducts. The entire circumference of large bronchi is supported by cartilage, whereas this support is only partial in the small bronchi; thus, large bronchi are rigid and stay patent in the face of massive collapse of the lung, whereas in the same conditions, small bronchi collapse also.

Bronchioles, the airways distal to the last plate of cartilage, have no mucous glands and few goblet cells. The mucosa of the bronchioles is lined by cuboidal epithelium with few or no cilia and is surrounded by smooth muscle and little fibrous tissue. Because of the absence of both glands and cilia, these small airways have little ability to drain themselves. These small bronchioles end in the terminal bronchioles, the most distal portion of the bronchus that preserves the characteristics of an air tract. The terminal bronchioles consist of smooth muscle with no supportive cartilaginous structure and are lined by a single layer of cuboidal epithelium. The average 5-liter human lung has been estimated to contain about 3×10^4 terminal bronchioles (Matsuba and Thurlbeck, 1971).

The ramification of the terminal bronchioles to respiratory bronchioles marks the beginning of the terminal respiratory units, the acini, which perform the lung's signal function, gas exchange. An acinus is the functional unit of the lung and consists of respiratory bronchioles, alveolar ducts, alveolar sacs, and alveoli (Fig. 1). Two to five respiratory bronchioles, lined by cuboidal epithelium and surrounded by smooth muscle, arise from each terminal bronchiole. One to five alveolar ducts diverge from each respiratory bronchiole. These ducts are relatively short and branch in rapid sequence with openings for 10–16 alveoli. The last duct terminates in three alveolar sacs that form irregular projections from the duct, each sac bearing numerous terminal alveoli. The human lung contains a total of about 300 million alveoli (Murray, 1976), and the mouse lung contains about 10^8 (Yeh et al., 1979).

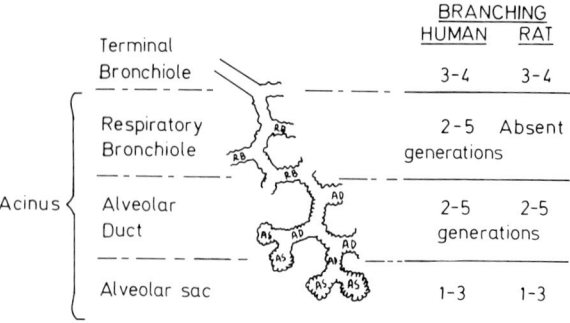

| | | BRANCHING | |
| | | HUMAN | RAT |

Terminal Bronchiole — 3–4 3–4

Respiratory Bronchiole — 2–5 generations / Absent

Alveolar Duct — 2–5 generations / 2–5

Alveolar sac — 1–3 1–3

FIG. 1. Schematic representation of the acinus, the smallest unit of pulmonary structure which carries out the cardinal function of the lung, respiration. Note that rat (and mouse) lungs have no respiratory bronchioles. (Redrawn with permission of Travis and Tucker, 1986.)

An alveolus is 250 μm in diameter and is made up of a continuous surface epithelium composed of only two cell types: Type 1 cells, which are only 0.2 μm thick and are sometimes referred to as the membranous pneumocyte, and Type 2 cells or granular pneumocytes, which are associated with synthesis, storage, packaging, and secretion of surfactant, the surface-active material in the lung. A basement membrane underlies the alveolar epithelium, which is continuous with that of the bronchioles and is separated from the capillaries by the interstitium of the lung, which contains reticulum and elastic fibers. Respiratory gases have to traverse 0.5–2.5 μm from the lumen of the capillaries to the alveoli and the air space. The capillaries are comprised of endothelium and basement membrane. The interstitial area of the lung consists of mesenchymal and inflammatory cells. Alveoli are connected by pores in the intraalveolar septa, "pores of Kohn," which are 10 μm in diameter and allow free intercommunication between alveoli as well as between different alveolar ducts or lobules. These pores become an effective bypass if a bronchiole is blocked.

The structure of the lung is remarkably similar among species as disparate in size as man and mouse (see Fig. 1 and Tables I and II), except that mouse lungs have no respiratory bronchioles; their terminal bronchioles ramify directly to alveolar ducts. Although this reduces the number of alveoli in mouse lung by a factor of 5 compared to man, the lungs of small animals still might be expected to respond to irradiation similarly to human lungs if the kinetics of the target cells in the two species are similar.

III. Cell Populations in the Lung

The lung consists of more than 40 different cell types. Because of its structure—it is mainly air (Table I)—it is difficult to identify different cells in the

TABLE I

TOTAL TISSUE VOLUMES AND SURFACE AREAS IN THE ALVEOLAR REGION OF NORMAL LUNGS

Parameter	Rat	Human
Body weight (kg)	0.3	79.0
Lung volume (ml)	8.6	4341.0
Total volume (cm³/both lungs)		
Air	5.980	3422.0
Capillary lumen	0.650	169.0
Type I	0.080	32.5
Type II	0.037	32.1
Endothelium	0.090	42.6
Interstitium	0.196	152.3
Surface area (m²/both lungs)		
Type I	0.390	89.0
Type II	0.015	7.0
Endothelium	0.383	91.0

lung, particularly in the alveolar regions. However, advances in morphometric techniques in the past few decades have allowed determination of cell numbers and cell characteristics in the alveolar regions of lung tissue (Crapo *et al.*, 1983). This section will focus only on those cells in the alveolar region, but not the cells in the airways because the former are probably more critical in the lung's response to irradiation.

TABLE II

CHARACTERISTICS OF CELLS FROM THE ALVEOLAR REGIONS OF NORMAL LUNGS

Characteristic	Mouse	Rat	Human
Body weight (kg)	0.03	0.3	79.0
Total cell number	10^7	6.7×10^8	2.3×10^{11}
Total lung cells (%)			
Type I	—	8.1	8.3
Type II	—	12.1	15.9
Endothelial	—	51.1	30.2
Interstitium	—	24.4	36.1
Alveolar surface covered (%)			
Type I	—	96.4	92.9
Type II	—	3.6	7.1
Average cell surface area (μm^2)			
Type I	—	7287.0	5098.0
Type II	—	185.0	183.0
Endothelial	—	1121.0	1353.0

The cells of a tissue are usually subdivided into two general categories: (1) parenchymal cells, which are characteristic of the organ and perform the organ's specific function, and (2) stromal cells, which comprise the supportive structure of the tissue, including vasculature. Because patent blood vessels are as important in gas exchange as an intact alveolar epithelium, both the epithelial cells lining the alveoli and the endothelial cells lining the blood vessels will be considered the pulmonary parenchyma; only the interstitial cells will be considered the stroma.

At least five major cell types have been identified in the alveolar regions of the lung: the epithelial cells lining the alveoli, Type 1 and Type 2 cells, endothelial cells, macrophages, and interstitial cells. Intestitial cells consist mostly of fibroblasts (50–60% in rat lung) as well as blood cells such as monocytes, lymphocytes, and plasma cells. An estimate of cell numbers and characteristics in the alveolar regions of normal lung from a variety of species including humans is given in Table II. Again, there are striking similarities between the species in the distribution and numbers of the various cell types despite the large variation in size and total cellularity.

A. Type 2 Cell

The Type 2 cells of the lung probably have been studied most extensively because they produce surfactant, the surface-active material responsible for preventing lung collapse. The Type 2 cell ($\sim$ 16% of all cells in the human lung; Table II) is a large cuboidal cell often seen at the junctions of alveolar septa. The exposed surface is covered with microvilli; lamellar inclusion bodies, organelles characteristic of secretory cells, are found in the cytoplasm. Extensive autoradiographic and electron microscopic studies indicate that the phospholipid and protein components of surfactant are synthesized in Type 2 cells, stored in the inclusion bodies, and secreted by exocytosis onto the alveolar surface. The Type 2 cell covers only 7 m^2 of the surface area of both lungs in humans (7%) and 0.01 m^2 of both lungs in the rat (3–6%) (Crapo et al. 1983) (Tables I and II). This cell also has been identified as a regenerative population in the lung and, after diffuse alveolar injury, generates a new alveolar epithelium by dividing and differentiating into Type 1 cells (Adamson and Bowden, 1974).

Because of its role as the stem cell of the alveolar epithelium and the necessity of surfactant for maintaining patent alveoli, the Type 2 cell has been considered a likely candidate for the role of target cell for radiation. Studies by Shapiro et al. (1982), Rubin et al. (1983), and Ahier et al. (1985) have indicated that changes in surfactant and its precursors may be the earliest indicators of radiation damage in the lung. These changes occur between 1 and 4 weeks after single doses of the same magnitude as those that cause death from radiation pneumonitis, albeit at later times, 4–6 months after exposure. Although these data do not necessarily indicate a causal relationship between these early cellular changes and the subsequent demise of the animal, they do indicate that these cells are damaged by radiation and manifest this injury at a time consistent with their turnover time.

B. Type 1 Cell

Type 1 cells are highly differentiated alveolar epithelial cells that do not divide. These cells cover the largest area of the lung (about 93% of the alveolar surface in humans), but because they are highly attenuated, they are difficult to identify by routine light microscopy (Crapo *et al.*, 1983; Gail and Lenfant, 1983). Although Type 1 cells are highly susceptible to injury by a variety of agents, because they are nondividing cells, they are less likely to be the putative cell for radiation damage.

C. Endothelial Cell

The endothelial cells have an average cell volume of about one-quarter of that of the alveolar Type 1 cells and cover about 91 m^2 of both lungs in humans (Table I) (Crapo *et al.*, 1983; Gail and Lenfant, 1983). The pulmonary capillary bed has a surface area in humans of about 70 m^2 (the size of a tennis court) and is the largest vascular bed in the body, allowing contact between the lung and the external environment 40 times greater than that of the skin. Endothelial cells form a continuous lining of the vascular lumen in the lung and account for about 30% of the cells in human lung parenchyma. The pulmonary capillary endothelium is a continuous, highly attenuated cell layer between blood and lung tissue. The cytoplasmic extensions (which may be less than 0.1 μm wide) lack specialized organelles, but contain numerous pinocytotic vesicles, many of which open directly to the vascular space. Thus, the endothelial cells are highly specialized for the transport of materials between blood and tissues and, in the lung, form a barrier that prevents leakage of excess water and macromolecules into the interstitium, which is due in part to the intracellular tight junctions joining these cells.

It has been known since 1925 that the lungs take up vasoactive substances and it is now known that the pulmonary circulation selectively processes many types of compounds (Gail and Lenfant, 1983). Studies of pulmonary endothelial cells *in vitro* have provided evidence that these cells perform a wide range of metabolic activities, including uptake of serotonin and norepinephrine, synthesis of angiotensin-converting enzyme (ACE) and prostacyclin, and metabolism of adenine nucleotides and drugs. Further work has shown that some of these compounds, e.g., ACE, are synthesized or inactivated on the luminal surface of the endothelial cell, whereas others such as serotonin and norepinephrine, are metabolized intracellularly. Such other substances as epinephrine and histamine are unaffected in the pulmonary circulation.

Based on histological observations of occluded and thrombosed small arterioles and capillaries (Phillips, 1966), it has long been suggested that damage to the vasculature plays a significant, if not the primary, role in radiation injury of the lung and that the endothelial cell could be the target cell. The depletion of these cells in the lung is almost impossible to study *in vivo*, and thus the vascular hypothesis has been difficult to investigate.

The metabolic activity of endothelial cells could, however, be a useful parameter for assessing damage to these cells. Prostacyclin is of particular interest because it is a potent vasodilator, and its release is stimulated by injury to the endothelium, e.g., pulmonary embolism, hypertension, and other factors. Ward and his colleagues (1983) and Ts'ao *et al.* (1983) showed that levels of prostacyclin and ACE do change after irradiation. Thus, techniques that identify changes in metabolic products of endothelial cells may be useful in defining the target cell if such changes exhibit a dose–response relationship.

D. Cell Kinetics

Knowledge of the kinetics of the various cell populations in the lung is a prerequisite to understanding of the response of this tissue to injury by irradiation and of the processes that initiate and control both pulmonary repair and regeneration. Precise information on the turnover time of various cell types in the lung is surprisingly sparse. Most of these studies have been done in rats and mice and there are few, if any, data available from human studies. One factor essential for determining any kinetic parameter in a tissue is the identification of a pure population of cells. In the lung, this can be accomplished for both the epithelial and endothelial cells of the alveoli, but only by adding electron microscopic techniques to standard light microscopic and autoradiographic methods.

The original studies of Bertalanffy and Leblond (1953) reported a turnover time of 8 days for the "nonvacuolated" cells and a turnover time of 29 days for the "vacuolated" cells of the lung. The vacuolated cells were correctly identified as Type 2 cells; however, the nonvacuolated cells, then believed to be Type 1 cells, subsequently were identified as macrophages. Later studies have not changed this early estimate of Type 2 turnover time, and it is generally accepted that the turnover time of an undisturbed population of Type 2 cells is between 28 and 80 days in the mouse. However, the long turnover time of these cells is decreased considerably after injury; e.g., after oxygen exposure, the entire population can be replaced in 3 days (Adamson and Bowden, 1974). When one considers that there are a total of 10^6 Type 2 cells in mouse lung, this is, indeed, very rapid proliferation and regeneration of these cells. After urethane administration, the number of Type 2 cells in the lung doubles between 1 and 28 days

(Kauffman, 1972). The subsequent differentiation of Type 2 cells into Type 1 cells remains a mystery, but it is known that during development, progressive waves of mitotic activity, followed by differentiation in the lung, are at least partially dependent on intimate contact between mesenchymal, endothelial, and epithelial cells (Vracko, 1974). If similar mechanisms are operational after radiation injury, control of Type 2 proliferation may be modulated by the adjacent parenchymal cells. It has been shown that an intact basal lamina is necessary for replacement of senescent cells as well as irreparably damaged cells.

Endothelial cells are a constantly renewing population with a daily turnover rate of about 1% of their number (Crystal, 1976). However, mitotic activity of the endothelial cells varies in different parts of the vascular tree; mitoses are most numerous at branching points or areas of turbulence and trauma (Meyrick and Reid, 1979). Like the Type 2 epithelium, the vascular endothelium has an immense capacity for regeneration after injury; at least 4% of the endothelial cells are actively synthesizing DNA within days after oxygen injury (Bowden and Adamson, 1974). Although there are no data available on the turnover time of the normal population of endothelial cells in the lung, studies in other tissues indicate that it is very long (Hirst et al., 1979).

IV. Phases of Lung Response

The clinical course and pathophysiological manifestations of lung response after irradiation have been well documented in experimental animals as well as in humans. There are a number of current reviews (Gross, 1977; Penney et al., 1982) in this area and only a brief summary will be given here to focus the problem.

Two syndromes, radiation pneumonitis and fibrosis, which are histologically distinct and separated in time, have been identified in the lung after irradiation. Radiation pneumonitis is an exudative reaction in the lung that occurs between 80 and 180 days after irradiation in experimental animals (Phillips, 1966; Travis et al., 1980; Siemann et al., 1982). This phase is characterized histologically by interstitial edema as well as edema in the air spaces (mainly after high doses that cause death), an infiltration of inflammatory cells (mostly macrophages), and an apparent desquamation of epithelial cells from the alveolar walls (Fig. 2). Measurement of hydroxyproline, an indicator of the collagen content of tissue, has shown that fibrosis is not a major component of this phase of pulmonary damage (Law, 1985; Travis, unpublished data). If the dose to both lungs is sufficiently high, the animals will succumb to this injury.

In animals surviving this first pneumonitis phase, 9 months after irradiation a second histologically distinct picture appears consisting of two main histological features, a diffuse interstitial fibrosis of the alveolar walls as well as focal scarring of the lung with loss of alveolar architecture resulting in contraction of

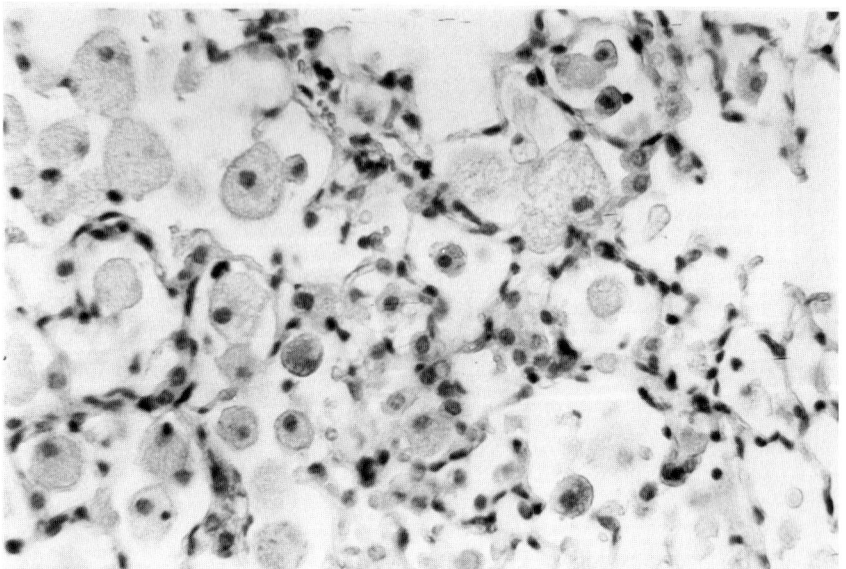

FIG. 2. Histological section of mouse lung at 16 weeks after 13 Gy, showing a fulminant radiation pneumonitis characterized by cellular infiltration in the alveolar walls and cells free in the alveoli. Hematoxylin and eosin (H and E) stain.

the lung tissue in that area (Fig. 3) (Travis, 1980; Travis *et al.*, 1980). These histological changes are often accompanied by the presence of pleural effusions found at necropsy. Recently, this lesion has been found to be accompanied by foamy cells and a granulomatous lesion with empty needle-like structures resembling cholesterol clefts (Fig. 4) (Travis *et al.*, 1985a). This lesion is similar to alveolar proteinosis, a nonspecific response of the lung to insult.

Although there is general agreement on the characteristics of the pneumonitis phase, the histological features of this second phase are controversial. Initially, Travis *et al.*, (1980) described an interstitial fibrosis detected by special stains for collagen. Pleural effusions were often found in these mice at the time of death or sacrifice. Travis *et al.* (1985a) recently described focal scarring and atelectasis, which are often accompanied by pleural effusions rather than an interstitial fibrotic process. Down *et al.* (1983, 1984) and Down and Steel (1983), however, observed only pleural effusions in their studies, using many strains of mice, and suggest that pulmonary fibrosis is not a significant factor in late radiation injury, at least after irradiation of mouse lung. Regardless of the exact nature of this late lesion, it is clear that none of these mice had histological signs of radiation pneumonitis.

Although it has been suggested that the later wave of injury is a secondary consequence of the pneumonitis phase, this later phase has been observed after

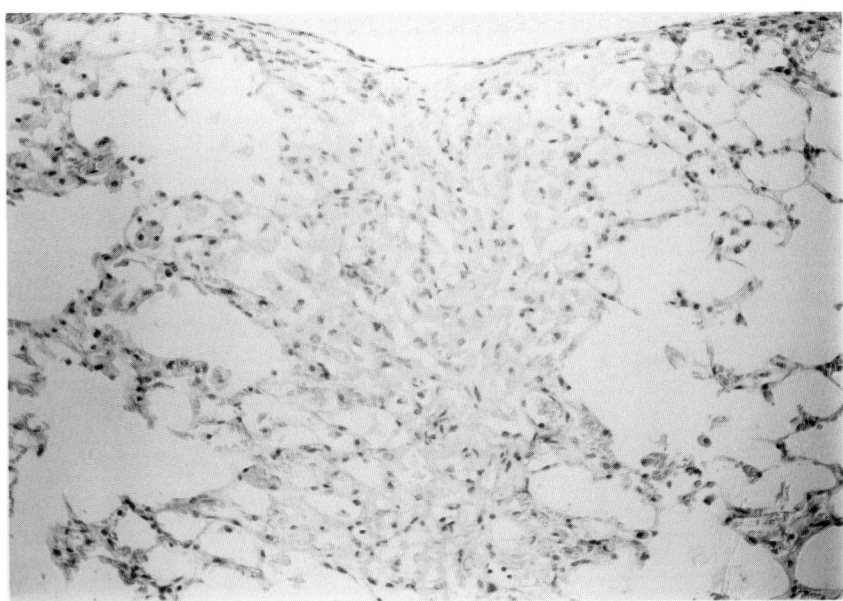

FIG. 3. Histological section of mouse lung 1 year after 11.75 Gy showing focal subpleural scarring and contraction of the lung. The adjacent lung appeared normal. H and E stain.

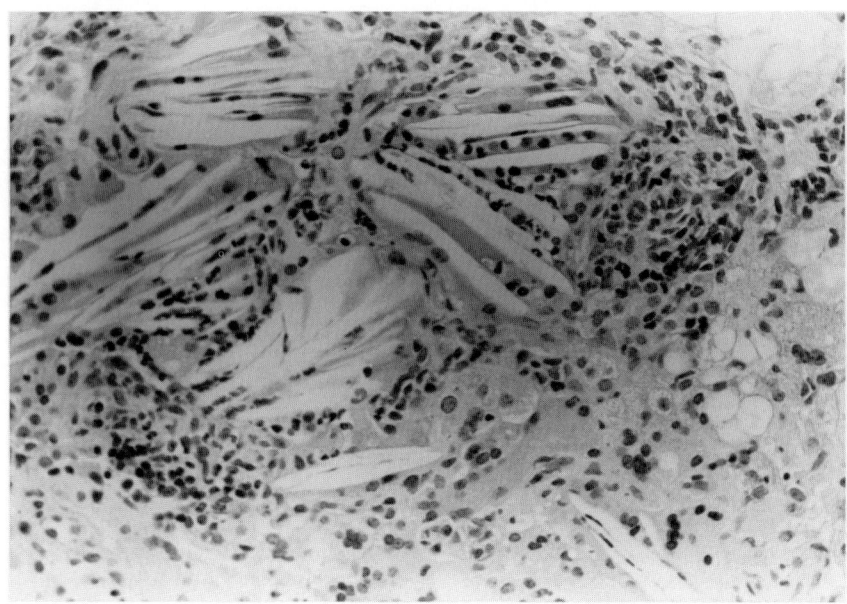

FIG. 4. Focal area of "foamy" cells accompanied by needle-like clefts found in mouse lung 1 year after 11.75 Gy to the whole thorax. H and E stain.

low radiation doses that caused no symptomatic signs of the first pneumonitis phase. In addition, the two phases can be dissociated; two doses split by 28 days prevent pneumonitis, but not the later phase (Fig. 5) (Travis and Down, 1981). Protection of the later phase of injury by the thiol radioprotector WR-2721 is greater than that of the pneumonitis phase (Fig. 6) (Travis *et al.*, 1985) and anesthetic drugs prevent the pneumonitis, but not the later wave of damage (Down *et al.*, 1983). In addition, one mouse strain, C57B1 exhibits only the later

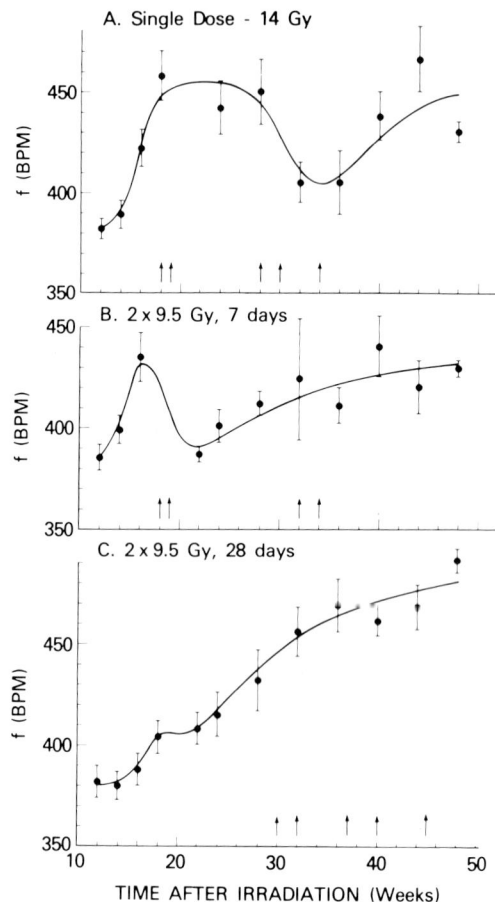

Fig. 5. Breathing rate (BPM) versus time after irradiation for two equal doses separated by 7 or 28 days. The pneumonitis phase (up to 28 weeks) is greatly reduced, at least symptomatically as measured by BPM when longer time intervals separate the two doses, indicating that the late "fibrosis" phase (after 36 weeks) is not a direct consequence of the first pneumonitis phase. (Reproduced with permission from Travis and Down, 1986.)

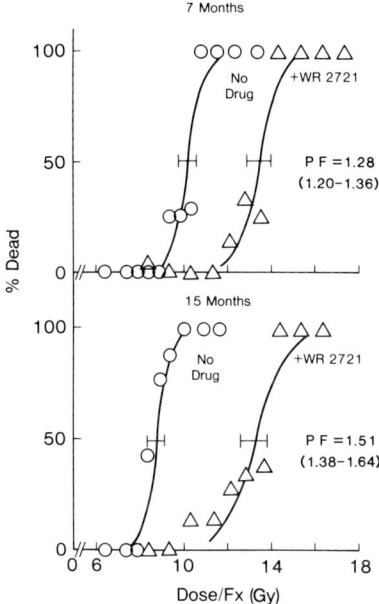

Fig. 6. Significantly greater protection against death is achieved at 15 months after whole lung irradiation than at 7 months by the thiol radioprotector WR-2721, again indicating that the two phases of damage in the lung can be dissociated. PF, Protection factor. (Reproduced with permission from Travis *et al.*, 1985a.)

wave of damage, with no preceding pneumonitis phase (Down and Steel, 1983). Although the interpretation of these data is open for discussion, certainly one possibility is that two different target cells with differing radiosensitivities are responsible for each of the two phases.

The two phases of radiation response in the lungs of small experimental animals are similar both pathologically and in time sequence of appearance to those observed in humans after irradiation of part or all of the lung. The first signs of pulmonary injury may be manifested between 2 and 4 months after irradiation, a time surprisingly similar to that observed in mice for the pneumonitis phase (Gross, 1977). Also, the histological changes in human lungs are strikingly similar to those observed in mice, edema of the alveolar septa, hyperplasia of Type 2 cells lining the alveolar walls, and an increase in alveolar macrophages (Jennings and Arden, 1962). As in small animals, neutrophils are not a prominent feature of this phase of injury. One difference between the lesions observed in human and mouse is that hyaline membranes lining both alveolar ducts and some alveoli are a prominent and common finding in human lungs (Jennings and Arden, 1962), but this lesion has not been observed in mice.

The pathological changes in human lungs after irradiation can be seen on radiographs, and they are often accompanied by dyspnea, the cardinal symptom of pneumonitis, as well as cough, fever, and rales (Gross, 1977). Dyspnea is also a symptom of pneumonitis in experimental animals (Travis *et al.,* 1979).

The second form of injury, chronic progressive pulmonary fibrosis, has been seen in all patients who recover from the first pneumonitis syndrome as well as in those who never manifest the first phase (Fajardo, 1982), which agrees with similar observations in mice; i.e., low doses that caused no histological or functional changes indicative of radiation pneumonitis still produced the later chronic, progressive radiation fibrosis (Travis and Down, 1981) (Fig. 5). Histologically, this phase is characterized by fibrosis of the alveolar septa and atypical alveolar epithelial cells. In the late stages, the alveoli collapse or are obliterated by collagen and connective tissue, resulting in atelectasis of that area of the lung and a scar, changes again remarkably similar to those seen in rodents (Gross, 1977). In mice, but not in humans, pleural effusions are a common finding of the late fibrotic phase, most probably because mouse data are usually obtained after irradiation of the whole lung, whereas data from clinical studies are obtained after only partial irradiation of the lung. Two exceptions are upper half body and total body irradiation, but the doses delivered with these techniques are significantly below those that would be expected to produce pulmonary complications. Although there may be a latent period between the termination of the first pneumonitis phase and the second chronic phase in humans, fibrosis generally is established by a year after irradiation and at this point is irreversible (Gross, 1977).

Most symptoms associated with radiation pneumonitis subside during the fibrosis phase and most patients with pulmonary fibrosis are asymptomatic (depending, of course, on the volume of tissue irradiated). The important functional alterations during this phase include a decrease in lung volume, arterial hypoxia, decreased compliance, and a reduction in maximum breathing capacity. Gross (1977) suggested that decreases in pulmonary blood flow can be demonstrated early and may be predictive of delayed changes like these.

V. Assays of Lung Damage in Humans and Experimental Animals

To assess sensitivity, one must be able to measure and then to quantify a response. The term radiosensitivity traditionally refers to the slope of the exponential portion of a radiation survival curve obtained from *in vitro* or *in vivo* assays of clonogenic cells. However, the use of clonogenic assays in organized tissues *in vivo* depends on identifying the target cells in the tissues and quantifying their response to radiation. Although such assays have been developed for some rapidly proliferating tissues *in vivo*—e.g., bone marrow and gut that have readily defined clonogenic cell compartments—the complexity and long cycle time of cells in slowly proliferating tissues such as the lung have made it difficult to establish such assays in these tissues.

Without the benefit of clonogenic assays to quantify the depletion of target cells in an organized tissue, other ways must be sought to determine the response of tissues to radiation if their sensitivity is to be determined. Determination of these tissues' radiosensitivity depends on quantifying organ response to radiation with nonclonogenic assays of damage, such as functional or histological changes, and determining isoeffect doses from the resultant dose–response curves. One assumption is, of course, that the shape of these dose–response curves is related to the depletion of the critical target cells in the tissue.

The data derived from various techniques of assessing injury can be described generally as measuring either a quantal "all-or-nothing" response (e.g., the standard lethality assay) or an ordinal, graded response. Although general models exist for describing the dose dependence of quantal responses, allowing the data to be analyzed by logit or probit analysis, no general model exists for describing the dose dependence of ordinal responses. Thus, estimates of effective doses (e.g., ED_{50}) for different levels of effect cannot be obtained from these data and different treatment schedules cannot be compared statistically by standard methods. One way of converting ordinal data into quantal data is by scoring the number of responders (either humans or animals) in each dose group that exceed a threshold response. This can be easily accomplished in most of the assays described below, at least for the data obtained from animal studies.

The large reserve capacity of the lungs results in an infrequent clinical occurrence of damage, particularly after irradiation of only part of the lung, the more common clinical situation. Nonetheless, many patients will exhibit at least some degree of pulmonary function changes if tested. There are many reports in the literature of physiological events that occurred after lung irradiation, perhaps the most meaningful data having been obtained from breast cancer patients in whom the lung was not compromised by disease prior to treatment. The functional parameters that appear to be sensitive indices of radiation damage in human lungs are lung volume, static lung compliance, blood gases, carbon monoxide diffusion capacity, and a change in the alveolar arterial oxygen gradient (Gross, 1977). Perhaps the most useful assay of lung damage after irradiation of the human lung and certainly the one closest to the experimental situation is the incidence of pneumonitis observed radiographically and at autopsy after irradiation of the whole thorax using upper half body (Prato *et al.*, 1976; Fitzpatrick and Rider, 1976) or total body irradiation techniques (van Dyk *et al.*, 1981; Fryer *et al.*, 1978). The data obtained from these latter studies are quantal, and thus dose–response curves can be generated and compared with those obtained from experimental animals.

The situation in experimental animals is somewhat simpler than in humans because most studies of radiation lung injury involve exposure of the whole thorax, at least in small experimental animals such as mice. Data derived from these studies are not complicated by compensating physiological factors which may themselves influence response when only part of the lung is irradiated.

A number of invasive and noninvasive techniques have been developed in the past 10 years for assessing the response of the lungs of small experimental animals, particularly mice, after irradiation. Most of these assays rely on restrictive patterns of ventilation and impairment of gaseous exchange believed to be associated with radiation pneumonitis and the later appearing injury, whether it be fibrosis or effusions.

The ultimate test of the function of a vital organ after treatment with any cytotoxic insult is survival. The LD_{50} assay, long used to assess radiation response of the lung of small animals, was at first limited to the time when pneumonitis was known to occur, i.e., 80–180 days (Phillips and Margolis, 1972; Wara et al., 1973). After the observation that animals surviving the first phase of injury exhibit a second wave of death, the LD_{50} assay for lung death was extended beyond 180 days to at least 1 year after irradiation (Siemann et al., 1982) (Fig. 7). The survival assay is a quantal response; thus the data can be fitted by logit or probit analysis and the LD_{50} determined with confidence limits. The single dose–response curves from all strains of mice have been found to be very steep, in the range of 2 Gy between LD_{10} and LD_{90} (Fig. 7).

Dyspnea, the cardinal symptom of radiation pneumonitis in humans, is the basis of a noninvasive assay of functional lung damage in mice, the breathing rate. This noninvasive functional assay is based on pulmonary function measurements in humans and was developed to assess both phases of response in mouse lung after irradiation (Travis et al., 1979, 1980). This system has been adapted for assessing the response of irradiated rat lung (Giri et al., 1985). The system consists of an airtight total body plethysmograph and an electrical capacitance manometer that records pressure changes in the chamber caused by the rate and amplitude of breathing. The animals can be tested without anesthesia or restraint and thus this noninvasive assay allows assessment of sequential changes in lung function in the same mouse after irradiation. One advantage of this assay is that it measures changes after sublethal doses to the lung (Fig. 8).

The increases in the rate of breathing measured with this technique are ordinal responses and, although these data give steep dose–response curves, they cannot be fitted by logit or probit analyses. However, the data can be transformed into a quantal response by scoring the number of animals that exceed a threshold breathing rate, usually 15%, above control values. These data, plotted

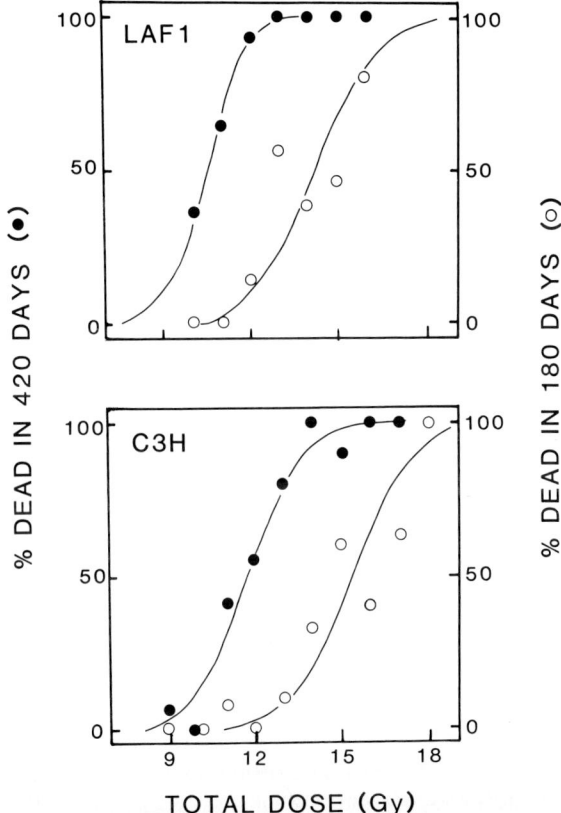

Fig. 7. Death as a function of dose two times after whole lung irradiation, 180 days (open circles) and 420 days (closed circles) for two mouse strains. (Redrawn with permission from Siemann *et al.*, 1982.)

as a function of dose and fitted by logit analysis, also give steep dose–response curves, as does the lethality assay (Fig. 9). The estimates of LD_{10} to LD_{90} range from 1 to 4 Gy. The increases in the rate of breathing have been shown to agree well with the severity of the histological changes scored from the same mice after irradiation (Travis *et al.*, 1980).

Abnormal blood gas values have been observed in some patients after lung irradiation, and it has been suggested that abnormal gaseous exchange also may occur in animals after irradiation of the thorax. Blood gas values are directly related to the ventilation and perfusion capacities of the lung. If the ventilation perfusion capacity is compromised by edema (during pneumonitis) or by col-

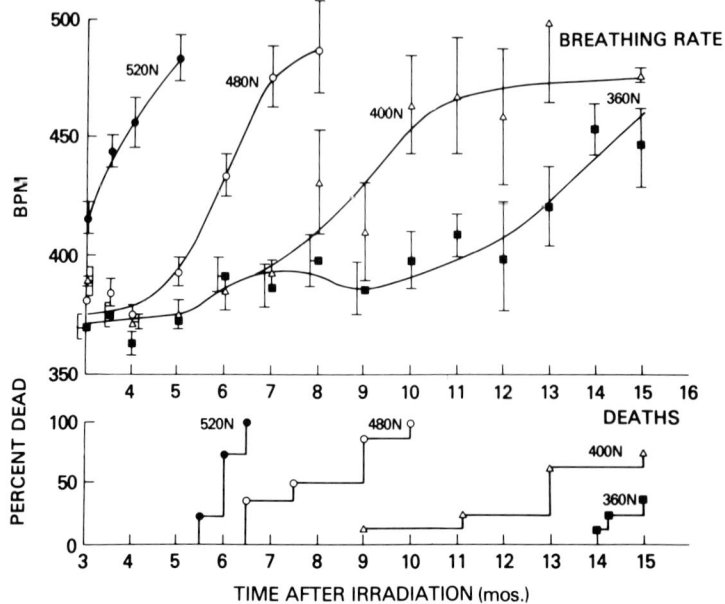

FIG. 8. Comparison of breathing rate and death as a function of time after whole thorax irradiation in mice showing that changes in breathing rate begin long before the death of the mouse because of pulmonary failure. (Redrawn with permission from Travis *et al.*, 1983a.)

lagen deposition (in fibrosis) or pleural effusions, resulting in inadequate ventilation of the blood, then a reduction in blood gas values could occur. Siemann *et al.* (1980) have used this clinical observation to develop a technique of measuring the partial pressure of oxygen (PaO_2) and carbon dioxide ($PaCO_2$) in mice. Their studies showed two reductions in PaO_2, one immediately before the mice succumbed to pneumonitis at 16 weeks and a second one in the few survivors at 27 weeks after the peak pneumonitis had subsided. Although this assay is not useful for predicting damage prior to death from pneumonitis, it may be useful for assessing late functional changes.

Another noninvasive assay of pulmonary function that is also based on physiological changes in the human lung after irradiation is carbon monoxide uptake (Depledge and Barrett, 1982). Like the rate of breathing, this technique also measures a response after sublethal doses. The method has been used to assess pulmonary function after low dose rate-total body irradiation (followed by bone marrow rescue), and it showed that gas exchange was impaired, but the rate of breathing was not elevated. These data indicate that gas transfer may be a more sensitive indicator of pulmonary damage than the rate of breathing.

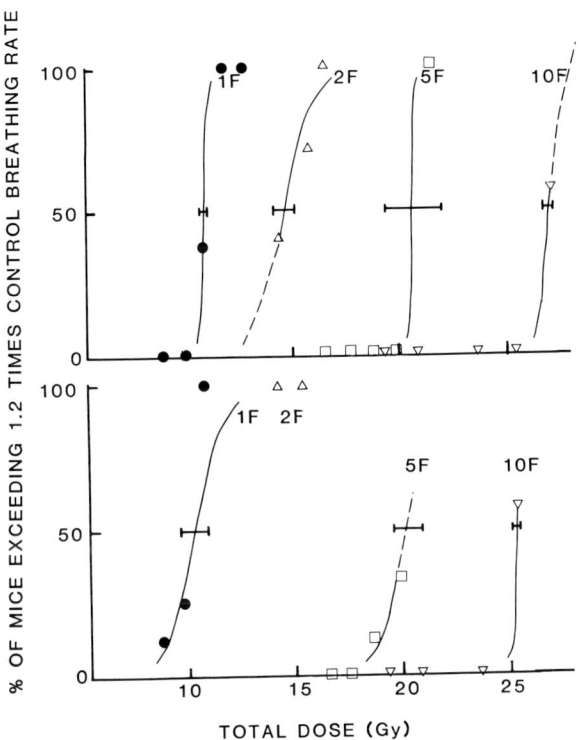

FIG. 9. The ordinal response, breathing rate, expressed as quantal data by scoring the number of mice exceeding a threshold BPM. Plotting these values as a function of dose yields steep dose–response curves. (Redrawn with permission from Parkins *et al.*, 1985.)

VI. Radiosensitivity of the Lung

A. Ways to Determine Sensitivity of Organized Tissues

As stated earlier, radiosensitivity generally refers to the slope of the terminal portion of a survival curve of cells irradiated *in vitro* or from *in vivo* survival curves of clonogenic cells. Although the available assays of functional pulmonary damage allow the determination of dose–response curves and isoeffect data, the lack of a defined target cell population and a method of studying its radiation response *in vivo* prevents the direct determination of survival parameters such as radiosensitivity of the putative target cell. The lung is only one of many tissues for which no direct assay of target cell response has been developed. In the absence of a direct assay, target cell response can be inferred from data on tissue dose–response, but only by assuming some relationship between cell killing and eventual tissue injury. Although the target cell for radiation damage in the lung is unknown, this target cell is assumed to be the same as that cell which, when depleted, causes death from lung damage. The above-mentioned relationship is most likely influenced by other physiological factors such as host response to injury.

An absolute requirement of methods for obtaining survival curve parameters from tissue dose–response curves is that these curves be derived from quantal all-or-nothing responses (e.g., death) or from ordinal (i.e., graded) responses that have been converted into quantal data (e.g., number of mice exceeding a threshold rate of breathing). The conversion of ordinal responses to quantal data requires that the number be derived from the monotonic portion of the ordinal dose–response curve.

One method for estimating target cell parameters such as D_0 from tissue-response data was introduced by Lange and Gilbert (1968). This approach is based on the Poisson distribution and requires quantal data. According to the Poisson model, the probability p of a response (e.g., lung death) after a dose regimen is given by

$$p = e^{-\lambda} \quad \text{or} \quad \lambda = -\ln p \tag{1}$$

where λ is the expected number of "death-preventing units" remaining in the tissue after irradiation. In Lange and Gilbert's model, the quantity λ is related to cell survival by assuming that

$$\lambda = r \cdot S \cdot N \tag{2}$$

where N is the number of target cells in the tissue, S is the fraction of surviving cells, and r is the probability that a surviving target cell rescues the tissue. The observed proportion of deaths in each dose group is an estimate of p, and the dependence of p on dose is modeled by replacing S with some model for cell survival. In this way, various parameters of the cell survival model can be estimated using data from the entire dose–response curve of the tissue.

In one analysis, cell survival after large single doses of radiation can be modeled by an exponential curve, i.e.,

$$\ln S = \ln E - D/D_0 \tag{3}$$

where E is the extrapolation number and D is dose. Substituting this model into the Poisson expression one obtains [Eqs. (1) and (2)]

$$\ln(-\ln p) = \ln(rN) + \ln(S) = \ln(rNE) - D/D_0 \tag{4}$$

Thus, the parameter D_0 can be estimated from data on responses to single doses using the exponential model (as well as the product rNE, which represents an upper bound on the number of death-preventing units in the tissue before irradiation).

The above method of estimating tissue radiosensitivity uses a maximum likelihood fit of the Poisson model to data on radiation responses and requires raw data, i.e., number of responders per total number of subjects in each dose group. However, the raw data are not always available for analyses, e.g., for previously published data. In these situations, D_0 can be estimated from the dose–response curve, again using the Poisson model, and can be shown to be one-third the dose required to increase response from 10% to 90%. This can be seen from the Poisson model where

$$\text{Probability (response)} = e^{-\lambda(D)\alpha_R} \tag{5}$$

where λ (D) is the number of target cells surviving a dose D and α_R represents the probability that a surviving target cell will rescue the tissue. Thus, $\ln \lambda_{90} - \ln \lambda_{10}$ is approximately equal to 3. {Note: $\ln\lambda_x = \ln[-\ln(x\%)] - \ln(\alpha_R)$.} In other words, the dose needed to change a response from 10% to 90% reduces survival by approximately three logs. Although a maximum likelihood fit of the Poisson model allows a more accurate estimate of D_0 from tissue response data, the ED_{10} to ED_{90} technique provides a reasonable estimate of D_0 when raw data are not available.

B. Sensitivity of the Lungs of Humans and Experimental Animals

1. Estimates of Sensitivity from Single-Dose Data

Using the first method of Lange and Gilbert (1968), data on lung lethality from several different experiments in our laboratory were analyzed using an exponential model for cell survival. In each case, maximum likelihood estimates of the appropriate parameter values were obtained as well as their 95% intervals.

Cell survival after large single doses of radiation was modeled by an exponential curve [Eqs. (3) and (4)] and estimates of D_0 were obtained from data on lung lethality at 28 weeks postirradiation derived from seven different single-dose experiments. Although the LD_{50} values from these seven experiments were quite consistent, ranging from 12.13 to 12.57 Gy (Table III), the slopes of the dose–response curves exhibited some variation due to binomial variability in the observed proportion of responders in each dose group. That the variable slopes did not, in all probability, reflect true changes in sensitivity from experiment to experiment is shown in Fig. 10 in which the combined data from all seven experiments yield a well-defined dose–response curve with relatively little scatter. Since D_0 is approximately equal to one-third the dose increment required to go from 10% to 90% response, varying estimates of D_0 were obtained from the seven experiments, ranging from 37.1 cGy for the steepest curve to 82.7 cGy for

TABLE IV

POISSON FIT OF EXPONENTIAL SURVIVAL MODEL
TO SINGLE-DOSE LETHALITY DATA

Experiment	D_0	Range
LFx 1	77.8	(−17.6, 173.3)
LR1.2	41.7	(22.0, 61.4)
LR1.1	82.7	(−207.4, 372.8)
LR1.52	37.1	(22.7, 51.5)
LR1.58	48.6	(20.4, 76.7)
LR1.3	49.9	(18.4, 81.3)
LP1.2	69.5	(−2.5, 141.6)
Combined	63.4	(44.6, 82.3)

TABLE III

SINGLE-DOSE LD_{50} VALUES (Gy) AT 28 WEEKS
(7 EXPERIMENTS)

Experiment	LD_{50}	95% CL
LFx 1	12.15	(11.63, 13.23)
LR1.2	12.57	(12.12, 13.07)
LR1.1	12.53	(12.12, 13.07)
LR1.52	12.43	(12.06, 12.75)
LR1.58	12.15	(11.67, 12.54)
LR1.3	12.27	(11.81, 12.84)
LP1.2	12.13	(11.33, 12.63)
Combined	12.34	(12.19, 12.49

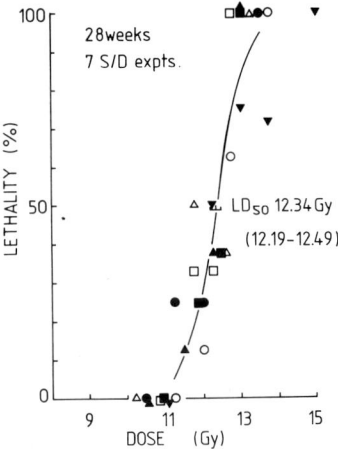

FIG. 10. Survival curves plotted at 28 weeks (196 days) after single radiation doses to the whole thorax of mice. Each symbol represents one of seven different experiments conducted over a period of 3 years. Note the steepness of the dose–response curve and the reproducibility of the data in one mouse strain in one laboratory. (Redrawn with permission from Travis and Tucker, 1986.)

the shallowest curve (Table IV). The estimate based on the combined data is D_0 = 63.4 cGy (95% confidence limits—44.6, 82.3 cGy).

This model was used also to analyze breathing rate data obtained at 28 weeks after irradiation in two experiments. The dose–response curves for these data are shown in Fig. 11, expressed as quantal data, i.e., 15% or 20% above the control breathing rate (BPM). Estimates of D_0 obtained from this assay are in good agreement with those obtained from lethality data in the same mice (Table V).

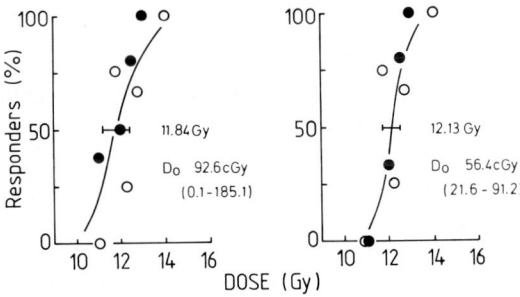

FIG. 11. BPM expressed as quantal data versus dose from mice from two of the experiments shown in Fig. 10. Note the good agreement between the isoeffect doses obtained from these data and the lethality data (Fig. 10). Open and closed symbols represent the two experiments. (Redrawn with permission from Travis and Tucker, 1986.)

TABLE V

Comparison of LD_{50} (Gy) and D_0 (cGy) Values from Death and BPM Curves (2 Experiments)

Experiment	Deaths		Breathing rate	
	LD_{50}	D_0	ED_{50} (1.2 × con)	D_0
LR1.52	12.43	38.5	12.11	35.0
	(12.06, 12.75)		(10.8, 12.9)	(10.3, 59.7)
LR1.58	12.15	48.6	12.15	73.5
	(11.67, 12.54)	(20.4, 76.7)	(11.5, 13.8)	(−29.3, 176.3)
Combined	12.30	62.7	12.13	56.4
	(12.06, 12.52)	(34.9, 90.5)	(11.75, 12.55)	(21.6, 91.2)

Although no direct estimates of D_0 for the target cells in lung *in vivo* are available for comparison, estimates of D_0 from *in vitro* colony assays of lung epithelial cells (Type 2) gave D_0 values higher by factors of 1.5–2 (Guichard *et al.*, 1980; Deschavanne *et al.*, 1981; Walker, 1986) than those obtained from the analysis of tissue response data, but because the *in vitro* estimates of D_0 were obtained after doses lower than the LD_{50}, these higher D_0 values are likely to reflect a more shallow slope of the dose–response curve, as might be expected over a lower dose range, and thus they would not be inconsistent with the D_0 values obtained with the Lange and Gilbert (1968) analysis.

Dose–response curves for lung damage (LD_{50}) for a number of different strains of mice from data published by different laboratories are shown in Fig. 12. The data from the seven single dose–response curves from our laboratory are shown as the dashed line. Estimates of the D_0 from each of these curves (using the ED_{10} to ED_{90} method) are given in Table VI and range from the lowest value of 0.63 Gy for C3H mice from our laboratory and CBA mice from the Gray Laboratory to 1.5 for C3H mice from Siemann *et al.*, (1982). Although the D_0 values vary by a factor of 2, these D_0 values for mouse lung are well within the range of published D_0 values for all mammalian cells *in vitro* and, more importantly, *in vivo*. Thus, using D_0 as a measure of sensitivity, the target cells of the lung would be expected to have a radiosensitivity similar to that of all other mammalian cells, at least after large single doses.

Also shown in Fig. 12 is the response curve for the incidence of pneumonitis in human lung obtained after whole thorax irradiation (van Dyk *et al.*, 1981). The D_0 for human lung (0.6 Gy) is in remarkably good agreement with the values obtained from mouse lung, which suggest that the sensitivity of the target cells in

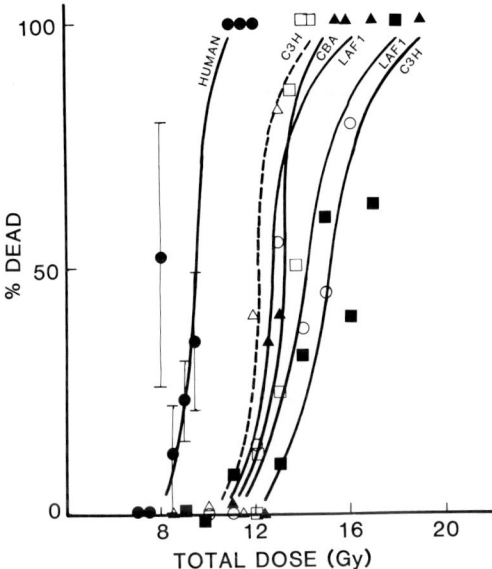

FIG. 12. Dose–response curves for pneumonitis in three mouse strains from different laboratories show good agreement between the slope of the curves, but the isoeffect doses (LD_{50}) vary among strains and among the different laboratories. Also plotted is the incidence of pneumonitis in human lungs after upper half body irradiation showing a similar shape of the human dose–response curve, but much lower isoeffect doses.

the lung in the two species is similar and that mouse lung is a good model for the response of human lung.

Although the cause of the later-appearing injury is unclear, mice surviving between 28 and 52 weeks from the seven experiments described above were analyzed using Eq. (4). The D_0 value obtained for this later-appearing injury was

TABLE VI

D_0 VALUES FOR LUNG ESTIMATED FROM DOSE–SURVIVAL CURVES OF RADIATION PNEUMONITIS

Species/strain	D_0 (Gy)	LD_{50} 80–180 days (Gy)	Reference
Human	0.60	9.30	van Dyk et al. (1981)
Mouse			
C₃Hf/Kam	0.63	12.34	Travis and Tucker (1986)
CBA/Ht	0.67	13.80	Travis et al. (1983)
LAF1/J	1.0	13.40	Travis and DeLuca (1985)
LAF1/J	1.5	14.30	Siemann et al. (1982)
C3/HeJ	1.5	15.40	Siemann et al. (1982)

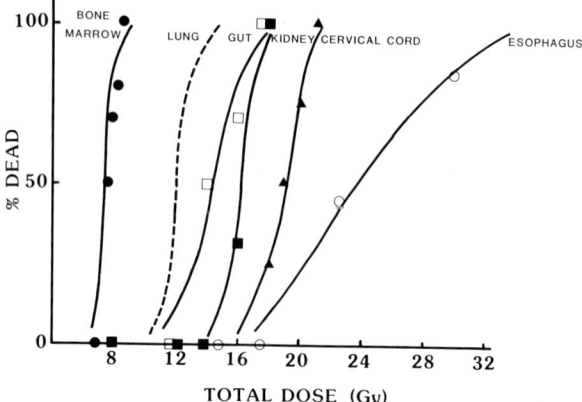

Fig. 13. Dose–survival curves for damage to a number of normal murine tissues. Except for esophagus, all have similar slopes, but the isoeffect dose (LD$_{50}$) varies with each tissue.

0.44 Gy (0.32–0.56 Gy), slightly less than the D_0 at 28 weeks. The value is in agreement with the steeper dose–response curves generally observed for this late wave of injury.

The question remains, however, whether and how estimates of target cell sensitivity from tissue response are related to the ultimate failure of the organ, resulting in death of the animal. Hendry *et al.* (1983) has obtained nearly identical D_0 values for the jejunum using an *in vivo* clonogenic assay, the Withers and Elkind (1970) microcolony assay for crypt cell survival, and an assay of tissue response, death from intestinal damage on day 7 after irradiation. Based on these and other studies, the Lange and Gilbert technique seems to be appropriate for assessing target cell radiosensitivity from tissue response data.

Clinically, radiosensitivity of the lung is more meaningful when it is compared with other tissues. Figure 13 shows a comparison of dose–response curves for various mouse tissues, including normal tissues that are critical when treating the thorax, e.g., spinal cord and esophagus. D_0 values estimated from these curves are shown in Table VII and indicate that the target cells for these different tissues have a sensitivity similar to those for the lung, with the exception of the esophagus, which appears to be less sensitive than the other tissues shown ($D_0 \sim$ 5 Gy) although this is based on very limited data.

All of these data suggest the following: (1) The radiosensitivity of the target cell for lung damage is similar to that of target cells in a number of other normal tissues; (2) the sensitivity of human lung is similar to that of mouse lung—mouse lung can be expected to be a good model for studying radiation damage; (3) the sensitivity of the target cell in the lung is similar to that of most mammalian cells. Therefore, if D_0 is used to assess sensitivity, the lung would not be expected to differ appreciably from most other tissues.

TABLE VII

D_0 Values Estimated from Dose–Response Curves for Radiation Damage to Various Normal Tissues in Experimental Animals

Tissue	Animal	Assay	ED_{50} (Gy)	D_0 (Gy)	References
Bone marrow	Mouse	Death, 30 days	7.20	0.34	Fang and Travis (unpublished data)
Lung	Mouse	Death, 80–100 days	12.34	0.63	Travis and Tucker (1986)
Gut[a]	Mouse	Death, 10–12 days	14.20	1.70	Travis (unpublished data)
Kidney	Mouse	Death, 500 days	16.00	0.47	Williams and Denekamp (1983)
Spinal cord	Rat	Paralysis, 7 months	19.00	1.10	van der Kogel (1979)
Esophagus	Mouse	Death, 20 days	24.80	4.27	Phillips and Margolis (1972)

[a] Total abdominal irradiation.

2. Sensitivity Estimated from Multifraction Data

If D_0 were used as an index of radiosensitivity, the lung would be expected to resemble most other tissues with respect to radiosensitivity. Clinically, however, the lung is more sensitive compared to other tissues, particularly spinal cord, esophagus, and heart, the other tissues considered dose-limiting when the whole thorax is treated. What the radiotherapist means in this case is that the limiting response occurs in the lung after lower doses than in the other tissues. Thus, the total dose that can be given to the thorax is not limited by the spinal cord or the esophagus, but by the lung.

This is seen clearly in Fig. 13. With the exception of death from bone marrow damage, death from pneumonitis occurs at much lower doses than death from damage in the other tissues. This does not indicate necessarily that the lung is more "radiosensitive," but rather that injury in the lung may be manifested when a smaller proportion of cells is damaged than in the other tissues. However, once this proportion is damaged, equal dose increments kill the same proportion of target cells in the lung as in the other tissues, i.e., the D_0 values are similar.

The same situation applies to the lung data for human and mouse; i.e., although the D_0 values are similar, pneumonitis in man occurs after lower doses (1.5–2 times less) than those after which pneumonitis occurs in the mouse (see Fig. 12), which implies that damage results when proportionally fewer cells are killed in the human lung than in mouse lung.

The discussion of pulmonary radiosensitivity up to this point concerns only large single doses of radiation, which are used clinically only in upper half body or total body irradiation techniques. The more common clinical technique is that of multiple fractions given daily or, in some situations, more than once a day. Therefore, a more meaningful clinical concept of tissue radiosensitivity is the effect of changing the size of dose per fraction on the response of tissues in multifractionated radiation schedules.

The most common technique of assessing target cell response from multifraction data is based on the assumption that equivalent levels of target cell killing are translated into equivalent levels of functional tissue injury, regardless of the dose regimen used to produce that level of injury. Based on this assumption, multifraction isoeffect data can be interpreted in terms of the dose–survival relationship of the putative target cells. The best known analysis of this sort is the estimation of the ratio α/β of the linear quadratic (LQ) cell survival parameters when reciprocal isoeffect dose is related linearly to dose per fraction.

The parameter α/β is related to the repair capacity of a tissue, which, although different from the classical concept of radiosensitivity, is ultimately important in tissue responses in clinical radiotherapy. The response of organized tissues to fractionated radiation can be estimated in several ways, e.g., by giving large numbers of small dose fractions and varying the number of fractions using a fixed dose per fraction, or, conversely, by fixing the number of fractions and varying the size of dose per fraction. Using the hypothesis that the fractionation response of acute and late effects tissues varies, Withers et al. (1983) found that late-responding normal tissues such as the lung are more sensitive to changes in size of dose per fraction than are acutely responding ones. When the total dose to produce the isoeffect as a function of the size of dose per fraction is plotted for acute- and late-responding tissues, the steepness of the plots of late-responding tissues indicates that the underlying survival curves for the target cells of these tissues are different from those of the acute-responding tissues. Because the slopes of the initial portions of the survival curves for the two different target cells within the clinical dose range are different, it was found that reducing the size of dose per fraction should spare late-responding tissues preferentially as compared with the acute-responding tissues. Conversely, increasing the size of dose per fraction has a relatively greater effect on the late-responding tissues than does the same increase in size of dose per fraction on the acute-responding tissue.

The Poisson model can also be applied to multifraction data, but the cell survival expression S in Eq. (2) must be replaced by the multifraction LQ model with parameters α and β:

$$S^n = e^{-n(\alpha x + \beta x^2)} \tag{6}$$

n represents the number of dose fractions in a multifraction regimen and x is the fractional dose. Cell survival is assumed to be reduced by the same proportion after each fractional dose. Substituting this survival expresson into Eq. (2), the following model results:

$$\ln(-\ln p) = \ln(rN) - \alpha D - \beta D^2/n \tag{7}$$

where $D = nx$ is total dose.

The response of mouse lung to multifractioned irradiation was analyzed using the Lange and Gilbert (1968) method to obtain estimates of α and β. Data on lung lethality from two different multifraction experiments were analyzed using Eq. (7). The number of dose fractions ranged from 1 to 10 for each experiment. For one experiment, the interval between dose fractions was 8 hr, and for the other 12 hr. Previous studies have shown that 8 hr is sufficient for complete repair of sublethal injury between fractions (Thames and Travis, 1985).

Values for the LD_{50} presented in Table VIII for each fractionation schedule in the two experiments show good agreement with only slight differences. Estimates of α/β obtained from Fe plots of reciprocal total dose versus dose per fraction yielded $\alpha/\beta = 0.407$ krad for the 8-hr interval and $\alpha/\beta = 0.312$ krad for the 12-hr interval (data not shown). Moreover, these and other plots of the data did not reveal any marked deviation from the multifraction LQ model. In particu-

TABLE VIII

LUNG LD_{50} VALUES AT 28 WEEKS (Gy)

Fx	LFx1 (T = 12 hr)		LR1.5 (T = 8 hr)	
	LD_{50}	95% CL	LD_{50}	95% CL
1	12.15	(11.53, 13.38)	12.43	(12.14, 12.71)
2	16.35	(15.35, 17.35)	16.40	(15.43, 17.07)
3	19.44	(18.89, 20.42)	19.11	(18.49, 19.59)
8	27.15	(25.79, 29.49)	27.31	—
10	30.53	—	29.57	(28.20, 31.00)

lar, the assumption of equal effect per dose fraction appears to be valid for these data.

Table IX shows the estimates of α,β that were obtained by fitting model Eq. (6) to the data on tissue-response data using the maximum "likelihood" technique. Although the confidence intervals for the estimates of α and β from the two data sets overlap, the estimates of the parameters themselves are rather different. In contrast, their ratio α/β is much more consistent between data sets and agrees well with the values obtained using the reciprocal dose (Fe) method.

It was not possible to obtain α/β ratios for human lung from any published data. In keeping with the previous discussion concerning placement of the dose–response curve on the dose axis, Table X presents a list of tolerance doses for a number of human organs when the whole organ is irradiated. The tolerance is considered in each case to be the total dose that will cause a 5% complication rate for each specific organ. The total doses were given in a conventional fractionation schedule, i.e., 2 Gy/day, 5 days/week. The data in this table show that if the criterion for radiosensitivity stated above is used, the lung again tolerates a lower dose of radiation than many other tissues.

TABLE IX

POISSON FITS OF LQ PARAMETERS TO FRACTIONATED LETHALITY DATA, 28 WEEKS

	Parameters			
	T = 8 hr		T = 12 hr	
α (krad^{-1})	2.403	(1.735, 3.071)	1.376	(.926, 1.827)
β (krad^{-2})	6.416	(4.841, 7.991)	4.154	(3.239, 5.069)
α/β (krad)	0.375	(.312, .442)	.331	(.236, 0.437)
SF at LD_{50}	2.5×10^{-6}		4.1×10^{-4}	
α/β (Fe)	0.407		.312	

TABLE X

RADIATION TOLERANCE DOSES AND COMPLICATIONS[a]

Organ	Injury at 5 years	1–5% TD$_{5/5}$ (Gy)
Oral mucosa	Ulcer, severe fibrosis	60
Esophagus	Ulcer, stricture	60
Stomach	Ulcer, perforation	45
Colon	Ulcer, stricture	45
Rectum	Ulcer, stricture	55
Kidney	Nephrosclerosis	23
Bladder	Ulcer, contracture	60
Testes	Permanent sterilization	2–3
Lung	Pneumonitis, fibrosis	
	lobe	40
	whole	25
CNS	Necrosis	50
Spinal cord	Necrosis, transection	50

[a] Adapted from Rubin and Casarett (1972).

VII. Whole versus Partial Lung Irradiation

All of the previous discussion deals with the radiosensitivity of the lung when the whole lung is irradiated. Although irradiation of the whole lung either as upper half body or total body irradiation has become a common treatment technique in the management of some diseases, the majority of patients still receive radiation to only part of the lung. Because of the large anatomical and functional reserve capacity of this tissue, much larger doses can be given to only part of the lung than to the whole lung. The isoeffect dose for pneumonitis when only part of the lung is irradiated is almost two times the dose that produces pneumonitis when the whole lung is irradiated. Thus, in this clinical situation, the lung would be considered less radiosensitive than the spinal cord, although the cellular radiosensitivity of the tissue remains unchanged.

VIII. Conclusions

The goal of this article was to assess the radiosensitivity of the human lung. However, before tissue sensitivity can be discussed, a common definition must be agreed upon because, as has been shown here, many factors will influence the assessment of a tissue's sensitivity. Using D_0 (derived from either survival curves or inferred from tissue response data) as a measure of sensitivity, the sensitivity of the target cells of the lung is within the range of that of most other mammalian cells. Thus, the lung may be assumed to respond to irradiation in a manner similar to most other normal tissues. Using the placement of the dose–response curve on the dose axis as a measure of sensitivity, the lung is more sensitive than most other normal tissues, with the exception of bone marrow.

Another factor influencing sensitivity is the reserve capacity of a tissue. The lung, which has a large functional reserve, can tolerate much higher doses when only part of the organ is irradiated. Under these conditions, partial lung irradiation, the lung appears less sensitive than the spinal cord as contrasted with irradiation of the whole thorax when the situation is reversed.

ACKNOWLEDGMENTS

The author thanks Dr. Susan Tucker for stimulating discussions and mathematical analyses, Dr. Thomas Barkley for helpful discussions concerning the clinical aspects of sensitivity, and Deborah Thomas Elum for preparation of this manuscript.

This work was supported by grants CA-06294 and CA-38106, awarded by the National Cancer Institute, United States Department of Health and Human Services.

REFERENCES

Adamson, I. Y. R., and Bowden, D. H. (1974). *Lab. Invest.* **30,** 35–42.
Ahier, R. G., Anderson R. L., and Coultas, P. G. (1985). *Radiother. Oncol.* **3,** 61–68.
Bertalanffy, J. D., and Leblond, C. P. (1953). *Anat. Rec.* **115,** 515–536.
Bowden, D. H., and Adamson, I. Y. R. (1974). *Lab. Invest.* **30,** 350–357.
Carmel, R. J., and Kaplan H. S. (1976). *Cancer* **37,** 2813–2825.
Crapo, J. D., Young, S. L., Fram, E. K., Pinkerton, K. E., Barry, B. E., and Crapo, R. O. (1983). *Am. Rev. Respir. Dis.* **128,** S42–S46.
Crystal, R. T. (1976). *In* "Lung Cells in Disease" (A. Bouhuys, ed.), pp. 17–38. Elsevier, Amsterdam.
Depledge, M. H., and Barrett, A. (1982). *Int. J. Radiat. Biol.* **41,** 325–334.
Deschavanne, P. J., Guichard, M., and Malaise, E. P. (1981). *Br. J. Radiol.* **54,** 973–977.
Down, J. D., and Steel, G. G. (1983). *Radiat. Res.* **96,** 603–610.
Down, J. D., Collis, C. H., Jeffery, P. K., and Steel, G. G. (1983). *Int. J. Radiat. Oncol. Biol. Phys.* **9,** 221–226.
Down, J. D., Laurent, G. J., McAnulty, R. J., and Steel, G. G. (1984). *Int. J. Radiat. Biol.* **46,** 597–607.
Fajardo, L. F. (1982). "Pathology of Radiation Injury". Masson, New York.
Fitzpatrick, P. J., and Rider, W. D. (1976). *Int. J. Radiat. Oncol. Biol. Phys.* **1,** 197–207.
Fryer, C. J. H., Fitzpatrick, P. J., Rider, W. D., and Poon, P. (1978). *Int. J. Radiat. Oncol. Biol. Phys.* **4,** 931–936.
Gail, D. B., and Lenfant, J. M. (1983). *Am. Rev. Respir. Dis.* **127,** 366–387.
Giri, P. G., Kimler, B. F., Giri, U. P., Cox, G. G., and Reddy, E. K. (1985). *Int. J. Radiat. Oncol. Biol. Phys.* **11,** 527–34.
Gross, N. J. (1977). *Ann. Intern. Med.* **86,** 81–92.
Guichard, M., Deschavanne, P. J., and Malaise, E. P. (1980). *Int. J. Radiat. Oncol Biol. Phys.* **6,** 441–447.

Hendry, J. H., Potten, C. S., and Roberts, N. P. (1983). *Radiat. Res.* **96**, 100–112.

Hirst, D. G., Denekamp, J., and Travis, E. L. (1979). *Radiat. Res.* **77**, 259–275.

Jennings, E. L., and Arden, A. (1962). *Arch. Pathol.* **74**, 351–360.

Kauffman, S. L. (1972). *Am. J. Pathol.* **68**, 317–323.

Lange, C. S., and Gilbert, C. W. (1968). *Int. J. Radiat. Biol.* **14**, 373–388.

Law, M. P. (1985). *Radiat. Res.* **103**, 60–77.

Matsuba, K., and Thurlbeck, W. M. (1971). *Am. Rev. Respir. Dis.* **104**, 516–524.

Meyrick, B., and Reid, L. (1979). *Am. J. Pathol.* **96**, 51–70.

Murray, J. F. (1976). "The Normal Lung: The Basis for Diagnosis in Treatment of Pulmonary Disease". Saunders, Philadelphia.

Nagaishi, C. (1972). "Functional Anatomy and Histology of the Lung". Univ. Park Press, Baltimore.

Parkins, C. S., Fowler, J. F., Maughan, R. L., and Roper, M. J. (1985). *Brt. J. Radiol.* **58**, 225–241.

Penney, D. P., Siemann, D. W., Rubin, P., Shapiro, D. L., Finkelstein, J., and Cooper, R. A. (1982). *Scanning Electron Microsc.* **I**, 413–415.

Phillips, T. L. (1966). *Radiology* **87**, 49–54.

Phillips, T. L., and Margolis (1972). *Front. Radiat. Ther. Oncol.* **VI**, 254–273.

Prato, F. S., Kurdyak, R., Saibil, E. A., Carruthers, J. S., Rider, W. D., and Aspin, N. (1976). *Cancer* **39**, 71–78.

Rubin, P., and Casarett, G. W. (1972). *Front. Radiat. Ther. Oncol.* **VI**, 1–16.

Rubin, P., Siemann, D. W., Shapiro, D. L., Finkelstein, J. N., and Penney, D. P. (1983). *Int. J. Radiat. Oncol. Biol. Phys.* **9**, 1669–1673.

Shapiro, D. L., Finkelstein, J. N., Penney, D. P., Siemann, D. W., and Rubin, P. (1982). *Int. J. Radiat. Oncol. Biol. Phys.* **8**, 879–882.

Siemann, D. W., Hill, R. P., and Bush, R. S. (1980). *Radiat. Res.* **81**, 303–310.

Siemann, D. W., Hill, R. P., and Penney, D. P. (1982). *Radiat. Res.* **89**, 396–407.

Thames, H. D., and Travis, E. L. (1985). *In* "Optimization of Cancer Radiotherapy" (B. R. Paliwal, E. D. Herber, and C. G. Orton, eds.), pp. 151–160. American Institute of Physics, New York.

Travis, E. L. (1980). *Int. J. Radiat. Oncol. Biol. Phys.* **6**, 345–347.

Travis, E. L., and Deluca, A. M. (1985). *Int. J. Radiat. Oncol. Biol. Phys.* **11**, 521–526.

Travis, E. L., and Down, J. D. (1981). *Radiat. Res.* **87**, 166–174.

Travis, E. L., and Tucker, S. L. (1986). *Br. J. Cancer* **53**, Suppl. VII, 304–319.

Travis, E. L., Vojnovic, B., Davies, E. E., and Hirst, D. G. (1979). *Br. J. Radiol.* **52**, 67–74.

Travis, E. L., Down, J. D., Holmes, S. J., and Hobson, B. (1980). *Radiat. Res.* **84**, 133–143.

Travis, E. L., Parkins, C. S., Down, J. D., Fowler, J. F., and Maughan, R. L. (1983a). *Int. J. Radiat. Oncol. Biol. Phys.* **9**, 691–699.

Travis, E. L., Parkins, C. S., Down, J. D., Fowler, J. F., and Thames, H. D., Jr. (1983b). *Radiat. Res.* **94**, 326–339.

Travis, E. L., Meistrich, M. L., Finch-Neimeyer, M. V., Watkins, T. L., and Kiss, I. (1985a). *Radiat. Res.* **103**, 219–231.

Travis, E. L., Peters, L. J., McNeill, J., Thames, H. D., Jr., and Karolis, C. (1985b). *Radiother. Oncol.* **4**, 341–351.

Ts'ao, C., Ward, W. F., and Port, C. D. (1983). *Radiat. Res.* **96**, 284–293.

van der Kogel, A. J. (1979). *In* "Late Effects of Radiation on the Spinal Cord". Radiobiological Institute of the Organization for Health Research TNO. Rijswijk.

van Dyk, J., Keane, T. J., Kan S., Rider, W. D., and Fryer, C. J. (1981). *Int. J. Radiat. Biol.* **7**, 461–467.

Vracko, R. (1974). *Am. J. Pathol.* **77**, 314–346.

Walker, H. C., Ahier, R. G., and Coultas, P. G. (1986). *Br. J. Cancer* **53,** Suppl. VII, 351–354.
Wara, W. M., Phillips, T. L., Margolis, L. W., and Smith, V. (1973). *Cancer* **32,** 547–552.
Ward, W. F., Solliday, N. H., Molteni, A., and Port, C. D. (1983). *Radiat. Res.* **96,** 294–300.
Warren, S., and Spencer, J. (1940). *Am. J. Roentgenol.* **43,** 682–701.
Williams, M. V., and Denekamp, J. (1983). *Radiat. Res.* **94,** 305–317.
Withers, H. R., and Elkind, M. M. (1970). *Int. J. Radiat. Biol.* **17,** 261–267.
Withers, H. R., Thames, H. D., Peters, L. J., and Fletcher, G. H. (1983). *In* "Biological Basis and Clinical Implications of Tumor Radioresistance" (G. H. Fletcher, C. Nervi, and H. R. Withers, eds.), pp. 139–153. Masson, New York.
Yeh, H. C., Schum, G. M., and Duggan, M. T. (1979). *Anat. Rec.* **195,** 483–492.

Relative Radiosensitivity of Fetal Tissues

R. L. BRENT, D. A. BECKMAN, AND R. P. JENSH

DEPARTMENTS OF PEDIATRICS AND ANATOMY
JEFFERSON MEDICAL COLLEGE OF THOMAS JEFFERSON UNIVERSITY
PHILADELPHIA, PENNSYLVANIA 19107

I. Introduction

The use of information obtained from animal experiments to estimate and predict human reproductive risks to environmental hazards has many difficulties. If, for example, one is evaluating the reproductive risks due to the exposure to drugs and chemicals, there are many examples of discrepancies between the effects in humans and animals. The most notorious example is the resistance of rat and mouse to thalidomide teratogenesis in comparison to the marked sensitivity of the human (Brent, 1964, 1972a, 1980b). Once it has been adequately demonstrated, however, that the effects are produced in humans and certain animal species, these concordant effects can be utilized to investigate risks and mechanisms.

For several reasons, high-energy radiation effects in animals often can be more readily compared to these effects in humans than drug and chemical effects (Brent, 1983, 1984; Brent and Gorson, 1972; Brent and Harris, 1976): (1) The effect of ionizing radiation on the developing mammalian embryo is due to its direct effect on the embryo. (2) Species differences in placental transport, maternal absorption, maternal metabolism, etc. have no impact on radiation embryological experiments, and therefore these species differences do not contribute to discordance between radiation effects in human and other mammalian embryos. (3) Many radiation reproductive effects have been documented to occur in *both* animals and humans and therefore their mechanisms and quantitative radiation risks can be pursued in animal models.

In establishing the presence ᵒf an environmental effect in the field of reproductive biology, there are a number of pitfalls. Frequently scientists jump to the conclusion that an exposure is responsible for a reproductive effect, based on

239

TABLE I

INCIDENCE OF REPRODUCTIVE FAILURES PER 1000
LIVE BIRTHS AND REPRODUCTIVE RISK

Conceptions necessary to attain 1000 live births	1350
Spontaneous abortions and failure to implant	350
Anatomical malformations observed at term	30
Genetic diseases and malformations diagnosed later in life	100
Total reproductive risk	35%
Recognized reproductive risk (with 15% spontaneous abortions)	28%

one case report or one epidemiological study (Brent, 1985a). In reality, the discovery and confirmation of a particular environmental reproductive risk should be based on a sound methodological approach (Brent, 1978, 1985b,c; Wilson and Brent, 1981): (1) A consistent, reproducible finding in human epidemiological studies; (2) development of an animal model at exposures comparable to the human exposure, either quantitatively or pharmacokinetically comparable; (3) demonstration that the reproductive effect is dose related and that there is a threshold dose; and (4) consistency with basic science embryological and teratological scientific knowledge and biological common sense.

Most reported effects of intrauterine radiation fit the criteria listed above. Reproductive effects in humans are difficult to evaluate in epidemiological studies because reproductive pathology occurs so commonly in the human (Brent, 1985b; Brent and Harris, 1976) (Table I). Thus, the magnitude of the reproductive problem is important to recognize because it is easy to associate reproductive risks with an environmental exposure if there are not adequate controls, since they occur so commonly. Furthermore, the human population is quite different from many experimental animal models because of genetic heterogeneity in the human and because animals are bred for optimal reproductive performance, which is not the situation in humans.

II. Radiation-Produced Reproductive Effects in Animals and Man

Let us examine the intrauterine effects of radiation in animals and man and determine where concordance and discordance occurs for the effect and/or the dose. The following effects will be discussed: (A) maternal effects of radiation on the developing embryo, (B) malformations, (C) growth retardation, (D) embryolethality; (E) sterility; (F) shortening of life span and tumorogenesis and (G) central nervous system effects.

A. Indirect Effects of Radiation on the Mother of the Developing Embryo

While one might intuitively conclude that maternal irradiation might not be a major component of the mechanism of the embryotoxic effects of irradiation, there is no need to hypothesize about the magnitude of the maternal component. With exposures of 4 Gy or less there is no measurable contribution to embryolethality or teratogenesis when pregnant rats were irradiated while shielding the embryos or zygotes on day 1, 10, or 13 of gestation (Brent, 1960a; Brent and Bolden, 1967a,b, 1968; Brent and McLaughlin, 1960). It is true that if the maternal dose is elevated, maternal effects may affect the embryo; therefore, it is possible that exposures of high-energy radiation which cause the maternal organism to be "sick" or decrease food and water intake could result in embryonic growth retardation or embryonic lethality (Brent and McLaughlin, 1960).

In any event, one can safely conclude that exposures in the pregnant woman from diagnostic radiation which do not expose the embryo present no measurable risk to the developing embryo, at least with regard to malformations, lethality, and growth retardation.

The literature pertaining to the following radiation effects is extensively reviewed in several articles (Brent, 1970, 1972b, 1976, 1977, 1980a, 1983, 1984; Brent and Gorson, 1972; National Academy of Science, 1980; Okada *et al.*, 1975).

B. Malformations

Major anatomical malformations have been produced by ionizing radiation in every mammalian species studied when the embryos are irradiated during the period of early organogenesis. The proportion of gestation sensitive to major malformations varies drastically in different species. While this sensitive period may account for 25% of mouse gestation, it represents only 6% of human pregnancy. Mole (1982) has suggested that the human embryo is less sensitive to major malformation induction during organogenesis than it is to central nervous system (CNS) cell depletion and disorganization during midgestation based on the epidemiological data from Hiroshima and Nagasaki. There is neither basic science data nor epidemiological information for this conclusion. Based on animal data, one would expect malformations to occur in the pregnant women in Hiroshima and Nagasaki with (1) embryos aged 18–40 days and (2) exposures of 0.5–1.5 Gy. The incidence of malformations would be expected to be only minimally increased in the 0.2–0.5 Gy group and not measurably increased in the group receiving less than 0.2 Gy. Any woman receiving substantial doses of radiation (more than 1.5 Gy) would have a great likelihood of aborting.

Thus, it is not unexpected that pregnant women exposed to the atomic bomb or receiving radiation therapy to the abdominal region during the period of organogenesis would not manifest a high incidence of major malformations in their exposed children. There are few, if any, clinical situations where a pregnant woman will receive a teratogenic dose of radiation during organogenesis. In Hiroshima and Nagasaki, one can estimate that the susceptible population for teratogenesis was extremely small and the bulk of the surviving patients were in mid- or late gestation, a time when the LD_{50} is much greater.

Therefore, it is appropriate to assume that the animal data are applicable to human embryos. Thus, one would conclude that (1) there is a threshold exposure for major malformations during the period of organogenesis that is $\sim$ 20 cGy; (2) as the exposure increases, the risk of malformation increases, reaching greater than 50% risk above 1 Gy, but the risk of embryonic loss is substantial in this dose range. Peak sensitivity occurs between days 18 and 30, falling rapidly over the next 10 days; and (3) with acute exposures above 2 Gy, death is the most likely result. As an added note, it is possible that the LD_{50} radiation dose may have been lowered substantially in Hiroshima and Nagasaki because of other burdens on the pregnant women (e.g., malnutrition, physical injury, typhoid fever).

There appears to be some controversy about the ability of the preimplanted embryo to be malformed by irradiation. The preimplanted embryo is easily killed by radiation, with a measurable increase in embryonic loss in the 10–15 cGy range. The embryonic loss is due to chromosomal damage, and therefore the embryos die from disorganization. Thus, it is not that the embryos are not malformed, but rather that the malformed embryos do not survive to term. Therefore, exposures of preimplanted embryos to any dose of irradiation present little risk of having a surviving malformed embryo at term.

C. Growth Retardation

There is unanimity of opinion that growth retardation can be produced over a broad range of mammalian gestation in experimental animals and humans. It was Rugh *et al.* (1964) who demonstrated that permanent growth retardation was most severe with a midgestation exposure in the mouse. This is also supported by the Hiroshima and Nagasaki data. Thus, while embryos during early organogenesis can be made smaller at term, they are not as growth retarded at maturity as are animals exposed to the same dose at midgestation (12–15 days in the mouse; 60–140 days in the human). Lastly, it is apparent that mammalian embryos that survive irradiation during the preimplantation period are not growth retarded at term or at maturity as are those irradiated at midgestation.

D. Embryolethality

Although irradiation of the preimplanted embryo does not result in either permanent growth retardation or malformations in surviving embryos at term, it does exhibit the lowest LD_{50}. In fact, an increase in mortality can be observed at term in the rat or mouse following 10–15 cGy of X-irradiation on the first day of gestation. Documentation of this phenomenon has not been accomplished in the human and it is very unlikely that it will ever be documented.

On the first few days of development, the most sensitive pregnancy test now available is not sensitive enough to consistently diagnose pregnancy. More importantly, most human exposures to diagnostic X ray do not exceed 10 cGy, and therefore one might not expect an increased mortality in a population exposed to diagnostic X ray during the preimplantation period. While we believe that the preimplanted human embryo is the stage most sensitive to the lethal effects of irradiation, we will probably never be able to document this conclusion in the human population.

E. Sterility

Laboratory investigations have been directed along two avenues. Some investigators have been interested in the functional capacity of an organism to sire or deliver offspring. Other investigators have attempted to predict the presence or absence of sterility on the basis of histologic interpretation, but this is difficult to accomplish. The response of the embryonic male and female gonad to irradiation will be discussed separately.

1. Males

Radiation exposure of the male rat embryo and fetus has yielded both interesting and unpredictable results. Exposure of the 9-day-old rat embryo to 1 Gy results in a low frequency of severe testicular hypoplasia or aplasia. Only about 3% of the animals will be so affected and these animals are sterile (Brent, 1960b). The mechanism of testicular aplasia is not known. Testicular hypoplasia or aplasia is not produced in the rat during the late organogenetic or early fetal period, but appears again following irradiation on days 17–22 of rat gestation (Erickson et al., 1963; Murphree and Pace, 1960). As little as 0.5 Gy diminished testicular size from 3.5 to 2.5 g, whereas 1.5 Gy resulted in testicles that weighed only 0.5 g. Although maximum reduction in testicular weight occurred in animals irradiated from days 19 to 22, minimal mating activity occurred in animals irradiated from days 18 to 20. Male fetuses that received 1.5 Gy on day 18, 19, or 20 were sterile as adults and made no attempt to mate. Brown (1964) exposed pregnant rats to various doses of daily radiation for 20 hr each day. The rats received 0.02–0.50 Gy/day throughout their intrauterine life. All male animals receiving 0.1, 0.2, or 0.5 Gy/day were sterile. A second study performed by Brown limited the daily irradiation from day 15 of pregnancy to day 23 postpartum to 0.01, 0.02, 0.05, 0.1, 0.2, and 0.4 Gy/day. The three highest dosage groups of male offspring were sterile from total doses of 3, 6, and 12 Gy fractionated irradiation. Ershoff and Brat (1960) also noted significant testicular injury following 3 Gy of acute radiation exposure on day 18 (17.92 cGy/min) and less damage from 3 Gy (0.3 mGy/min) administered from days 13 to 20.

Radiation exposure of neonatal male rats also resulted in permanent sterility, provided that the exposure of acute radiation was 3 Gy or greater (Petrosyan and Pereslegin, 1962). If the young rat is irradiated on day 18 or 17 postnatally with 0.5, 1.5, 3, or 5 Gy (1 Gy/min), testicular damage can be produced in only 3% of the seminiferous tubules of 18-day-old animals given 3 Gy (Leonard *et al.*, 1964). It appears, therefore, that following parturition, the rat testes rapidly lose the radiosensitivity demonstrated during the last few days of pregnancy.

Erickson *et al.* (1963) exposed pregnant pigs at various stages of gestation to 4 Gy. Organization of the sex cords does not begin until day 50 of gestation in the male pig fetus, yet 12% of male fetal pigs irradiated with 4 Gy prior to this time were sterile as adults. After day 50 of gestation, the incidence of sterility rose sharply (animals irradiated at 50 days had a 50% incidence of sterility). Sterility rose to 70% and 75% after irradiation of 55- and 75-day-old male pig fetuses, respectively. Pregnant goats were exposed on days 13, 40, and 50 of gestation to 3 Gy. The male offspring were not tested for fertility, but there was a 32% reduction in sperm count, 26% less sperm activity, and an increase of 133% in abnormal sperm (O'Brien *et al.*, 1966).

Radiation-induced sterility in the female mouse has been studied more than male sterility. Rugh and Wohlfromm (1964a,b) exposed 10.5- to 12.5-day-old mice to 1 Gy, but did not observe sterility in the males. Although there was some reduction in the number of matings, the litters sired were normal in size. Petrosyan and Pereslegin (1962) reported that 3 Gy administered to neonatal male rats sterilized them, but Rugh and Wohlfromm (1964a,b) found that 4 Gy to the neonatal male mouse did not permanently sterilize. In fact, they administered 1 Gy on each day of gestation or 4 Gy to sexually maturing mice and concluded that these doses ''are not sufficient to seriously affect the reproductivity of the male over their normal reproductive period.'' The specific state of maximal sensitivity for sterilization occurs during the latter part of mouse gestation on day 16 (Rugh and Jackson, 1958). Laskey *et al.* (1973) observed that male rats exposed to up to 3 cGy/day from tritiated water from the time of conception were fertile as adults, although they exhibited small testes.

2. Females

The effects of radiation on the developing ovary are more fascinating and even less predictable than those reported for the male embryo and neonate.

Beaumont (1962) exposed pregnant rats at 13.5, 15.5, 17.5, or 19.5 days postcoitum to 1 Gy and then determined the population of oocytes in the female offspring at 100 days of age. Animals irradiated at the 15.5-day stage with 1 Gy had the greatest depopulation of oocytes, namely, 241 rather than the 5025 in control rats.

These findings help explain the results of Ershoff and Brat (1960). They reported that 3 Gy protracted radiation exposure from days 13 to 20 was more effective than one acute dose of 3 Gy on day 18. This is understandable, since day 18 of gestation is not the day when the ovary is the most sensitive to radiation. The course of protracted radiation exposure straddled the most sensitive period for the ovary in the rat, day 15.5 (Beaumont, 1962). Although sterility was not tested, there were histologic changes in the gonad (Beaumont, 1962).

In pregnant Wistar rats exposed to tritiated water and tritiated thymidine continuously through pregnancy, total postnatal oocyte counts were reduced to 50% by 1450 μCi of tritiated water (Haas et al., 1973). Total aplasia of the ovaries was produced by 5800 mCi. Tritiated thymidine was 10 times more effective than tritiated water in its effect on the developing ovary. Brown (1964) compared the effects of low-dose protracted radiation exposure in the male and female rat. After 11 generations of irradiation, the rats exposed to 2 cGy/day (20 hr/day) exhibited a decrease in litter size. The male rat was more sensitive than the female predominantly due to the highly radiosensitive period in the testes during the last few days of gestation (Langendorff and Neumann, 1972).

Irradiation of the ovaries has resulted in some interesting discoveries. If large doses of radiation (2–4 Gy) are administered to one ovary 3–4 days before conception, the unirradiated ovary will ovulate an increased number of eggs to make up for the reduction in number of eggs ovulated from the irradiated ovary (Hahn and Feingold, 1973). If one ovary is irradiated with 45 Gy and the other ovary removed on the fourth day of pregnancy, the size of the implantation sites and the number of live embryos will be reduced. This effect can be reversed by administration of 17 β-estradiol and progesterone (Gibbons and Chang, 1973).

The mouse presents a different picture. First of all, the female mouse can be sterilized by a dose of radiation below 1 Gy and is therefore much more readily sterilized than the females of most other species or the male mouse. Second, the mouse ovary has its greatest sensitivity in the neonatal period rather than *in utero*. Rugh and Wohlfromm (1964a,b) tested the fertility of mice exposed to 1 Gy from the time of conception to the age of 2 months postpartum. Exposure to 1 Gy on any day of mouse gestation from fertilization to day 18 did not result in sterile females, although the litter size of adult females that were irradiated on day 13 of gestation was smaller (Rugh and Wohlfromm, 1966). An increase in ovarian cyst formation was noted in adult mice irradiated on days 12 and 13 of gestation. Only 44% of the mice irradiated as newborns were fertile. Mice 1-, 2-, and 3-weeks old receiving 1 Gy were sterile, whereas those receiving the same exposure at 1 and 2 months had 15% and 14% fertility, respectively. Sterility was not produced by 30 cGy given to the newborn mouse or 1- to 3-week-old mice. In fact, 30 cGy to the newborn mouse did not decrease fertility, but the fertility of 1- to 3-week-old mice was decreased (Rugh and Wohlfromm, 1964b).

The explanation for these findings was provided by Peters (1961), Peters and Levy (1964), Russell and Oakberg (1963), and others. On the one hand, mice receiving 20 cGy on the day of birth had 50% of their small oocytes destroyed. On the other hand, the 1-, 2-, and 3-week-old mice that received 20 cGy had 95%, 98%, and 100%, respectively, of their small oocytes destroyed. The exact mechanism of oocyte killing is not understood, but the cells die shortly after radiation exposure, resulting in partial destruction of the pool supplying future mature ova. Although 50% of the small oocytes are killed in the neonatal mouse, there are 85% of the normal number of oocytes in the ovary at maturity. On the other hand, a 3-week-old mouse that receives 20 cGy will have only 1% of the normal number of oocytes at maturity. The change in oocyte sensitivity is believed to be due to the synchrony of mouse oocyte development and the transformation of oocytes of the newborn in the early diplotene stage to the highly sensitive late diplotene and pachytene stages in the 1- to 3-week-old mice.

During this sensitive period, the oocyte also is very sensitive to changes in dose rate. Female mice 10-days old exposed to 25 cGy at a dose rate of 0.09 mGy/min had 10 times as many surviving oocytes as those at 2.85 Gy/min (Rugh and Wohlfromm, 1965). This indicates that there are repair processes in the oocyte. Russell *et al.* (1959) exposed pregnant mice to 2 Gy protracted over the entire period of gestation and noted a slight reduction in adult female fertility (Table II).

Pregnant goats exposed to 3 Gy on day 13 of gestation had a normal number of oocytes as adults (O'Brien *et al.*, 1966). Fractionation of the 3-Gy exposure, with exposures at 30, 40, and 50 days, resulted in a decrease in the number of follicles in the ovaries of the surviving goats to 59% of the controls.

Apparently, the female rhesus monkey is more resistant to the sterilizing effects of radiation than other mammalian species, since the minimal dose to sterilize the monkey fetus is ~ 20 Gy. Although the primary oocytes disappeared within 2 days following 1.4–3 Gy in the mouse and rat at the primordial stage, the monkey needed 7 Gy to produce the same results (Baker, 1970; Baker and Beaumont, 1967). The radiation resistance of the monkey oocytes may be due to the asynchronous development of the germ cells.

There are few data on the effect of radiation *in utero* on the fertility of the offspring in the human. The large number of survivors in Hiroshima and Nagasaki exposed *in utero* have not lived long enough to have complete fertility histories. In 2345 women who had been exposed to irradiation at Hiroshima and Nagasaki, there was no observable reduction in fertility, even when the dose was

TABLE II

Appropriateness of the Utilization of Animal Models (Primarily Rodents) to study *in Utero* Radiation Effects

Effect	Qualitative evaluation	Quantitative evaluation
Embryonic lethality		
Preimplantation	Yes	Yes
Organogenesis	Yes	Yes
Fetal period	Yes	Yes
Malformations		
Organogenesis	Yes	Yes
Permanent growth retardation		
Preimplantation	Yes	Yes
Organogenesis	Yes	Yes
Fetal period	Yes	Yes
Sterility		
Female	No	No
Male	Yes	?
Indirect effects of irradiation	Yes	?
Tumorigenesis	Yes	No
Shortening of life span	Yes	No
Central nervous system		
Microcephaly	Yes	Yes
Mental retardation	No	No
Neurophysiological measurements	Yes	?
Threshold exposures	Yes	Yes and no, depending on parameter

greater than 1 Gy (Blot and Sawada, 1972). This population included children and adult women, and none were *in utero* at the time of exposure.

Mondorf and Faber (1968) studied the fertility of 180 females who had been examined fluoroscopically for intussusception during their infancy. They found no difference between the number of children born and the age distribution of births in the controls and the exposed girls. The gonadal exposure was estimated to be 1–5 cGy. Tabuchi (1964) and Tabuchi *et al.* (1967) reviewed the data of the Atomic Bomb Casualty Commission in Japan and found that menarche was delayed 1–3 months in girls exposed to high doses of radiation while *in utero*.

It is evident that it is quite difficult to generalize about the effects of radiation *in utero* or during the neonatal period on subsequent sterility. There are marked sex and species differences in the response to radiation (Baker, 1970). However, from the data on both humans and animals we have accumulated to date, we can tentatively conclude that acute doses below 25 cGy absorbed by the fetus during gestation will not result in sterility in the human male or female.

F. Shortening of Life Span and Tumorigenesis

Stewart *et al.* (1956, 1958) suggested that intrauterine radiation exposure from diagnostic radiation is associated with an increase in leukemia. Numerous reports and analyses of this association have been reported (Brent, 1983, 1984). While it is a relatively low-risk association (5×10^{-4}/cGy), it is substantially greater than the risk of leukemia in the population of irradiated adults (10^{-6}/cGy/year or 10^{-5}/cGy/10 years). If this association of *in utero* irradiation and leukemia is both accurate and causal, then it means that the embryo is 50 times more sensitive than the adult to the leukemogenic effects of X rays. We may never be able to resolve whether this is a selection problem (a higher risk population is exposed to diagnostic radiation) or whether the human embryo is truly very sensitive to the leukemogenic effects of X rays.

Animal experiments have not revealed that the embryo is more sensitive than the adult to X ray-induced tumorigenesis (Brent and Bolden, 1961; Rugh *et al.*, 1966; Brent, unpublished). Because of the differences between carcinogenesis in the mouse and human, animal models may be much less appropriate to investigate this question than other radiation embryological questions. The mouse has proven viral etiologies for some of its tumors, has vertical transmission of these agents, and usually consists of experimental groups that are genetically homogenous. All we can say is that animal models, to this date, do not support the allegation that the embryo is more sensitive to the leukemogenic effects of X rays than the adult. The only longevity study dealing with life-span analysis and intrauterine irradiation of which we are aware was recently completed in our laboratory (Brent, unpublished). This study does not indicate that mouse embryos which survive the neonatal period have a decreased life expectancy.

In spite of the controversy in this area and the inconsistency between human and animal data, the application of this information does not present a problem when establishing maximum permissible exposures for women of reproductive age because the risk under the worst situation is relatively small. In fact, the radiation risk is actually smaller than Stewart *et al.* (1956, 1958) reports (5×10^{-4}/cGy), since this risk is actually the spontaneous risk of leukemia combined with the radiation-associated risk. Since the spontaneous risk is 3.3×10^{-4}/cGy per 10 years (childhood), the radiation-associated risk is $\sim 1.7 \times 10^{-4}$/cGy per 10 years (childhood), or 1/6000. Thus, the maximum risk for leukemia from *in utero* irradiation is extremely small and would not change the present maximum permissible exposure for pregnant women or the parameters for counseling women exposed to diagnostic irradiation.

G. Central Nervous System Effects

The CNS exhibits a long period of sensitivity to ionizing irradiation from conception into the neonatal period. Even during the neonatal period, very high exposures (4–5 Gy) can result in mental retardation. It has been known for a long time that exposures above 50 cGy present a significant risk to the developing nervous system (Blot and Miller, 1973; Dekaban, 1968; Goldstein and Murphy, 1929; Miller and Mulvihill, 1976; Wood et al., 1967; Yamazaki et al., 1954; Zappert, 1926). The recent work of Otake and Shull (1984) indicates that the most sensitive period for severe mental retardation is from week 8 to week 15 of gestation, a time of neuronal cell proliferation and migration.

In reality, each period of gestation manifests CNS pathology: (1) During the preimplantation period, irradiation results in a high mortality, but the survivors have normal growth potential and appear morphologically normal in rodent species (Brent and Bolden, 1967a,b, 1968; Russell, 1954; Schlesinger and Brent, 1978). Yet, if one examines the dead and dying zygotes, death is primarily the result of chromosome damage. Furthermore, these embryos die primarily during early organogenesis. Analysis of embryos dying early from unbalanced translocations reveals that most of them die from CNS maldevelopment and disorganization. Therefore, it is very likely that although preimplantation radiation does not produce a high incidence of malformations at term, it is likely that most of the early deaths are due to CNS disorganization. (2) While the risk of mental retardation is low during the period of early organogenesis, the risk of severe anatomical malformations is high in the 0.50–1.5 Gy range and during the period of 18–30 days of human gestation (embryonic age). The malformations most likely to be produced at this stage are hydrocephaly, anencephaly, encephalocele, and spina bifida. As mentioned previously, these are unlikely malformations to occur with a single random exposure occurring during pregnancy because no increase in malformations would be expected below 0.2 Gy, and exposures above 1.5 Gy are most likely to kill the embryo. Therefore, epidemiological studies dealing with diagnostic radiation or therapeutic radiation would be unlikely to reveal these effects in surviving human embryos. (3) Severe anatomical malformations are less likely to occur in the period from days 30 to 50 because the neural tube is closed and the embryo still maintains a significant capacity to replace damaged cells. It has not been determined whether the transition into the period of radiation sensitivity with regard to interference with cell proliferation and cell migration is gradual or abrupt. However, from ~60 to 130 days, the embryo's ability to repair radiation damage which interferes with these processes is reduced and, therefore, severe microcephaly and mental retardation can result (Mole, 1982; Otake and Schull, 1984).

Although animal experiments can be utilized to corroborate the midgestation effects in the human, it is unlikely that we would have been able to predict the nature of these effects in the human from animal experiments. On the other hand, we have *no* human data to support the animal findings pertaining to the pre-implantation period and the period of early organogenesis. However, since the nature of early human embryonic development is more similar to other mammals than is its fetal development, it is very likely that the mouse and rat data can be applied qualitatively and quantitatively to the preimplanted and embryonic phases of the human.

There are many methods available to study behavior and mentation following the irradiation of rodents and monkeys, but it is difficult to apply these results to the human species. We have studied nine behavioral and developmental parameters in postnatal rats that had been irradiated *in utero* on day 9 or 17 of development (Jensh and Brent, unpublished): (1) pinna detachment, (2) eye opening, (3) testes descent, (4) vaginal opening, (5) surface righting, (6) negative geotaxis, (7) auditory startle, (8) air righting, and (9) visual placing.

As expected, rats irradiated on day 9 (prior to organogenesis) exhibited no developmental abnormalities, whereas rats irradiated on day 17 manifested some developmental abnormalities (Table III). While it was possible to ascertain some developmental delay in rat embryos irradiated on day 17, the radiation effect on reflexes and development was no more sensitive than the radiation effect on growth in postpartum animals. Furthermore, growth as well as developmental effects exhibited a threshold dose below which no effect was observed.

The work currently being done at the Radiation Effects Research Foundation dealing with the effect of midgestation exposure on neurophysiological parameters (Otake and Schull, unpublished) will be watched with great interest, since it may give us an indication of the range of neurological deficits that occur in adult humans who have been irradiated *in utero,* although other environmental factors

TABLE III

EFFECT OF *in Utero* IRRADIATION ON GROWTH, REFLEXES, AND DEVELOPMENTAL PARAMETERS

| | Gestational day of irradiation | | | | | | | |
| | Day 9 | | | | Day 17 | | | |
Effect	0.1[a]	0.2	0.4	0.6	0.1	0.2	0.4	0.6
Growth retardation at term	−	−	−	−	−	−	−	−
Growth retardation postpartum	−	−	−	−	−	+	+	+
Developmental parameters (4)	−	−	−	−	−	−	+(2)[b]	+(2)
Reflexes (5)	−	−	−	−	−	+(1)	+(1)	+(1)

[a] Dose of X ray (grays).

[b] Numbers in parentheses are the number of parameters tested having a positive effect.

coexisted with this radiation. This work may also answer an important question regarding the existence of a threshold dose for these neuropathological effects, since all teratological and embryopathological effects that have been studied in the past have been threshold phenomena.

III. Appropriateness of Animal Studies for Determining Human Intrauterine Radiation Risks

As stated in the original thesis of this article, animal studies can be used to confirm a finding in humans, to study mechanisms, and to determine the quantitative risk of the intrauterine radiation exposure. Table II lists the radiation risks that are known or alleged and the appropriateness of animal studies for studying these effects. Most effects can be readily studied and applied, but there are exceptions: sterility, shortening of life span, tumorigenesis, and many aspects of CNS development. This is not to say that important information cannot be gained from these studies, but there is a greater potential for both qualitative and quantitative errors because the above-mentioned processes and functions are significantly different in the rodent and in man. In some instances, the use of primates as experimental subjects may solve the problem.

IV. The Risk of Intrauterine Irradiation in the Human and the Contribution of Animal Studies: Summary

While it is true that animal studies cannot be used by themselves to predict human disease following *in utero* radiation, they can be utilized for studying the mechanisms of underlying radiation effects in embryos in humans and the risk associated with some of these effects at various exposures. There are certain radiation effects on embryos that cannot be evaluated in these ways because of the marked discordance in development or function between the rodent and human. The effects in question pertain to intrauterine radiation effects dealing with sterility, tumor development, life-span studies, and certain aspects of CNS functions.

Although some investigators emphasize that one period of gestation is more sensitive than another, this is probably not the case, at least with regard to the first two-thirds of the gestation period. All stages of gestation exhibit pathological effects following intrauterine radiation. While these effects are markedly different, every stage will be affected by 0.5 Gy: death in the preimplantation period, major malformations during early organogenesis, minor malformations during later organogenesis, histogenic depletion (growth retardation) and disorganization and cell depletion in midgestation, and cell depletion during the later part of gestation. Surprisingly, the threshold dose for these irreparable effects is quite similar, namely, about 0.2 Gy except for late gestation when permanent effects may not be produced at this low dose. Finally, since all radiation effects in embryos are multicellular phenomena and can reach a frequency of 100%, it is unlikely that they are stochastic phenomena (Brent, 1983, 1984). This means that the risks are not linearly related to radiation dose. The only possible exception is the lethal effect produced on the first day of gestation. If, however, death at this stage is due to multiple chromosomal breaks and failure of repair, this effect may not be linearly related to dose.

Based on the information available, it should be apparent that the present maximum permissible exposure for women of reproductive age in the work place (0.5 cGy maximum permissible exposure) is quite appropriate. The risk of 0.5 cGy per year for women working in the radiation industry presents a risk several orders of magnitude below the expected human risks of reproduction (Tables I, IV–VII). These risk estimates are based on available human experience of intra-

TABLE IV

A COMPILATION OF THE EFFECTS OF 1 Gy OF ACUTE RADIATION ON EMBRYONIC DEVELOPMENT AT VARIOUS STAGES OF GESTATION IN RAT AND MOUSE

		Stage of gestation (days)			
		Organogenesis			Fetal stages
Feature	Preimplantation	Early	Late	Histogenesis	
Mouse	0–4.5	6.5–8.5	8.5–12	12–15	15–18
Rat	0–5.5	8–10	10–13	13–17	18–22
Corresponding human gestation period	0–9	18–36	36–60	60–130	130–280
Lethality	+++[a]	++	+−	+−	−
Growth retardation					
At term	−	+++	++	++	+
As adult	−	+	+	+++	++
Gross malformations	−	+++	+	−	−
Cell depletions	−	+−	++	+++	+
Sterility	−	+−	−	+−	++
Increased germ cell mutations	+−	+−	+−	+−	+−
Cytogenetic abnormalities	+−		+		+
Cataracts	−	+	+	+	+
Neuropathology	−	+	+	+++	++
Tumor induction	−	+−	+−	+−	+−
Behavior disorders	−	+−	+−	+−	+−
Reduced life span	−	−	−	−	−

[a] −, No observed effect; +−, questionable but reported or suggested effect; +, demonstrated effect; ++, readily apparent effect; +++, occurs in high incidence.

TABLE V

Estimation of the Risks of Radiation in the Human Embryo Based on Human
Epidemiological Studies and Mouse and Rat Radiation Studies

Embryonic age (days)	Minimum lethal dose (cGy)	Approximate LD$_{50}$ (cGy)	Minimal dose[a] (cGy) for		
			Permanent growth retardation	Mental retardation	Gross malformation
1–5	10	<100	No effect in survivors		
18–36	25–50	140	20–50	—	20
36–50	50	200	25–50	—	50
50–150	>50	>100	25–50	50[b]	—[c]
To term	>100	Same as mother	>50	100	—[c]

[a] The minimal dose to induce genetic, carcinogenic, and cell depletion phenomena is unknown.

[b] Otake and Shull suggest an increased risk at lower exposures.

[c] Gross malformations cannot be produced this late in gestation except in the genitourinary system and as tissue hypoplasia in specific organs such as brain and testes as well as growth retardation.

uterine irradiation and the appropriate use of animal studies, permitting us to conclude that, while it is impossible to state categorically that 0.5 cGy or 5 cGy represents absolutely no risk, the estimated risks are at least several orders of magnitude below the spontaneous or usual reproductive risks (Table VI). Likewise, exposures from diagnostic radiation below 5 cGy present such a small or nonmeasurable risk that radiation risk counselors can support the continuation of wanted pregnancies.

Lastly, inadvertent diagnostic exposures during the menstrual cycle present no greater concern whether in the first or second half of the cycle. Planned,

TABLE VI

Risk of 0.5 cGy (0.5 rem) Maximum Exposure for Women
Radiation Workers with Reproductive Potential

Risk	Background exposure	0.5 cGy exposure
Spontaneous abortion	150,000/10^6	0
Major anatomical malformation	30,000/10^6	0
Severe mental retardation	5000/10^6	0
Childhood leukemia in first 10 years of offspring's life	350/10^6	
Based on Stewart *et al.* (1956, 1958)		166/10^6
Based on ABCC data		2.5/10^6

TABLE VII

Estimate of Risks of 1 cGy Exposure (Low LET) to the Developing Human Embryo

Age (days)	Mutagenic effect[a]	Childhood carcinogenic effect		Gross malformations, death, growth retardation	Permanent cell depletion
		Stewart[b]	ABCC[c]		
1	No data	No data	No data	?[d]	No effect
18 to term	10^{-7} per locus	3.2×10^{-4}	5×10^{-6}	Same as controls	_?

[a] Based on the adult risk and as estimated doubling dose for mutagenesis of 1 Gy, assuming a linear dose–response curve and no threshold for mutagenic effects.

[b] The data of Stewart *et al.* (1956, 1958) would indicate that the embryo is more sensitive than the adult, but this association may be other than a radiation effect.

[c] The Atomic Bomb Casualty Commission data on carcinogenesis indicate no increased risk to the developing human, so the risk presented is the adult risk, assuming a linear dose–response curve.

[d] With the possible exception of preimplantation embryolethality, present data indicate a stochastic relationship with a threshold of $\sim$ 20 cGy.

medically necessary radiographic examinations should be performed whether the woman is in the first or second half of the cycle, since preovulation exposures or postconception exposures before the first missed menstrual period represent similar minimal risks following an exposure to 5 cGy or less. It follows that this continuation of the 10-day rule (confining all medical diagnostic exposures in women of reproductive age to the first portion of the menstrual cycle) is inappropriate, since this would tend to concentrate exposures of women in the preovulatory phase, thus increasing the population's total genetic exposure to radiation just before conception (National Council on Radiation Protection and Measurement, 1977).

ACKNOWLEDGMENT

Supported by NIH HD/CA 19165.

REFERENCES

Baker, T. G. (1970). *Mutat. Res.* **11,** 9–22.
Baker, T. G., and Beaumont, H. M. (1967). *Nature (London)* **214,** 981.
Beaumont, H. (1962). *J. Endocrinol.* **24,** 325–339.
Blot, W. J., and Miller, R. W. (1973). *Radiology* **106,** 617–619.
Blot, W. J., and Sawada, H. (1972). *Am. J. Hum. Genet.* **24,** 613–622.
Brent, R. L. (1960a). *Am. J. Dis. Child* **100,** 103–108.
Brent, R. L. (1960b). *Clin. Obstet. Gynecol.* **3,** 928–950.
Brent, R. L. (1964). *J. Pediatr.* **64,** 762.

Brent, R. L. (1970). *In* "Late Effects of Radiation" (R. M. Fry, D. Grahn, and M. L. Griem, eds.), pp. 23–60. Taylor & Francis, London.

Brent, R. L. (1972a) *J. Clin. Pharmacol.* **12,** 61–70.

Brent, R. L. (1972b). *In* "Davis' Gynecology and Obstetrics" (J. J. Sciarra, ed.), Vol. 2, pp. 1–32. Harper & Row, New York.

Brent, R. L. (1976). *In* "Prevention of Embryonic, Fetal and Perinatal Disease" (R. L. Brent and M. I. Harris, eds.), pp. 211–218. DHEW, (NIH) 76–853, Bethesda, Maryland.

Brent, R. L. (1977). *In* "Handbook of Teratolgy" (J. G. Wilson and F. C. Fraser, eds.), Vol. 1, pp. 153–223. Plenum, New York.

Brent, R. L. (1978). *Teratology* **17,** 183.

Brent, R. L. (1980a). *Teratology* **21,** 281–298.

Brent, R. L. (1980b). *In* "Controversies in Therapeutics" (L. Lasagna, ed.), pp. 134–150. Saunders, Philadelphia.

Brent, R. L. (1983). *Clin. Obstet. Gynecol.* **26,** 484–510.

Brent, R. L. (1984). *Teratology* **29,** 20A.

Brent, R. L. (1985a). *Teratology* **31,** 429–430.

Brent, R. L. (1985b). *In* "Prevention of Physical and Mental Congenital Defects, Part A: The Scope of the Problem" (M. Marois, ed.), pp. 55–68. Liss, New York.

Brent, R. L. (1985c). *In* "Prevention of Physical and Mental Congenital Defects, Part C: Basic and Medical Science, Education, and Future Strategies" (M. Marois, ed.), pp. 191–195. Liss, New York.

Brent, R. L., and Bolden, B. T. (1961). *Radiat. Res.* **14,** 453–454.

Brent, R. L., and Bolden, B. T. (1967a). *Radiat. Res.* **30,** 759–773.

Brent, R. L., and Bolden, B. T. (1967b). *Proc. Soc. Exp. Biol. Med.* **125,** 709–712.

Brent, R. L., and Bolden, B. T. (1968). *Radiat. Res.* **36,** 563–570.

Brent, R. L., and Gorson, R. O. (1972). *In* "Current Problems in Radiology" (R. D. Moseley, Jr., D. H. Baker, R. O. Gorson, A. Lalli, H. B. Latourette, and J. Quinn, III, eds.), Vol. 2, pp. 1–48. Medical Publ., Chicago.

Brent, R. L., and Harris, M. (1976). "The Prevention of Embryonic, Fetal, and Perinatal Disease." U.S. Gov. Printing Office, Washington, D.C.

Brent, R. L., and McLaughlin, M. M. (1960). *Am. J. Dis. Child.* **100,** 94–102.

Brown, S. (1964). *Genetics* **50,** 1101–1113.

Dekaban, A. (1968). *J. Nucl. Med.* **9,** 471–477.

Erickson, B., Murphree, R., and Andrews, J. (1963). *Radiat. Res.* **20,** 640–648.

Ershoff, B., and Brat, V. (1960). *Am. J. Physiol.* **198,** 1119–1122.

Gibbons, A. F. E., and Chang, M. C. (1973). *Biol. Reprod.* **9,** 343–349.

Goldstein, L., and Murhpy, D. P. (1929). *Am. J. Obstet. Gynecol.* **18,** 189–195, 281–283.

Haas, R. J., Schreml, W., Fliedner, T. M., and Calvo, W. (1973). *Int. J. Radiat. Biol.* **23,** 603–609.

Hahn, E. W., and Feingold, S. M. (1973). *Radiat. Res.* **53,** 267–272.

Langendorff, H. U. M., and Neumann, G. K. (1972). *Strahlentherapie* **144,** 324–337.

Laskey, J. W., Parrish, J. L., and Cahill, D. F. (1973). *Radiat. Res.* **56,** 171–179.

Leonard, A., Imbaud, F., and Maisin, J. (1964). *Br. J. Radiol.* **37,** 764–768.

Miller, R. W., and Mulvihill, J. J. (1976). *Teratology* **14,** 355–358.

Mole, R. H. (1982). *Int. J. Radiat. Biol.* **42,** 1–12.

Mondorf, L., and Faber, M. (1968). *J. Reprod. Fertil.* **15,** 165–169.

Murphree, R., and Pace, H. (1960). *Radiat. Res.* **12,** 495–504.

National Academy of Science (1980). "National Academy of Science Committee on the Biological Effects of Ionizing Radiation." National Research Council Publication, Washington, D.C.

National Council on Radiation Protection and Measurements (1977). Report No. 54: Medical radiation exposure of pregnant and potentially pregnant women, pp. 1–21. NCRP Publ., Washington, D.C.

O'Brien, C. A., Hupp, E. W., Sorensen, A. M., and Brown, S. O. (1966). *Am. J. Vet Res.* **27,** 711–721.

Okada, S., Hamilton, H. B., Egami, N., Okajima, S., Russell, W. J., and Takeshita, K. (1975). *J. Radiat. Res. Tokyo* **16**(Suppl.), 1.

Otake, M., and Schull, W. J. (1984). *Br. J. Radiol.* **57,** 409–414.

Peters, H. (1961). *Radiat. Res.* **15,** 582–593.

Peters, H., and Levy, E. (1964). *J. Reprod. Fertil.* **7,** 37–45.

Petrosyan, S., and Pereslegin, I. (1962). *Med. Radiol.* **5,** 38–45 (in Russian).

Rugh, R., and Jackson, S. (1958). *J. Exp. Zool.* **138,** 209–221.

Rugh, R., and Wohlfromm, M. (1964a). *Atompraxis* **10,** 511–518.

Rugh, R., and Wohlfromm, M. (1964b). *Atompraxis* **10,** 33–42.

Rugh, R., and Wohlfromm, M. (1965). *Radiat. Res.* **26,** 493–506.

Rugh, R., and Wohlfromm, M. (1966). *Fertil. Steril.* **17,** 396–410.

Rugh, R., Duhamel, L., Osborne, A. W., and Varma, A. (1964). *Am. J. Anat.* **115,** 185–198.

Rugh, R., Duhamel, L., and Skaredoff, L. (1966). *Proc. Soc. Biol. Med.* **121,** 714–718.

Russell, L. B. (1954). *In* "Radiation Biology" (A. Hollaender, ed.), Part II. McGraw-Hill, New York.

Russell, L. B., Badgett, S. K., and Saylors, C. L. (1959). *In* "Immediate and Low Level Effects of Ionizing Radiation" (A. A. Buzzati-Traverso, ed.), pp. 343–359. Taylor & Francis, London.

Russell, W. L., and Oakberg, E. F. (1963). *In* "Cellular Basis and Aetiology of Late Somatic Effects of Ionizing Radiation" (R. J. C. Harris, ed.), pp. 224–232. Academic Press, New York.

Schlesinger, D. M., and Brent, R. L. (1978). *Radiat. Res.* **75,** 202–216.

Stewart, A., Webb, D., Giles, D., and Hewitt, D. (1956). *Lancet* **2,** 447.

Stewart, A., Webb, D., and Hewitt, D. (1958). *Br. Med. J.* **1,** 1495–1508.

Tabuchi, A. (1964). *Hiroshima J. Med. Sci.* **13,** 125–173.

Tabuchi, A., Nakagawa, S., Hirai, T., Sato, H., Hori, I., Matsuda, M., Yano, K., Shimada, K., and Nakao, Y. (1967). *Hiroshima J. Med. Sci.* **16,** 49–66.

Wilson, J. G., and Brent, R. L. (1981). *Am. J. Obstet. Gynecol.* **141,** 567–580.

Wood, J., Johnson, K., Omori, Y., Kawamoto, S., and Keehn, R. (1967). *Am. J. Public Health* **57,** 1381–1390.

Yamazaki, J., Wright, S., and Wright, P. (1954). *Am. J. Dis. Child.* **87,** 448–463.

Zappert, J. (1926). *Monoschr. Kinderheilkd.* **34,** 490–493.

Tolerance of the Central and Peripheral Nervous System to Therapeutic Irradiation

STEVEN A. LEIBEL AND GLENN E. SHELINE

DEPARTMENT OF RADIATION ONCOLOGY
UNIVERSITY OF CALIFORNIA, SCHOOL OF MEDICINE
SAN FRANCISCO, CALIFORNIA 94143

I. Introduction

The major limitation to treatment of neoplasms of the brain with radiation therapy is the occurrence of various reactions of injury resulting from the radiosensitivity of normal brain tissue encompassed in the treatment volume. These reactions have been classified into three groups according to the time of appearance: (1) acute reactions occurring during the course of treatment; (2) early delayed reactions occurring a few weeks to 2–3 months after irradiation; and (3) late delayed reactions developing several months to years after treatment (Kramer, 1968; Sheline *et al.*, 1980). The presence and severity of these reactions are governed by a number of factors, including the total radiation dose, the radiation fraction size, and the volume of tissue irradiated. The age of the patient, the presence of preexisting brain injury by tumor or surgery, infection, vascular disease due to hypertension or diabetes mellitus, and chemotherapy also may influence the susceptibility of the patient to injury.

The tolerance of the spinal cord frequently limits the total radiation dose and treatment technique used in the management of extracranial head and neck and thoracic neoplasms as well as in that of tumors involving or surrounding the spinal cord. Radiation myelopathy may take the form of an early delayed reaction (transient radiation myelopathy) or late injury (Kramer and Lee, 1974).

The optic nerves are an extension of the white matter of the central nervous system (CNS). Excessive doses to the optic nerve and retina may result in loss of vision. The cranial and peripheral nerves generally are considered more radioresistant than the brain and spinal cord, and their presence in the treatment field

257

does not restrict the delivery of high-dose irradiation to the same extent as does the presence of the brain and spinal cord. However, injury to these structures may occur, and when it does it is manifested in progressive, late complications (Kinsella *et al.*, 1980).

This article will review the clinical syndromes, pathological features, and time–dose relationships of radiation injury to the brain, spinal cord, optic nerves and chiasm, retina, and to the peripheral nervous system. The clinical features of injury from therapeutic irradiation will be emphasized. Although an extensive literature exists on the effects of ionizing radiation on the nervous systems of experimental animals, most of this information is of little use to clinicians. The doses used in these studies are frequently far below those used in clinical practice or consist of large, single doses which bear little relationship to the fractional doses used in clinical radiation therapy. Because the literature on the tolerance of the central and peripheral nervous system to therapeutic irradiation is largely fragmentary, anecdotal, and often conflicting, firm inferences are difficult to derive. The observations and time–dose relationships presented in this article are based on conventional photon or γ irradiation and are not applicable to other forms of ionizing radiation such as neutrons or heavy particles.

II. Radiation Injury to the Brain

A. Acute Reactions

Acute reactions occur during the course of radiation therapy and are characterized by symptoms suggesting increased intracranial pressure or exacerbation of preexisting neurological symptoms or signs. The pathogenesis of the acute syndrome is thought to be radiation-induced edema (Sheline *et al.*, 1980). However, an increase in peritumoral edema has not been confirmed by computerized tomographic (CT) scans obtained during the course of treatment (Deck, 1980). Acute symptoms may occur, but are uncommon with the conventional dose-fractionation schedules (180–200 rad/day, given 5 days/week to a total dose of 6000 rad to a portion of or to the whole brain) employed in the radiation therapy of brain tumors. Using fraction sizes of 200 rad/day, Salazar *et al.* (1976) delivered total doses of 7000–8000 rad to a limited portion of the brain for the treatment of malignant gliomas. Dexamethasone was administered to the patients throughout the course of treatment, and no significant, acute radiation complications were observed.

Clinical experience using larger daily dose fractions has been derived from the treatment of patients with brain metastases. Over 2000 patients were entered into two multiinstitutional Radiation Therapy Oncology Group (RTOG) studies comparing different time–dose fraction schemes for the palliation of brain metastases. Whole brain irradiation was administered using five different dose schedules, including 5 fractions of 400 rad in 1 week; 10 fractions of 300 rad in 2 weeks; 15 fractions of 267 rad in 3 weeks; 15 fractions of 200 rad in 3 weeks; and 2 fractions of 600 rad in 2 days. Corticosteroid therapy was administered when clinically indicated. An increased incidence of acute reactions was not observed (Kramer et al., 1977; Borgelt et al., 1980). More recently, the RTOG completed studies employing fraction sizes of 500–600 rad given twice weekly for a total dose of 3000 rad with or without the hypoxic cell radiosensitizer, misonidazole (Phillips et al., 1984). Each of these fraction schemes was well tolerated, suggesting that the acute tolerance of larger daily fractions is acceptable provided that total doses are appropriately reduced coincident with the increase in fraction size.

Acute complications have been observed with fraction sizes larger than 600 rad. Young et al. (1974) treated 83 patients with brain metastases using 1500 rad delivered in two fractions of 750 rad each over 3 days. Of the patients, 49% developed acute complications including headache, nausea, vomiting, and pyrexia, and 6 patients developed cerebral herniation. The more severe complications occurred in patients exhibiting signs of increased intracranial pressure prior to the beginning of treatment. Hindo et al. (1970) reported the outcome in 54 patients with cerebral metastases treated with a single fraction of 1000 rad to the whole brain. Four patients died within 48 hr and the fifth died within 7 days of treatment. Those patients who died shortly after treatment had severe neurological impairment before therapy and the contribution of irradiation to the cause of their deaths is uncertain. Borgelt et al. (1981) found no increase in treatment morbidity in 26 patients receiving 1000 rad in one fraction compared with those receiving more protracted schedules of radiation.

B. Early Delayed Reactions

The early delayed reaction occurs a few weeks to a few months after irradiation and is analogous to transient radiation myelopathy following spinal cord irradiation (Kramer and Lee, 1974). This reaction may appear as somnolence or an accentuation of preexisting signs and symptoms in patients with intracranial neoplasms. The somnolence syndrome was first described by Druckmann (1929) who reported marked, but transient somnolence in 3% of 1100 children irradiated for trycophytosis of the scalp. This syndrome is now recognized as a common side effect occurring in 58–78% of children with acute lymphoblastic leukemia (ALL) undergoing CNS prophylactic irradiation (Freeman et al., 1973; Littman et al., 1984). The syndrome is clinically characterized by somnolence, anorexia, and irritability. Diffuse slowing may be seen on EEG and cerebrospinal fluid (CSF) pleocytosis or mild protein elevation may be present (Bleyer and Griffin, 1980). Symptoms characteristically develop 24–56 days after the completion of cranial irradiation and spontaneously resolve in 10–38 days. There are no accompanying focal neurological abnormalities. The incidence of somnolence is similar in those patients receiving and not receiving methotrexate. Symptoms are more severe in children less than 3 years of age; there is no increase in CNS relapse or survival difference in children who do or do not develop this syndrome. Freeman *et al.* (1973) proposed that the somnolence syndrome was secondary to a temporary disruption of myelin synthesis.

Parker *et al.* (1978) prospectively followed 27 children with ALL who received CNS prophylaxis consisting of cranial radiation to a dose of 2400 rad. An increased incidence of somnolence (75%) was found in children receiving fractions of 150 rad compared with those treated at 125 rad per fraction (40%). Although Cecchetti and Brandoli (1979) suggested that the incidence of somnolence was related to dose, Littman *et al.* (1984) were unable to confirm that a lower total dose of cranial irradiation and lower daily fraction size influenced the development of the somnolence syndrome. The incidence (58%) and severity of somnolence were equal in children receiving 1800 rad in 18 fractions and 1800 rad in 10 fractions. The inconsistency in these observations may be related to the different chemotherapeutic agents used together with cranial irradiation.

Rider (1963) reported two adult patients who received incidental cranial irradiation to doses of 5500 rad in 16 or 27 fractions while undergoing therapy for extracranial tumors. Approximately 10 weeks after the completion of therapy, both patients developed nausea, vomiting, ataxia, dysarthria, dysphagia, horizontal nystagmus, and a positive Romberg sign. Recovery began after 4 weeks and was complete in 6–8 weeks. Both patients were alive and well 6 years later. Rider (1963) proposed this to be a new syndrome which should be differentiated from delayed radiation necrosis. Boldrey and Sheline (1967) described in 8 patients the development of transient signs and symptoms which appeared during the first 10 weeks after radiation therapy for low-grade gliomas, meningiomas, and pituitary adenomas. The reactions could not be ascribed to tumor progression and the symptoms resolved within 6 weeks. Hoffman *et al.* (1979) studied 51 patients with malignant gliomas treated with 5000 rad (170–180 rad/fraction) to the whole brain and a 1000-rad boost to a reduced field with concomitant 1,3-bis(2-chloroethyl)-1-nitrosourea (BCNU) given at 8-week intervals. Detailed neurological examinations, CT scans, and radionuclide scans were performed at 8-week intervals. Changes suggesting progression of tumor occurred in 49% (25/51) of the patients within 18 weeks after radiation therapy. Spontaneous improvement was seen over the next 16 weeks in 28% (7/25) of these patients without modification in treatment.

The early delayed reaction is transient and usually associated with an uneventful recovery. Rarely, however, this reaction may be fatal (Lampert *et al.*, 1959; Lampert and Davis, 1964). Autopsy studies have revealed extensive plaques of demyelination with central necrosis, absence of oligodendroglia, and gliosis accompanied by minimal vascular change and preservation of neurons and axons. These changes were limited to the tissues encompassed in the radiation beam (Lampert and Davis, 1964).

The pathogenesis of the early delayed reaction is thought to be a temporary inhibition of myelin synthesis due to radiation-induced effects on the oligodendroglial cells. The turnover time of myelin is 5 weeks to 2 months which corresponds to the latency and recovery time of the syndrome (Jones, 1964; Hoffman *et al.*, 1979; Norton, 1972; Edwards and Wilson, 1980). CT scans may show transient demyelination during the period of deterioration (Groothuis and Vic, 1980).

C. Late Delayed Reactions

The late delayed reactions, particularly in the form of radiation necrosis, constitute the major hazard of exposure of the brain to therapeutic radiation. Late brain injury may develop from several months to years following irradiation. The clinical presentation of radiation necrosis depends on the site and volume of the brain irradiated. This type of injury is often progressive, irreversible, and potentially fatal.

1. Pathology

Late delayed brain injury is mainly restricted to the white matter (Burger *et al.*, 1979; Manz *et al.*, 1979; Martins *et al.*, 1977; Fike *et al.*, 1984). Radiation necrosis may appear grossly as an ill-defined, firm, space-occupying mass with a yellow waxy color, or the white matter may be dry and granular in appearance. Cystic cavitation and peticheal hemorrhages may be present (Rubinstein, 1972). Microscopically, there may be minimal gliosis to extensive coagulation necrosis. Oligodendroglial cells are absent (Groothuis and Vick, 1980). The most characteristic findings are vascular changes with hyalin thickening of vessel walls, endothelial proliferation, fibrinoid necrosis, perivascular fibrosis, and thrombosis (Burger *et al.*, 1979; Mandybur and Gore, 1969; Manz *et al.*, 1979; Martins *et al.*, 1977; Fike *et al.*, 1984; Groothuis and Vick, 1980). The breakdown of white matter with release of necrotic cellular material (i.e., unsaturated fatty acids) may result in cerebral edema and mass effect (Edwards and Wilson, 1980).

There are multiple theories as to the pathogenesis of late delayed brain injury, including (1) vascular injury (Martins *et al.*, 1977; Yoshii and Phillips, 1982), (2) a direct effect on glial cells (Manz *et al.*, 1979), or (3) autoimmune demyelination (Lampert *et al.*, 1959). It is probable that multiple mechanisms are involved and the importance of each depends on the radiation dose and latent interval. Evidence suggests that demyelination is important in the early delayed reaction and that the effect on the small and medium arterioles becomes progressively more important with time (Sheline *et al.*, 1980).

2. Time–Dose Relationships

Radiation tolerance doses have not been precisely defined. Published reports have been largely anecdotal without a description of the population at risk. Pertinent information such as total dose at the site of necrosis, size and number of treatment fractions, overall treatment time, radiation field arrangement, and volume treated is frequently omitted. Until recently, few data have been published on the relationship of radiation dose, treatment time, and fractionation on the development of radiation necrosis.

Sheline *et al.* (1980) reviewed over 100 cases published in the literature in which radiation was thought to be the cause of subsequent brain necrosis. In 80 of these cases, an estimate of total radiation dose, treatment time, and fractionation could be analyzed: 28 patients had undergone irradiation for extracranial tumors, of which 11 were cancer of the skin and 17 represented other extracranial sites; 20 patients were treated for either pituitary tumors or craniopharyngiomas; 32 patients, 4 of whom also received chemotherapy, were treated for primary brain tumors. The interval from irradiation to injury ranged from 4 months to $7\frac{1}{2}$ years. Orthovoltage doses were converted to megavoltage rad equivalent (MRE) using the appropriate physical conversion factors and differences in relative biological effectiveness (RBE) of the quality of the radiation. In 26 patients, the total dose was 7000 rad MRE or greater. Twenty patients were treated with a total MRE dose of 5000 or less, which is generally considered safe. However, 17 of these 20 patients (84%) were treated with relatively large individual fraction sizes ranging from 250 to 3750 rad. The remaining 3 patients received 170–220 rad/fraction. Sheline *et al.* (1980) plotted total megavoltage rad dose against the number of treatment fractions for each case on a double logarithmic scale. An isoeffect line with a slope of 0.44 defined the lower limits of tolerance, implying that brain necrosis is largely dependent on the number of treatment fractions (fraction size) as well as the total dose. A dose of 6000 rad in 30 fractions over ~ 6 weeks approached the threshold for radionecrosis in the brain.

The Ellis formula (Ellis, 1968), $NSD(ret) = D \times N^{-0.24} \times T^{-0.11}$ (where D is the total dose in MRE, N is the number of fractions, and T is the total time in days), was modified and another equivalent single-dose quantity, termed "neuret," was derived. The formula, neuret $= D \times N^{-0.41} \times T^{-0.03}$, was based on the slope of the isoeffect line of 0.44 and the assumption that the exponent of T was -0.03. Considerable support exists in the literature that a formula with very close to these exponents is applicable to the entire CNS. Hornsey *et al.* (1981) established an isoeffect curve for X ray-induced brain damage in the rat, with the end point being death within 1 year. A modified Ellis formula was derived, with the exponent of N being -0.38 and the exponent of T -0.02. Pezner and Archambeau (1981) developed a mathematical formula for brain tolerance based on a series of dose schedules reported in the literature and derived a formula for human brain tolerance with an exponent of N of -0.45 and an exponent of T of -0.03. They concluded that this formula more accurately related dose schedule to risk of brain injury than the NSD formula. Van der Kogel (1979) reviewed the literature on the radiation tolerance of the human spinal cord. A double logarithmic plot of total dose against number of fractions gave an isoeffect curve with a slope of 0.4. From his review, he concluded that in humans the time factor was negligible for a treatment period of 6 weeks. Wara *et al.* (1975) derived an exponent of N of -0.377 and an exponent of T of -0.058 for the human spinal cord. These exponents suggest that the tolerance of the CNS is highly dependent on the number of fractions and minimally dependent on overall treatment time (at least up to 6 weeks). The formula, however, may not be valid for very small fractions, and if multiple fractions are given per day (hyperfractionation), the relative importance of N and T may differ from that for single daily fractions (Ang *et al.*, 1985). By plotting the incidence of necrosis versus dose in ret and neuret, Sheline and colleagues (1980) found that the threshold for necrosis was about 1700–1800 ret and 1000–1100 neuret. The neuret formula appeared not only to describe better the effect of fractionation, but also to delineate the threshold dose below which radiation necrosis was not observed. No relationship between neuret dose and the interval to necrosis was observed.

Accurate incidence rates of cerebral necrosis are difficult to establish because autopsies are performed infrequently and because most reports of incidence fail to include the number of exposed patients who have survived a sufficient length of time to develop the reaction. Marks *et al.* (1981) reported a series of 139 patients treated with a minimum of 4500 rad in daily increments of 180–200 rad. The patients were followed until death or for a minimum of 3 years. Treatment was given through equally or unequally weighted parallel opposed fields, with one field treated each day. Of the 139 patients, 7 (5%) developed pathologically documented cerebral necrosis. Two additional patients had clinically suspected necrosis. The incidence figure was admittedly minimal, as only 17% of their patients had reoperation or autopsy. The majority of patients received 5000–6000 rad in 5–7 weeks. With use of the equivalent dose (ED) formula derived for the spinal cord by Wara *et al.* (1975), $ED = D \times N^{-0.377} \times T^{-0.058}$, no case of necrosis occurred in 51 patients with an ED of less than 1250 ($\sim$ 6000 rad in 35 fractions); 2 of 60 patients developed necrosis with an ED $\geq 1250 \leq 1330$ ($\sim$ 6000 rad in 29 fractions); and 5 of 28 patients developed necrosis with ED $\geq$ 1330–1460. There were no cases of radionecrosis with doses of 5400 or less given in 30 fractions over 42 days. The lowest dose associated with necrosis was 5500 rad in 30 fractions over 44 days. If both portals had been treated at each treatment session, the incidence of necrosis probably would have been lower for any given dose level. In each case of radiation injury, the dose-fractionation schedule exceeded the lower limits of tolerance described by Sheline *et al.* (1980).

3. Diagnosis of Radiation Necrosis

A clinical diagnosis of radiation necrosis is difficult to establish in patients with intracranial neoplasms who deteriorate neurologically after receiving a substantial dose of cranial irradiation. Radiation-induced changes may be confused with tumor recurrence, since both conditions may present similar but inconstant and nonspecific clinical and radiographic findings (Mikhael, 1980). Neurodiagnostic studies, including EEG, radionuclide brain scan, angiography, and CT, generally fail to differentiate radiation necrosis from recurrent tumor (Edwards and Wilson, 1980). Mikhael (1980) studied radiation-induced lesions on CT scans and was able to correlate radiographic changes with radiation dose. A dose of 5550 rad in 180–200 rad per fractions delivered over 36–42 days resulted in the appearance of diffuse, noncontrast-enhancing, low-density lesions in the white matter without a mass effect. These lesions appeared after a latent period of 18–28 months. In patients receiving 6000–7000 rad in 30–35 fractions over 46–72 days, CT scans revealed either a localized mass lesion with peripheral or central contrast enhancement or diffuse low- to high-density lesions in the white matter, some of which also showed contrast enhancement. These changes occurred after a latent period of 9–20 months. Although a CT scan may suggest radiation-induced changes, if the injury occurs at the tumor site, only biopsy can confirm the diagnosis.

New diagnostic methods to differentiate recurrent tumor from necrosis are being investigated. Positron emission tomography (PET) with [^{18}F]fluorodeoxyglucose (FDG) is a method which allows the determination of cerebral glucose metabolism *in vivo*. Using PET, Patronas *et al.* (1982) studied 5 patients who had undergone radiation therapy for cerebral tumors and who were clinically deteriorating. All 5 patients had similar clinical and CT findings. PET scanning with FDG was able to distinguish two cases of radiation necrosis from three cases with recurrent tumor. In the case of radiation necrosis, the rate of glucose utilization in the lesion was markedly less than in normal brain parenchyma. In areas of recurrent glioma, the rate of glucose metabolism was elevated. All five diagnoses were confirmed by biopsy or autopsy. Fike and Cann (1984) have attempted to differentiate canine gliomas from radiation necrosis by quantitative CT scan using the kinetics of contrast uptake and washout. Their results indicate that there may be a difference in contrast uptake between tumor and irradiated brain. The role of magnetic resonance imaging in distinguishing radiation necrosis from recurrent tumor remains to be determined.

4. Treatment of Radiation Necrosis

Surgical resection of favorably situated focal lesions may result in considerable improvement or in complete recovery from the secondary effects of the brain injury. After surgery, improvement most commonly occurs in patients with focal necrosis who have been irradiated for primary extracranial lesions. Resection is of little value in patients with diffuse lesions or involvement of the brain stem or optic nerves. Surgical procedures involving less than total or subtotal resection of the necrotic tissue (i.e., biopsy) provide little benefit. Corticosteroids may result in improvement or stability of neurological deficit when surgery is not possible (Edwards and Wilson, 1980). Rizzoli and Pagnanelli (1984) recently reported clinical and CT scan-documented improvement in 2 patients with radiation necrosis of the brain who were treated with anticoagulant therapy. This observation remains to be confirmed.

5. The Effects of Chemotherapy

The extent to which chemotherapeutic agents influence the production of radiation necrosis is poorly documented. Evidence exists, however, suggesting that some of these agents add t�📍 the risk of injury. The most dramatic example occurs in children with ALL treated with aggressive chemotherapy, both intravenous and intrathecal, and prophylactic brain irradiation. Brain injury has occurred in these children with doses as low as 2400 rad in 150- to 200-rad fractions which, without chemotherapy, would be well below normal tolerance levels. Two late delayed syndromes, necrotizing leukoencephalopathy and mineralizing microangiopathy, have been recognized in children with ALL treated with prophylactic cranial irradiation and aggressive intravenous and intrathecal methotrexate (Bleyer, 1981).

Necrotizing leukoencephalopathy characteristically develops 4–12 months following CNS prophylactic therapy for ALL. Its onset is heralded by the insidious development of dementia, drooling, dysarthria, or dysphagia. The majority of patients survive with some degree of permanent neurological impairment, although a few patients may recover completely. In others, the outcome is fatal (Bleyer and Griffin, 1980; Price and Jamieson, 1975; Fusner et al., 1977). This syndrome has not been reported in patients receiving radiation therapy alone in the range of 1800–2400 rad (Bleyer and Griffin, 1980). Necrotizing leukoencephalopathy occurs in less than 1–2% of patients receiving intrathecal methotrexate or high-dose intravenous methotrexate alone (Bleyer and Griffin, 1980). The risk increases with combination therapy; the addition of intrathecal methotrexate (<50 mg), high-dose intravenous methotrexate (>40–80 $mg/m^2/$ week), or both, to cranial irradiation raises the incidence of injury (Bleyer and Griffin, 1980; Price and Jamison, 1975). With cranial irradiation and intrathecal methotrexate, 2–10% of patients may become symptomatic (Bleyer and Griffin, 1980; Aur et al., 1978). The addition of high-dose intravenous methotrexate to cranial irradiation also increases the risk up to 15%. The incidence may be as high as 45% in patients treated with all three modalities. Methotrexate is most neurotoxic when given during or after radiation therapy; leukoencephalopathy has not been reported for patients exposed to methotrexate before radiation therapy (Bleyer and Griffin, 1980).

This syndrome is thought to result from demyelination of the cerebral hemispheric white matter. The histopathological changes are characterized by demyelination and coagulative necrosis with gliosis. Initially, the lesions appear as multifocal areas of coagulation necrosis; in advanced stages the white matter may be reduced to a gliotic calcified layer (Bleyer and Griffin, 1980). CT scans show noncontrast-enhancing paraventricular hypodensity. In the later stages, ventricular dilatation and widening of the sulci, with or without calcification of the residual cerebral white matter, occur (DiChiaro *et al.*, 1979). Radionuclide brain scans may show increased deposition of the radionuclide in the paraventricular areas (Aur *et al.*, 1978; Liu *et al.*, 1978; Bleyer and Griffin, 1980). Subclinical leukoencephalopathy may be demonstrated by serial CT scanning (Ochs *et. al.*, 1983b). Peylan-Ramu *et al.* (1978) reported abnormal CT findings in 8 of 14 (57%) asymptomatic patients studied 19–67 months after the initiation of CNS prophylaxis. These patients had received 2400 rad concomitantly with intrathecal methotrexate (5 doses, 12 mg/m²) followed by monthly maintenance intrathecal methotrexate for 30 months. Nonsymptomatic CT abnormalities have not been seen with less intensive prophylaxis (Bode, 1980).

Mineralizing microangiopathy may develop in 25–30% of children who survive more than 9 months after CNS therapy for ALL (Bleyer and Griffin, 1980). This syndrome is nonfatal and neurological dysfunction is generally mild, subclinical, and transient. Neurological signs and symptoms may include headaches, focal seizures, ataxia, behavior disorders, or perceptual motor disability (Price, 1979; McIntosh *et al.*, 1977). Predisposing factors include age less than 10 years at the time of treatment, long duration of survival after radiation, and the number of CNS leukemic relapses (Price and Birdwell, 1978). CT scans characteristically show calcification of the basal ganglia.

Mineralizing microangiopathy affects the gray matter of the CNS and is characterized by dystrophic calcification of the basal ganglia and subcortical zones. Histopathologically there is deposition of calcium within the walls of small blood vessels. The smaller blood vessels are often occluded by precipitated calcific material. The brain tissue surrounding the diseased vessels may be mineralized and necrotic (Price and Birdwell, 1978). The incidence of intracranial calcifications increases with the use of large cumulative doses of intravenous methotrexate (>4.5 g/m²) and cytosine arabinoside (0.4–5.5 g/m²) during maintenance chemotherapy.

The toxicity observed in patients receiving CNS therapy for ALL suggests that chemotherapeutic agents interact with radiation in an additive or synergistic manner. It has been hypothesized that (1) methotrexate and radiation may share common mechanisms of neurotoxicity; (2) methotrexate may sensitize the brain to irradiation; or (3) CNS irradiation may modify the distribution of methotrexate, exposing portions of the brain to increased concentrations of the drug (Bleyer and Griffin, 1980).

Both early and late neuropsychological deterioration have been recognized in long-term surviving patients with small cell carcinoma of the lung who received prophylactic cranial irradiation with doses of ~3000 rad in 10 fractions (Volk *et al.*, 1983; Licciardello *et al.*, 1983; Ellison *et al.*, 1982). As with ALL, these patients received a variety of chemotherapeutic agents which may have enhanced the effect of radiation on the CNS. Komaki *et al.* (1985) did not observe late neurological sequelae after decreasing the dose to 2500 rad in 10 fractions.

Burger *et al.* (1979) reported 4 cases of radiation necrosis in 24 brains examined at postmortem following 6000 rad in 6–6½ weeks for malignant gliomas. Of the 4 patients, 3 had had multiple courses of chemotherapy which included procarbazine, BCNU, and dibromodulcitol. There was a predilection for necrosis in the white matter adjacent to the persistent neoplasm, suggesting that the tumor played an important role in causing the injury. Neurotoxicity may result from chemotherapy alone (Weiss *et al.*, 1974), and it is probable that in addition to methotrexate, chemotherapeutic agents such as BCNU, procarbazine, and vincristine may also enhance the radiation effects on the CNS. Prospective studies are needed to establish the true incidence of injury and the causal relationships of the various agents used.

6. Other Risk Factors

Other factors which may predispose patients to the development of late radiation injury include the patient's age and the presence of other underlying disease. Although no controlled studies have been published comparing the risk of radiation-induced brain damage in children versus adults, it is generally believed that children, especially under the age of 2–3 years, are more susceptible to such injury (Sheline *et al.*, 1980). Myelination is not complete until 2 or 3 years of age and, before this time, high doses of radiation may be associated with an increased risk of consequences such as intellectual impairment (Kornblith *et al.*, 1982). Patients with underlying vascular disease from diseases such as diabetes mellitus, Cushing's disease, and acromegaly may have an increased susceptibility to radiation injury (Smith *et al.*, 1979; Bloom and Kramer, 1984; Aristizabal *et al.*, 1979). Underlying CNS infection may also predispose the patient to radiation neurotoxicity (Cumberlin *et al.*, 1979).

D. Impaired Intellectual Function

A decreased level of intellectual function has been recognized in children after radiation therapy for brain tumors and ALL. The specific cause of this impairment is difficult to determine; radiation therapy, surgery, chemotherapy, the tumor and its location, increased intracranial pressure, reduced school attendance, psychological stress, and the adequacy of rehabilitation may all contribute (Eiser, 1978). A marked paucity of information exists on the effects of irradiation on intellectual function in adults.

Danoff *et al.* (1982) retrospectively reviewed 36 children previously operated and irradiated for brain tumors. The mean age at diagnosis was 7 years and the mean follow-up time 9.3 years. Tumor doses ranged from 4000 to 6500 rad at 180–200 rad/day. Of the patients, 82% received a tumor dose of 5000–5600 rad. No patient received chemotherapy. Intelligence testing documented that 6 patients (17%) were mentally defective [intelligence quotient (IQ) $\leq$ 69)], 28% were dull/normal or borderline (IQ of 70–89), and 20 patients (55%) were of normal or above normal intelligence ($\geq$90). Of patients who were older than 3 years at the time of treatment, 10% (3/31) were mentally defective compared with 60% (3/5) of children treated before the age of 3 years.

Mental retardation is more common in patients with tumors involving the hypothalamic or thalamic regions (Denoff *et al.*, 1982) or brain stem (Hirsch *et al.*, 1979). Hirsch *et al.* (1979) observed that 56% (5/9) of patients with medulloblastoma or brain stem involvement were mentally defective (IQ < 70) compared with 18% (3/17) without brain stem involvement. An association between the presence of hydrocephalus and IQ has not been identified (Danoff *et al.*, 1982).

Bamford *et al.* (1976) and Danoff *et al.* (1982) were unable to identify a correlation between the volume of brain irradiated and mental impairment. Kun *et al.* (1983), however, found intellectual deficits (verbal or memory IQ < 80) in 53% of children with supratentorial or posterior fossa tumors who, following resection, were treated with subtotal supratentorial or whole brain irradiation compared with 17% of patients whose treatment was restricted to the posterior fossa. The authors concluded that intellectual dysfunction was more likely to occur in patients receiving irradiation to the supratentorial region. In a later report, however, a statistically significant correlation could not be made between full-scale IQ and treatment volume (Kun and Mulhern, 1983).

Most of the patients reported have been treated within a narrow dose range, and little information has been published relating dose of radiation to the development of intellectual deficit. Spunberg *et al.* (1981) found that the 4 patients in their series receiving doses below 4000 rad scored within the borderline or high IQ range, whereas the 2 patients receiving more than 5000 rad had moderate mental retardation.

Intellectual function has been studied in patients treated with surgery and radiation therapy (with or without chemotherapy) and compared with that in patients receiving surgery alone. Hirsch *et al.* (1979) compared the IQs of children with medulloblastoma receiving surgery, craniospinal irradiation, and chemotherapy with patients with cerebellar astrocytoma treated with surgery alone. Of patients with medulloblastoma, 11% had IQ levels of 90 or greater compared with 62% of patients with cerebellar astrocytomas. When patients with brain stem involvement were excluded, the incidence of IQ levels of less than 70 was similar in both groups. The role of radiation as a cause of intellectual disability is clouded by the addition of chemotherapy. Raimondi and Tomita (1979) retrospectively studied 16 children 3–12 years following treatment for medulloblastoma. All patients received craniospinal irradiation and 4 also received systemic chemotherapy. The mean full-scale IQ level was normal in 4 patients, borderline in 7 patients, and 5 patients were mentally defective (IQ ≤ 69). These patients were compared with 6 patients with cerebellar astrocytoma who underwent posterior fossa craniotomy, but did not receive postoperative irradiation; one patient had a borderline IQ score while the IQs of the remainder of the patients were within the normal range.

Packer *et al.* (1983), however, found that patients treated with surgery alone also expressed neuropsychological dysfunction. In a study of 94 children previously treated for brain tumors, neuropsychological impairment occurred in 44% of children receiving craniospinal irradiation, 41% of patients receiving local irradiation, and in 31% of children receiving no radiation therapy. The incidence of dysfunction was relative to location of the tumor; 25% of children with suprasellar tumors exhibited neuropsychological injury compared with 15% of those with other supratentorial primary sites and 8% of patients with posterior fossa lesions. This study suggests that the tumor, its location, and surgical treatment may all contribute to the development of injury.

Hochberg and Slotnick (1980) studied 132 adult patients with high-grade astrocytomas after a combination of surgery, external beam irradiation (4500 rad to the whole brain with a 1500-rad boost), and lomustine (CCNU). Patients selected for study had survived at least 1 year after treatment and had failed to return to pretreatment educational or work levels. Testing revealed difficulties in problem solving or coping with newly learned tasks requiring attention. Previously acquired ability, overlearned information, and judgment ability were not impaired.

Although Hochberg and Slotnick (1980) found progressive deterioration in 3 adults undergoing serial testing after irradiation and chemotherapy, Kun and Mulhern (1983) observed no significant deterioration in IQ with time after treatment with radiation in their children. Serial improvement occurred in 4 children. The possibility that this improvement was due to recovery from injury or special educational programs was suggested.

Soni *et al.* (1975) found no evidence of neurological or psychological impairment in children with ALL who had received 2400 rad to the whole brain plus chemotherapy. Other reports, however, suggest that treatment with prophylactic CNS irradiation and chemotherapy may induce a small but significant reduction in IQ scores accompanied by perceptual and learning disabilities (Eiser, 1979; Moss *et al.*, 1981) (Table I).

TABLE I

IQ FOLLOWING CNS PROPHYLAXIS FOR ALL

| Author | Mean full-scale IQ | | | p value |
	Cranial RT	No cranial RT	Controls	
Eiser (1978)	89 <2 mos[a]	102	—	—
	109 ≥ 6mos[a]			
Moss et al. (1981)	99	—	112	<0.001
	—	103	99	—
Meadows et al. (1981)	89[b]	109[b]	114[b]	—
Esseltine et al. (1983)	95	105	106	<0.03

[a] Interval between diagnosis of ALL and CNS irradiation.
[b] Median IQ scores.

Meadows et al. (1981) prospectively studied 31 children with ALL who received cranial prophylaxis with whole brain irradiation (2400 rad) plus intrathecal methotrexate. Controls included 6 children diagnosed 7–13 years earlier who did not receive cranial irradiation, and 6 children with recently diagnosed Wilms' tumor treated with multiple drug chemotherapy alone. Patients were studied during the first month after diagnosis and 12–34 months after the first study. A subgroup of 18 patients were reevaluated at a minimum of 3 years after diagnosis. The first and second test showed no variation in IQ score. However, 11 of 18 patients with ALL who received cranial irradiation showed a decrement in IQ of 10 or more points at the time of the third test. The mean full-scale IQ of this group was 89 compared with 109 and 106 at the time of the first and second test, respectively. Unfortunately, the patients with Wilms' tumors did not have the third examination, and there was no control group of concurrently treated ALL patients not receiving CNS irradiation. A decrease in IQ was related to age at the time of treatment and the initial IQ score. Children 2–5 years of age when treated showed a greater decrement in IQ score than those who were 6 years of age or older at the time of diagnosis. Children whose initial IQ score was 110 or greater averaged 23 IQ points less by the third study, while those whose initial score was less than 110 had an average drop of 8 IQ points.

Ochs et al. (1983a) compared 55 patients with ALL receiving 1800 rad to the brain plus intrathecal methotrexate with 53 patients receiving moderate dose intravenous methotrexate and intrathecal methotrexate without irradition. Compared with the "normal" population, the verbal and performance IQ scores were decreased in 40% of both groups. The authors concluded that both forms of prophylaxis result in neuropsychological changes. Tamaroff et al. (1985) performed retrospective neuropsychological testing on 91 patients treated with intrathecal methotrexate alone or in combination with 1800 or 2400 rad cranial irradiation. The mean full-scale IQ of the patients treated at each dose level was similar and was significantly lower than that of the nonirradiated group ($p <$ 0.01). Performance IQ scores were 8–9 points lower in the irradiated patients. No difference was found in verbal scores of the three groups studied. All patients displayed deficits in attention, concentration, and short-term memory, whereas visual motor integration was impaired only in patients receiving irradiation.

The decrements in IQ score that have been observed in patients receiving brain irradiation for ALL and primary brain tumors generally reflect a reduction in the performance score, although in some series a decrease in verbal scores has also been recorded (Kun *et al.*, 1983; Moss *et al.*, 1981; Duffner *et al.*, 1983; Spunberg *et al.*, 1981; Eiser, 1973; Hochberg and Slotnick, 1980; Esseltine *et al.*, 1983; and Tamaroff *et al.*, 1985). The discontinuity between performance and verbal scores is consistent with nonspecific cerebral injury (Reitan and Davison, 1974). A similar pattern also has been seen in children with learning disabilities (Duffner *et al.*, 1983).

III. Radiation Injury to the Spinal Cord

Radiation myelopathy was first described by Ahlbom (1941) in a report of the results of radiation therapy for hypopharyngeal cancer. Since then, numerous reports have appeared of its occurrence in the cervical (Boden, 1948; Abbatucci *et al.*, 1978), thoracic (Locksmith and Powers, 1968; Reinhold *et al.*, 1976; Dische *et al.*, 1981; Hatlevoll *et al.*, 1983), and lumbar (Maier *et al.*, 1969) spinal cord. The incidence of radiation myelopathy is difficult to determine, since most patients receiving large doses of irradiation to the spinal cord have had carcinoma of the pharynx, esophagus, or lung, which are associated with a poor prognosis, and survival has been too short for the development of injury. Reported incidence figures are dependent on the distribution of case material, duration of survival following irradiation, the spinal cord dose (including overall time and fractionation), and the completeness of follow-up (Dische *et al.*, 1981). Radiation myelopathy may appear as a transient syndrome (early delayed) or as a more serious late reaction.

A. Transient Radiation Myelopathy

This syndrome is characterized by momentary, transient, electrical shock-like paresthesias or numbness radiating from the neck to the extremities precipitated by neck flexion (Lhermitte's sign). The sensation may radiate only down both legs or involve the upper extremities as well. The symptoms are symmetrical and do not conform with the distribution of a specific dermatome. There are no associated abnormal neurological findings. The syndrome characteristically develops after a latent period of 1–6 months (average 3–4 months) following irradiation and gradually abates over the next 2–9 months (average 3–6 months) (Jones, 1964; Boden, 1948; Dynes and Smedal, 1960).

Jones (1964) observed that this syndrome, including the latent period and duration, is similar to that produced after traumatic injury to the head and neck, vertebral subluxation, or cervical disc protrusion. He attributed the syndrome to transient demyelination of the ascending sensory neurons, most probably in the posterior columns, lateral spinothalamic tracts, or both. The latent period is consistent with interference in the synthesis and turnover of myelin, and it is hypothesized that radiation inhibits the normal proliferation of myelin-producing oligodendroglial cells in the irradiated region. Because of the turnover time of myelin, the effect is not immediately apparent. As the oligodendroglial cells recover, myelin synthesis is resumed. The induction of symptoms by neck flexion results from the elongation or stretching of the denuded, hypersensitive posterior fibers of the cord.

This syndrome is spontaneously reversible and requires no specific therapy. As a rule, it does not portend the development of later progressive myelopathy. However, occasionally, especially if it appears after a long latent period (9–12 months), radiation myelitis may follow (Jones, 1964).

B. Delayed Radiation Myelopathy

Delayed radiation myelopathy occurs several months to years after irradiation, with most cases appearing 12–28 months after exposure (Kramer and Lee, 1974). About 75% of cases occur within 18 months of completion of radiation therapy (Lambert, 1978). The latent period decreases with increasing dose (Schultheiss *et al.*, 1984).

Neurological signs and symptoms may be abrupt or insidious in their onset, but typically become chronic or progressive over a period of several months. The neurological sequelae depend on the level of involvement and include the development of paresthesias followed by sensory and motor dysfunction, bowel and bladder disturbance, and paraplegia or quadraplegia. A partial Brown-Séquard syndrome, characterized by motor weakness and pyramidal tract signs in one lower extremity and alteration in sensory perception in the opposite extremity, may develop below the level of injury. There may be a partial loss of function that does not progress to complete paralysis, or complete loss of function may ensue (Atkins and Tretter, 1966; Maier *et al.*, 1969). The neurological changes associated with radiation myelopathy are irreversible, and about one-half of the patients die from secondary complications (Lambert, 1978).

Less frequently, radiation myelopathy is manifested by the acute onset of paraplegia or quadraplegia which develops over several hours or a few days and results from infarction of the cord secondary to radiation-induced vascular changes. Myelopathy also may be heralded by signs of lower motor neuron disease in the upper or lower extremities due to selective injury to anterior horn cells (Reagan *et al.*, 1968).

The diagnosis of radiation myelopathy requires a history of radiation therapy in doses sufficient to result in cord injury. The portion of cord irradiated must be at least slightly above the dermatome level of expression of the lesion, and other possible causes must be excluded (Rubin, 1975). The latent period from the completion of treatment to the onset of symptoms must also be considered (Margolis *et al.*, 1981). There are no confirmatory laboratory tests to differenti- ate radiation myelopathy from other spinal cord lesions, and the diagnosis is one of exclusion. The myelogram may be normal or show an area of expansion or attenuation at the level of irradiation (Atkins and Tretter, 1966; Kramer and Lee, 1974; Reagan *et al.*, 1968). CSF pressure is normal and no abnormal cells are seen. Although the protein concentration in the CSF is usually within normal limits, mild elevations have been reported (Boden, 1948; Pallis *et al.*, 1961).

The site of radiation myelopathy is the white matter. The pathogenesis has been attributed to either a direct effect on oligodendroglial cells or vascular damage. Because of the long latent period, vascular injury is considered the primary cause of radiation myelopathy (Kramer and Lee, 1974; Lambert, 1978). Histopathological findings include patchy demyelination with circumscribed areas of necrosis surrounded by gliosis and lipid-laden macrophages. Vascular changes include endothelial and medial smooth muscle swelling in small arteries and arterioles, fibrinoid and hyalin degeneration, endothelial proliferation, acute and chronic vasculitis, intimal and medial necrosis, and vascular occlusion. Wallerian degeneration is seen above and below the necrotic zones. The neurons are spared (Rubin, 1975).

Radiation tolerance levels of the spinal cord are difficult to establish. The number of cases of radiation myelopathy reported is small, and often there is little information regarding the size of the exposed population. Radiation equip- ment has ranged from orthovoltage to megavoltage. A wide variety of time–dose fractionation schedules have been used in the reported cases. From the available data, however, it may be concluded that an increased incidence of myelopathy occurs with large fraction sizes (fewer fractions), shorter treatment time, and the delivery of high total doses (Lambert, 1978).

Since little information is available on incidence, investigators attempted to establish tolerance levels by plotting cases on a log-log graph of treatment time against dose and drawing sloping lines separating cases which developed my- elitis from those that did not. These lines were considered isoeffect lines. Boden (1948) developed such an isoeffect line for the human spinal cord based on a plot of dose versus overall treatment time in days; the line had a slope of 0.26 (Atkins and Tretter, 1966). Boden concluded that the tolerance limits were 3500 R delivered over 17 treatment days for large fields and 4500 R administered over

17 days for small fields. Pallis *et al.* (1961) observed cases of radiation myelopathy at doses below the levels suggested by Boden (1948) and proposed an isoeffect line paralleling Boden's, but with a 20% shift to lower dose levels. Pallis *et al.* (1961) felt that the tolerance limits for the cord were 3300 R in 42 days for long segments and 4300 R in 42 days for short segments.

Atkins and Tretter (1966) plotted their cases of radiation myelopathy and those previously reported on a log-log scale of dose versus the number of fractions and generated a regression line with a slope of 0.38. The slope of the regression line was thought to be an effect of fractionation. They suggested that the isoeffect lines of Boden (1948) and Pallis *et al.* (1961) did not adequately reflect the effect of fractionation, since it overestimated the tolerance of the spinal cord for small numbers of fractions and underestimated it for 20 or more fractions. Phillips and Buschke (1969) also concluded that the tolerance of the spinal cord appeared to increase more rapidly with the number of fractions and that the number of fractions may be more important than overall time.

Wara *et al.* (1975) reviewed cases of thoracic cord injury at the University of California, San Francisco (UCSF), and those cases reported in the literature and compared the Ellis NSD formula to an equivalent dose (ED) formula, $ED = D \times N^{-0.377} \times T^{-0.058}$, in which D is the total dose in rad, N is the number of fractions, and T is the overall treatment in days. The ED dose successfully distinguished patients with and without injury. The 1% incidence level for thoracic myelopathy occurred at an ED of 1015 ret and the 50% incidence at 1476 ret. Using a double logarithmic scattergram comparing dose with number of fractions, it was demonstrated that 2000 rad in 1 week, 3000 rad in 2 weeks, and 5000 rad in 5 weeks (delivered in 5 fractions/week) appeared safe.

Although the evidence is not conclusive, it is generally believed that the tolerance dose of the cervical spinal cord is greater than that of the thoracic cord and that tolerance is inversely related to the length of the cord irradiated. Kramer and Lee (1974) considered the tolerance dose of the cervical and lumbar spinal cord to be 5000 rad given in 5 weeks and the tolerance of the thoracic spinal cord to be 4500 rads given in 5 weeks. He recommended lowering the dose by 10% if long segments of the cord were treated. Abbatucci *et al.* (1978) found that there was little risk of developing cervical spinal cord myelopathy at doses lower than 5500 rad in 27 fractions over 37 days and that injury was almost inevitable with a dose greater than 7000 rad in 35 fractions over 49 days. These authors concluded that a dose of 5000 rad delivered to 3–5 vertebrae would not induce radiation myelopathy. It was suggested that for larger volumes (6–7 vertebrae), a dose of 4500 rad represented the upper limit of tolerance. Lambert (1978) found no evidence of thoracic cord injury in 23 patients with carcinoma of the lung and esophagus who received 5000 rad or less and who survived at least 18 months. Of 35 patients receiving greater than 5000–6300 rad to the thoracic cord, 2 patients (5.7%, or 3.5% of all 58 long-term survivors) developed myelopathy.

One patient received a dose of 5600 rad in 27 fractions and developed a mild foot drop and gait disturbance. The second patient had a progressive myelopathy after receiving 5300 rad to a major portion of the cord with a small region boosted to 5800 rad in 30 fractions. Lambert (1978) concluded that the tolerance of the thoracic cord to well-fractionated radiation may be underestimated.

Individual factors, in addition to fractionation, may be related to late radiation damage and need to be considered. Hypertension has been reported experimentally to exacerbate the effects of radiation to the spinal cord (Asscher and Anson, 1962). However, an increased risk of myelopathy in patients with hypertension has not been observed clinically (Reinhold et al., 1976). The radiosensitivity of the spinal cord appears to be unaffected by the presence of a tumor in the cord (Kopelson, 1982). An increased incidence of radiation myelopathy has not been reported in patients with head and neck cancer receiving concurrent chemotherapy with such agents as methotrexate, bleomycin, 5-fluorouracil (5-FU), hydroxyurea, cis-platinum, vincristine, actinomycin D, and cyclophosphamide used alone or in combination (Fu, 1979).

Van der Kogel and Barendsen (1977) found two types of injury after radiation to the cervical spinal cord in male rats. The early reaction, which appeared within 7 months, was thought to be the result of demyelination, while the more delayed injury with an onset of 7–18 months was thought due to vascular damage. The end point used in their studies was the development of paralysis. When the ED_{50} (the dose to produce paralysis in 50% of animals) for paralysis due to demyelination with white matter necrosis was plotted against fraction number, the slope was 0.46. For the more delayed syndrome, the slope was 0.42. The time factor up to 112 days was felt to be zero. For changes in the lumbar area, the slope was 0.40. Van der Kogel (1979) noted that the pathological syndromes, latent periods, and tolerance in rat and man were similar. Although man might be expected to exhibit more variability, the similarities suggested that the rat spinal cord may be a good model for the establishment of time–dose relationships relevant to man. Van der Kogel (1979) plotted dose against fraction number on a double logarithmic scale for reported cases of human spinal cord injury and found a slope of 0.4. He concluded that the isoeffect formula relating tolerance to fraction number is $NSD = D \times N^{-0.4}$, with the time factor being negligible for overall treatment times up to 6 weeks. A dose of 5000 rad in 25 fractions was thought to be near the tolerance of the cord.

These isoeffect formulas predict that tolerance of the spinal cord continues to increase with decreasing fraction size. Recent experiments by Ang et al. (1985), however, have demonstrated that reduction in dose per fraction from 200 rad down to 130 rad does not result in a further increase in the tolerance dose. When isoeffective doses for inducing spinal cord injury were plotted against dose per fraction on a double logarithmic scale, the curve was linear, with fraction sizes between 1100 and 200 rad, but the curve flattened at fraction sizes of 200 rad or less. Thus, while these observations need confirmation, one should be cautious in applying isoeffect formulas to predict tolerance using fraction sizes smaller than 200 rad.

IV. Radiation Injury to the Optic Nerves and Chiasm

Radiation-induced visual changes may occur after radiation of the optic chiasm or optic nerves, an extension of the white matter of the CNS, and have been reported after the treatment of head and neck tumors, pituitary adenomas, and craniopharyngiomas. Radiation injury of the optic nerve can be divided into two clinical groups: (1) ischemic optic neuropathy from injury to the distal end of the nerve (optic disc), and (2) retrobulbar optic neuropathy from more proximal nerve injury.

Ischemic optic neuropathy results from occlusion of the posterior ciliary arteries that supply the optic nerve and choroid and is manifested by progressive visual loss which occurs from several months to 4 years after treatment. Opthalmoscopic changes include disc pallor, papilledema, and splinter hemorrhages on or near the disc. Retrobulbar optic neuropathy occurs as a result of infarction due to small vessel injury. Patients with preexisting small vessel occlusive disease may be at an increased risk of developing this type of injury. Retroorbital optic nerve neuropathy is characterized by sudden visual loss which may occur 1–8 years after treatment. Opthalmoscopic examination reveals optic atrophy without disc edema or hemorrhage (Parsons *et al.*, 1983).

Parsons *et al.* (1983) evaluated 92 optic nerves in 55 patients who were treated for head and neck carcinoma and who were followed for a minimum of 3 years. There were no cases of injury at doses of less than 5000 rad. The risk of injury correlated with fraction size in patients receiving greater than 5500 rad. Of patients receiving doses of 6000–7300 rad at 165–190 rad per fraction, 8% developed optic nerve injury. The risk of injury increased to 41% in those patients receiving similar total doses at daily fractions of 195 rad or greater.

The importance of fraction size is best illustrated in cases of optic nerve or chiasmal injury resulting from treatment of patients with pituitary adenomas (Table II). Harris and Levine (1976) reported 55 patients undergoing radiation therapy for pituitary adenomas or craniopharyngiomas. After megavoltage irradiation, 5 patients sustained visual impairment ranging from visual field changes to total blindness. The patients with pituitary adenomas received 4500–5000 rad in 4–5 weeks, and those with craniopharyngioma received 5500–7000 rad. Optic chiasm injury occurred only in those patients receiving daily doses of 250 rad or more (250–300). The total dose did not appear to be as significant a factor as fraction size, although the number of patients receiving greater than 5000 rad was small. NSD values, calculated according to the Ellis formula, did not appear to be a good discriminant of visual complications, and it was thought that the

TABLE II

RELATIONSHIP OF FRACTION SIZE TO OPTIC NERVE AND CHIASMAL INJURY
FOLLOWING RADIATION THERAPY OF THE PITUITARY REGION

Author	Total dose (rad)	Daily fraction size (rad)	Incidence of injury
Harris and Levine (1976)	4500–7000	≤200	0/27
		≥250	5/28
Aristizabal et al. (1977)	3400–5200	<200	0/7
		200–220	2/99
		>220	2/16
Sheline and Tyrell (1984)	~4500	≤200	0/180
		225	1/1

Ellis formula underestimated the importance of fraction size for radiation damage to the optic nerves. Cases observed during surgical exploration or at postmortem examination revealed gross atrophy and pallor, with yellowish discoloration of the optic nerves and chiasm. Histopathological evaluation showed severe demyelination and gliosis of the optic nerves.

Aristizabal et al. (1977) reported 4 patients who developed blindness related to radiation injury to the optic pathways among 122 patients receiving pituitary irradiation. Complications were not observed in patients receiving less than 4600 rad. Of the patients receiving at least 5000 rad, injury occurred in 2 of 99 patients treated with daily doses of 200–220 rad and 2 of 16 patients treated with fraction sizes of greater than 220 rad. The complication rate rose progressively from 0 to 25%, with an increasing ret dose from less than 1500 to greater than 1700. In a subsequent literature review of 14 patients treated with moderate doses of radiation (≤5000 rad) who had evidence of injury to the brain or optic chiasm, 7 (50%) were noted to have Cushing's disease (Aristizabal et al., 1979). Aristizabal et al. (1979) suggested that the vascular and hormonal changes associated with Cushing's disease may predispose the patient to radiation injury. However, 2 of these patients were treated with high daily fraction sizes, and 2 received orthovoltage irradiation. The remaining patients were treated with multiple field techniques which could have resulted in high daily doses to the region of the chiasm.

Atkinson et al. (1979) reported progressive visual failure in 4 of 23 patients receiving pituitary irradiation for acromegaly. Treatment was delivered at a dose of 4200–4500 rad in 18–22 days. In the 4 patients developing injury, the daily dose ranged from 283 to 313 rad. Bloom and Kramer (1984) reported visual complications in 5 of 40 patients treated for acromegaly with doses of 4500–

5000 rad at 196–217 rad per fraction. Visual injury was not seen in 140 patients with chromophobe adenoma treated with the same dose or fraction size or in 43 patients with craniopharyngioma treated with doses of 5500–7000 rad. The authors suggest that small vessel disease associated with acromegaly may make these patients more susceptible to injury. No cases of injury were found where the dose was reduced to 4600 rad at 180 rad per fraction. The incidence of chiasmal injury following irradiation of pituitary tumors will be a fraction of 1% if the total dose is limited to 4500 rad and the daily fraction size does not exceed 200 rad (Sheline and Tyrell, 1984).

V. Radiation Retinopathy

The retina is a specialized neural end organ which is supplied by an end arterial system, is very sensitive to vascular injury, and has little ability for repair. Depending upon dose, radiation-induced injury to the retina is characterized by a latent period of 1.5–6 years after exposure. Visual deterioration ensues as a result of progressive obliteration of small retinal vessels. During the latent period visual acuity may remain normal. Clinical findings include focal retinal capillary nonperfusion or occlusion of branch arterioles, edema, capillary microaneurysm formation, flame hemorrhages, cotton wool exudates, hard exudates, and retinal neovascularization. Retinal detachment, vitreous hemorrhage, rubeosis iridis, and neovascular glaucoma may complicate radiation retinopathy. Pathological findings in the retina vary from atrophy and disorganization to full thickness necrosis (Parsons et al., 1983).

Shukovsky and Fletcher (1972) found that each patient in their series with retinal injury received a dose of greater than 2000 ret (Ellis formula). An isoeffect line ranging from 6300 rad in 33 days to 7000 rad in 44 days seemed to represent the threshold of tolerance.

The retina may be more sensitive to irradiation than previously believed. Wara et al., (1979) reported 4 patients who developed retinopathy after retinal doses of 4640–4920 rad at 180 rad per fraction. Parsons et al. (1983) reviewed data on 28 eyes of 27 patients who received radiation to all or nearly all of the retina. Retinopathy was not observed when treatment was limited to the medial third of the eye. There was one instance of retinal injury at 4500 rad in a patient who also received chemotherapy. All but one patient who received doses of 6000 rad (170–200 rad per fraction) or more developed retinal injury, and retinopathy occurred in 3 of 4 patients receiving doses of 5000–5500 rad. One of the latter patients received chemotherapy in addition to radiation, and the remaining 2 were treated with large daily fraction sizes (305–325 rad). Parsons et al. (1983) concluded that the risk of retinal injury increased with doses exceeding 5000 rad at 180–200 rad per fraction. The concomitant use of chemotherapy may increase the risk of retinal damage (Chan and Schukovsky, 1976), and other conditions such as hypertension, diabetes mellitus, and autoimmune vasculitis may predispose the patient to injury (Wara *et al.*, 1979).

VI. Cranial and Peripheral Nerve Injury

Radiation-induced cranial nerve injuries have been reported, but are rare. The largest experience of cranial nerve injuries is that of Berger and Bataini (1977) of the Foundation Curie who reported 35 cranial nerve palsies in 25 patients irradiated for head or neck cancer. All but 2 patients were treated with megavoltage radiation. Doses to the injured nerves ranged from 6250 rad in 41 days (NSD = 1808 ret) to 10,000 rad in 43 days (NSD = 2840 ret) given in 6 fractions totaling 1100 rad/week. The hypoglossal nerve was injured alone in 16 patients and in combination with other nerves in 3 additional patients. The vagus and spinal accessory nerves were the next most frequently involved. Cranial nerve injuries were noted from 12 months to 172 months after treatment and the latent interval was inversely related to dose. Four cases of hypoglossal nerve injury were observed with ret doses ranging from 1800 to 2000 rad. The remainder of injuries occurred at doses of 2150 rets or above.

Cheng and Schultz (1975) reported 4 cases of unilaterial hypoglossal nerve injury with hemiatrophy of the tongue occurring 3–9 years after radiation therapy for carcinoma of the head and neck. Injuries occurred with doses ranging from 6000 rad in 16 fractions over 30 days to 8000 rad in 32 fractions over 47 days (1198–2150 ret). Injury to the abducens nerve (Lampert *et al.*, 1959; Parsons, 1984) and recurrent laryngeal nerves (Parsons, 1984) have also been described. The pathogenesis of radiation-induced cranial nerve injury has been attributed to perineural fibrosis. Injury of the hypoglossal nerve and of cranial nerves X and XI may be due to entrapment in the lateral pharyngeal space. Hypoglossal nerve dysfunction has also been reported in association with soft tissue necrosis of the glossotonsillar sulcus (Parsons, 1984).

The most common radiation-induced peripheral nerve injury is that of the brachial plexus in patients treated for breast carcinoma. Fortunately, with proper techniques the incidence of radiation brachial plexopathy is low. Montague *et al.* (1983) reported an incidence of 0.4% in patients treated at the M.D. Anderson Hospital with 5000 rad in 25 fractions to the supraclavicular region and axilla.

Radiation brachial plexopathy develops 5–30 months following irradiation. The most common presenting symptom is paresthesias in the fingers accompanied by hypoesthesias and weakness of the hand or fingers. Some patients develop only transient paresthesias or slight paresis of a muscle or muscle group. Others may have complete or nearly complete paralysis. Motor lesions are often progressive during the first few months, with a later tendency to stabilize. Partial recovery may occur. On rare occasions, bracheal plexus injury is transient and reversible (Salner *et al.*, 1981). The latency period tends to be inversely related to the degree of paresis. Electromyography (EMG) may reveal partial to total denervation (Stoll and Andrews, 1966; Westling *et al.*, 1972).

Total dose and, more importantly, daily fraction size are important factors in the development of brachial plexus injury (Table III). Stoll and Andrews (1966) reported 117 consecutive patients who were given radiation to the supraclavicular, axillary, and scalene regions after radical mastectomy. A single, large irregularly shaped field was irradiated using a 4-MeV linear accelerator. The minimum plexus dose was 5500 rad in 11–12 fractions over 25–26 days in 33 patients and 5100 rad in 11–12 fractions over 25–28 days in the other 84. Brachial plexopathy developed in 73% (24/33) of those receiving 5500 rad. The incidence decreased to 15% (13/84) in patients receiving 5100 rad. Postmortem examination of the brachial plexus region was performed in 2 patients. In 1 patient extensive fibrosis surrounded the nerves of the brachial plexus. Microscopic examination revealed varying degrees of fibrous thickening of the neurilemma sheath, demyelination, and fibrous replacement of some nerve fibrils. Distal to the fibrotic area of the median nerve in the upper arm, extensive myelin loss and atrophy and fibrous replacement of nerve fibrils were observed. In the second patient who had only minimum signs and symptoms, examination of involved nerves showed some loss of myelin, but minimal fibrosis.

Westling *et al.* (1972) reported 171 patients who received postoperative radiation therapy for breast carcinoma using separate fields to the internal mammary and supraclavicular regions and a third medial-angled (30°) axillary field. Cobalt-60 radiation was used and two of the three fields were treated each day. A "peak" dose of 400 rad was given daily to each field, 5 days/week, to a total of 4400 rad. A study of the isodose distribution revealed that this technique resulted in a 30% greater dose in the region of the brachial plexus in about 70% of the patients. The total calculated average dose to the brachial plexus was 5400 ±

TABLE III

RELATIONSHIP TO TOTAL DOSE AND DAILY FRACTION SIZE
TO THE DEVELOPMENT OF BRACHIAL PLEXOPATHY

Author	Number of patients	Total dose to plexus (rad)	Dose per fraction (rad)	Incidence of injury (%)
Stoll and Andrews (1966)	33	5500	450–500	73
	84	5100	425–460	15
	139	4350	435	10
Westling *et al.* (1972)	71	5400	490	60
Basso-Ricci *et al.* (1980)	490	6000	≤200	3.2
	200	4900	≤200	0

1000 rad delivered in 11 fractions over 21–26 days. A total of 60% developed brachial plexopathy. In a group of patients treated by a different technique, the incidence of injury was 14–16% for doses of 5100–5200 rad in 15–20 days, and with a total dose of 4400 rad there was no case of nerve injury in 20 patients.

Svensson et al. (1975) reanalyzed the data of Westling et al. (1972) using a formula in which the cumulative radiation effect (CRE) was defined as $CRE = d \times q \times N^{0.65}$, where d is the delivered dose to the plexus in rad, $q = (T/N)^{-0.11}$, T is the total time in days, and N is the number of fractions. Of the patients, 8% developed brachial plexopathy with a CRE of 1700–1800. The incidence increased to 57% for patients receiving a CRE of 2100–2230.

Rawlings et al. (1983) emphasized the importance of fraction size as an antecedent to plexus injury and recommended that fraction sizes no greater than 250 rad be used and that radiation therapy in the completely dissected axilla should be avoided. Basso-Ricci et al. (1980) reported a 3.2% incidence of brachial plexopathy in 490 patients receiving 6000 rad to the plexus at ~ 200 rad per fraction, whereas no injuries occurred in 200 patients receiving 4900 rad with the same daily fractionation schedule ($p < 0.009$—Fisher's test). Permanent brachial plexus injury does not appear to be enhanced by chemotherapy regimens used concomitantly or sequentially with radiation therapy (Danoff et al., 1983).

It may be extremely difficult to distinguish brachial plexopathy caused by metastatic tumor from that caused by radiation-induced injury. There is no single differentiating sign or symptom at the time of presentation. Painless upper plexus lesions with lymphedema suggest radiation injury whereas painful lower plexus lesions with Horner's syndrome are more suggestive of tumor infiltration (Kori et al., 1979). Stabilization and occasional improvement are found with radiation-induced plexopathy in contrast to the progressive downhill course of carcinomatous plexopathy (Thomas and Colby, 1972). Surgical exploration and biopsy are indicated when the etiology of neuropathy is not clear and when no other evidence of metastatic disease is present (Bagley et al., 1978). However, surgical exploration may fail to detect tumor infiltration (Kori et al., 1979). Radiation-induced plexopathy is a difficult management problem. Exploration and neurilysis have been advocated in the European literature, but this procedure has resulted in little benefit in this country (Match, 1975).

Radiation injury to the lumbosacral area may occur. Maier et al. (1969) reported the development of symmetric lower extremity paralysis accompanied by a loss of deep tendon reflexes and muscle fasciculations in 15 of 343 patients after radiation therapy to the pelvis and paraaortic regions for testicular tumors. No sensory deficits resulted and bladder and rectal sphincter functions were preserved. The patients received 3510–5680 rad over 26–87 days (mean NSD 1366 ret). The daily fraction size was not presented; one field was treated each day. Two explanations of the observed pure lower motor neuron signs were proposed: (1) a selective injury to anterior horn cells within the lumbar cord, and

(2) an injury to the lumbosacral plexus. A small number of patients have developed similar injury following treatment to the pelvis only, suggesting that the site of injury is the lumbosacral plexus (Ashenhurst *et al.*, 1977; Aho and Sainio, 1983).

Although the mechanism of late nerve injury has not been clearly defined, these injuries are generally accompanied by dense fibrosis. It has been hypothesized that interference with blood supply by direct pressure or by obliterative small vessel changes in the nerve fibers may play a role. An impaired response of the Schwann cell to postradiation injury (lack of remylination) may contribute to late nerve damage (Kinsella *et al.*, 1980).

VII. Summary

The clinical syndromes and time–dose relationships of radiation injury to the central and peripheral nervous system have been reviewed. Tolerance levels have been primarily based on anecdotal cases of patients who have developed injury, usually without reference to the population at risk. Few data are available on the incidence of injury under any given set of conditions. It is clear, however, that the daily fraction size in addition to total dose has a substantial effect on the risk of the development of injury to both the central and peripheral nervous systems.

Radiation injury to the brain can be divided into acute, early delayed, and late delayed injuries. Clinical experience suggests that daily doses of 180–200 rad to all or a portion of the brain are well tolerated and that total doses of 6000–8000 rad can be administered without appreciable *acute* side effects. Daily doses of up to 600 rad are acutely well tolerated, but higher daily dose fractions should be administered with caution. Acute reactions, when they occur, are usually mild and transient. Corticosteroids may be required and provide symptomatic relief.

Early delayed reactions occur in 20–25% of patients receiving cranial irradiation. Following treatment for intracranial neoplasms, this syndrome may go unrecognized or be attributed to the recently treated tumor. Although the severity of signs and symptoms are usually mild and require no specific therapy, in rare instances the severity of the reaction may require intensive medical support. Corticosteroids may be useful in relieving signs and symptoms. It is important to recognize that the early delayed syndrome is generally transient and does not necessarily indicate a failure of the treatment or a need for a change in therapy.

The major risk of high-dose radiation to the brain is late delayed injury. There is little information on the incidence of late irradiation injury, and time–dose relationships have been based on case reports collected from the literature. The available data suggest that the risk of injury increases with doses in excess of 6000 rad delivered in 30 fractions over ~ 6 weeks. Prospective studies are needed to establish the influence of chemotherapy, age, surgery, and vascular disease on the incidence of injury.

The literature on the neuropsychological effects of brain irradiation is difficult to interpret. It appears that a decreased intellectual capacity (especially the cognitive aspects) can occur with 2400 rad at 200 rad per fraction in children with ALL and with doses of $\sim$ 6000 rad in adults, at least when radiation therapy is accompanied by aggressive chemotherapy. In patients with brain tumors, it is difficult to assess the role of the tumor (including site, direct brain injury from the tumor, and increased intracranial pressure). The influence of surgery and chemotherapy requires further evaluation. Prospective serial studies of both children and adults with brain tumors or ALL are needed with proper controls (including those treated with and without irradiation and chemotherapy) in order to assess more rigorously the effect of the tumor and its treatment and the effect of host factors on the brain. By careful monitoring of the neuropsychological effects of the disease and treatment, the precise mechanisms of impairement can be established and corrective measures instituted. More studies are needed to establish the incidence of injury and to identify those groups of patients who might benefit from special education and psychological intervention. The early identification of neuropsychological changes may prevent long-term deleterious effects.

Radiation injury to the spinal cord may appear as an early delayed or a late reaction. Daily radiation fraction size appears to be a significant factor in the development of late spinal cord injuries. A dose of 5000 rad in 180- to 200-rad fractions approaches the threshold for tolerance.

Optic nerve and chiasmal injury is a major risk in patients with pituitary or parapituitary tumors who are treated with high daily dose fractions. The observation that injury is less frequent in patients with primary brain tumors who receive incidental irradiation to the optic chiasmal areas than in patients with pituitary tumors suggests that a mass effect from compression of the optic nerves may increase the risk of injury. In patients with pituitary tumors, the use of fraction sizes of 180–200 rad, with the total dose limited to 4500 rad, minimizes the risk of such injury.

Clinical studies indicate that the retina is sensitive to irradiation and that clinically evident injury may occur with doses as low as 4600 rad when the whole retina is included in the radiation field. Every effort should be made to spare the posterior retina whenever possible.

Injuries to cranial and peripheral nervous systems are rare. The risk of brachial plexus injury increases with total dose in excess of 6000 rad, with daily fraction sizes of 200 rad, and with the use of large daily fraction sizes.

In each of the central and peripheral nervous system sites, carefully designed studies are needed to establish the true risk of injury under various clinical conditions. Large animal studies are needed to confirm the dose–response relationships which have primarily been based on anecdotal reports in the literature and dose–response curves based on small animal models. These studies should include the effect of small fraction sizes (hyperfractionation) on the risk of injury. Animal studies will also help to establish the tolerance of peripheral nerves which may be exposed to high single-dose fractions through the use of interoperative irradiation.

ACKNOWLEDGMENTS

The authors gratefully acknowledge Ms. Colleen Condon and Ms. Cindy Nakada for their secretarial assistance, and Ms. Eleanor Haas for her editorial assistance.

REFERENCES

Abbatucci, J. S., Delozier, T., Quint, R., Roussel, A., and Brune, D. (1978). *Int. J. Radiat. Oncol. Biol. Phys.* **4,** 239–248.
Ahlbom, H. C. (1941). *Acta Radiol.* **22,** 155–171.
Aho, K., and Sainio, K. (1983). *Neurology* **33,** 953–955.
Ang, K. K., Van der Kogel, A. J., and van der Schueren, E. (1985). *Int. J. Radiat. Oncol. Biol. Phys.* **11,** 105–110.
Aristizabal, S. A., Boone, M. L., and Laguna, J. (1979). *Int. J. Radiat. Oncol. Biol. Phys.* **5,** 349–353.
Aristizabal, S., Caldwell, W. L., and Avila, J. (1977). *Int. J. Radiat. Oncol. Biol. Phys.* **2,** 667–673.
Ashenhurst, E. M., Quartey, G. R. C., and Starrveld, A. (1977). *J. Can. Sci. Neurol.* **4,** 259–263.
Asscher, A. W., and Anson, S. G. (1962). *Lancet* **29** December.
Atkins, H. L., and Tretter, P. (1966). *Acta Radiol.* **5,** 79–94.
Atkinson, A. B., Allen, I. V., Gordon, D. S., Hadden, D. R., Maguire, C. J. F., Trimble, E. R., and Lyons, A. R. (1979). *Clin. Endocrinol.* **10,** 469–479.
Aur, R. J. A., Simone, J. V., Verzosa, M. S., Hustu, H. O., Pinkel, D. P., and Barker, L. F. (1978). *Sangre* **23,** 1–12.
Bagley, F. H., Walsh, J. W., Cady, B., Salzman, F. A., Oberfield, R. A., and Pazianos, A. G. (1978). *Cancer* **41,** 2154–2157.
Bamford, F. N., Jones, P. J., Pearson, D., Ribeiro, G. G., Shalet, S. M., and Beardwell, C. C. (1976). *Cancer* **37,** 1149–1151.
Basso-Ricci, S., della Costa, C., Viganotti, G., Ventafridda, V., and Zanolla, R. (1980). *Tumori* **66,** 117–122.
Berger, P. S., and Bataini, J. P. (1977). *Cancer* **40,** 152–155.
Bernath, A., Kane, R., and Porter, P. (1982). *Proc. ASCO* **1,** 149.
Bleyer, W. A. (1981). *Cancer Treat. Rep.* **65,** (Suppl. 1) 89–98.
Bleyer, W. A., and Griffin, T. W. (1980). In "Radiation Damage to the Nervous System" (H. A. Gilbert and A. R. Kagan, eds.), pp. 155–174. Raven, New York.
Bloom, B., and Kramer, S. (1984). In "Secretory Tumors of the Pituitary Gland" (P. McL. Black, N. T. Zervas, E. C. Ridgway, and J. B. Martin, eds.), pp. 179–190. Raven, New York.
Bode, U. (1980). *Am. J. Pediatr. Hematol. Oncol.* **2,** 21–24.
Boden, G. (1948). *Br. J. Radiol.* **24,** 464–469.
Boldrey, E., and Sheline, G. E. (1967). *Acta Radiol.* **5,** 5–10.
Borgelt, B., Gelber, R., Kramer, S., Brady, L. W., Chang, C. H., Davis, L. W., Perez, C. A., and Hendrickson, F. R. (1980). *Int. J. Radiat. Oncol. Biol. Phys.* **6,** 1–9.

Borgelt, B., Gelber, R., Larson, M., Hendrickson, F., Griffin, T., and Roth, R. (1981). *Int. J. Radiat. Oncol. Biol. Phys.* **7,** 1633–1638.

Burger, P. C., Mahaley, M. S., Jr., Dudka, L., and Vogel, F. S. (1979). *Cancer* **44,** 1256–1272.

Cecchetti, E., and Brandoli, V. (1979). *Int. J. Radiat. Oncol. Biol. Phys.* **5,** 367–371.

Chan, R. C., and Shukovsky, L. J. (1976). *Radiology* **120,** 673–675.

Cheng, V. S. T., and Schultz, M. D. (1975). *Cancer* **35,** 1537–1544.

Cumberlin, R. L., Luk, K. H., Wara, W. M., Sheline, G. E., and Wilson, C. B. (1979). *Cancer* **43,** 1014–1020.

Danoff, B. F., Cowchock, S., Marquette, C., Mulgrew, L., and Kramer, S. (1982). *Cancer* **49,** 1580–1586.

Danoff, B. F., Goodman, R. L., Glick, J. H., Haller, D. G., and Pajak, T. F. (1983). *Int. J. Radiat. Oncol. Biol. Phys.* **9,** 1625–1630.

Deck, M. D. F. (1980). *In* "Radiation Damage to the Nervous System" (H. A. Gilbert and R. A. Kagan, eds.), pp. 107–127. Raven, New York.

Di Chiro, G., Arimitsu, T., Brooks, R. A., Morgenthaler, D. G., Johnson, G. S., Jones, A. E., and Keller, M. R. (1979). *Radiology* **130,** 661–666.

Dische, S., Martin, W. M. C., and Anderson, P. (1981). *Br. J. Radiol.* **54,** 29–35.

Druckmann, A. (1929). *Strahlentherapie* **33,** 382–384.

Duffner, P. K., Cohen, M. E., and Thomas, P. (1983). *Cancer* **51,** 233–237.

Dynes, J. B., and Smedal, M. I. (1960). *Am. J. Roentgenol* **83,** 78–87.

Edwards, M. S. B., and Wilson, C. B. (1980). *In* "Radiation Damage to the Nervous System" (H. A. Gilbert and A. R. Kagan, eds.), pp. 129–143. Raven, New York.

Eiser, C. (1978). *Arch. Dis. Child.* **53,** 391–395.

Ellis, F., (1968). *Curr. Top. Radiat. Res.* **4,** 357–397.

Ellison, N., Bernath, A., Kane, R., and Porter, P. (1982). *Proc. ASCO* **1,** 149.

Esseltine, D. W., Tarshish, E., Schulpen, A. E., Chevalier, L., and Whitehead, V. M. (1983). *Proc. ASCO* **2,** 74.

Fike, J. R., and Cann, C. E. (1984). *Radiology* **151,** 115–120.

Fike, J. R., Sheline, G. E., Cann, C. E., and Davis, R. L. (1984). *Prog. Exp. Tumor Res.* **28,** 136–151.

Freeman, J. E., Johnston, P. G. G., and Voke, J. M. (1973). *Br. Med. J.* **4,** 523–525.

Fu, K. K. (1979). *Front. Radiat. Ther. Oncol.* **13,** 113–132.

Fusner, J. E., Poplack, D. G., Pizzo, P. A., and Di Chiro, G. (1977). *J. Pediatr.* **91,** 77–79.

Groothuis, D. R., and Vick, N. A. (1980). *In* "Radiation Damage to the Nervous System" (H. A. Gilbert and A. R. Kagan, eds.), pp. 93–106. Raven, New York.

Harwood-Nash, D. C. F., and Reilly, B. J. (1970). *Am. J. Roentgenol.* **108,** 392–395.

Harris, J. R., and Levine, M. B. (1976). *Radiology* **120,** 167–171.

Hatlevoll, R., Host, H., and Kaalhus, O. (1983). *Int. J. Radiat. Oncol. Biol. Phys.* **9,** 41–44.

Hindo, W. A., DeTrana, F. A., III, Lee, M.-S., and Hendrickson, F. R. (1970). *Cancer* **26,** 138–141.

Hirsch, J. F., Renier, D., Czernichow, P., Benveniste, L., and Pierre-Kahn, A. (1979). *Acta Neurochir.* **48,** 1–15.

Hochberg, F. H., and Slotnick, B. (1980). *Neurology* **30,** 172–177.

Hoffman, W. F., Levin, V. A., and Wilson, C. B. (1979). *J. Neurosurg.* **50,** 624–628.

Hornsey, S., Morris, C. C., and Myers, R. (1981). *Int. J. Radiat. Oncol. Biol. Phys.* **7,** 393–396.

Jones, A. (1964). *Br. J. Radiol.* **37,** 727–744.

Kinsella, T. J., Weichselbaum, R. R., and Sheline, G. E. (1980). *In* "Radiation Damage to the Nervous System" (H. A. Gilbert and A. R. Kagan, eds.), pp. 145–153. Raven, New York.

Komaki, R., Cox, J. D., Byhardt, R. W., Holoye, P. Y., Libnoch, J. A., Hansen, R., and Anderson, T. (1985). *Am. J. Clin. Oncol.* **8,** 20–21.

Kopelson, G. (1982). *Int. J. Radiat. Oncol. Biol. Phys.* **8,** 925–929.

Kori, S. H., Foley, K. M., and Posner, J. B. (1979). *Neurology* **29,** 583.

Kornblith, P. L., Walker, M. D., and Cassady, J. R. (1982). *In* "Principles and Practice of Oncology" (V. T. DeVita, Jr., S. Hellman, and S. A. Rosenberg, eds.), pp. 1181–1253. Lippincott, Philadelphia.

Kramer, S. (1968). *Clin. Neurosurg.* **15,** 301–318.

Kramer, S., and Lee, K. F. (1974). *Semin. Roentgenol.* **9,** 75–83.

Kramer, S., Hendrickson, F., Zelen, M., and Schotz, W. (1977). *Natl. Cancer Inst. Monogr.* **46,** 213–221.

Kun, L. E., and Mulhern, R. K., Jr. (1983). *Am. J. Clin. Oncol.* **6,** 651–656.

Kun, L. E., Mulhern, R. K., and Crisco, J. J. (1983). *J. Neurosurg.* **58,** 1–6.

Lambert, P. M. (1978). *Cancer* **41,** 1751–1760.

Lampert, P. W., and Davis, R. L. (1964). *Neurology* **14,** 912–917.

Lampert, P., Tom, M. I., and Rider, W. D. (1959). *Am. Med. Assoc. Arch. Pathol.* **68,** 90/322–98/330.

Licciardello, J., Bromer, R., Karp, D., Hoffer, S., Paquette, D., and Hong, W. (1983). *Proc. ASCO* **2,** 189.

Littman, P., Rosenstock, J., Gale, G., Krisch, R. E., Meadows, A., Sather, H., Coccia, P., and DeCamaryo, B. (1984). *Int. J. Radiat. Oncol. Biol. Phys.* **10,** 1851–1853.

Liu, H. M., Maurer, H. S., Vongsrivut, S., and Conway, J. (1978). *Hum. Pathol.* **9,** 635–648.

Locksmith, J. P., and Powers, W. E. (1968). *Am. J. Roentgenol.*

McIntosh, S., Rothman, S. G., Lobel, J. S., and O'Brien, R. T. (1977). *J. Pediatr.* **91,** 909–913.

Maier, J. G., Perry, R. H., Saylor, W., and Sulak, M. H. (1969). *Am. J. Roentgenol.* **93,** 153–160.

Mandybur, T. I., and Gore, I. (1969). *Neurology* **19,** 983–992.

Manz, H. J., Woolley, P. V., and Ornitz, R. D. (1979). *Cancer* **44,** 473–479.

Margolis, L., Smith, M. E., Fortuin, F. D., Chin, F. K., Leibel, S. A., and Hill, D. R. (1981). *Cancer* **48,** 1680–1683.

Marks, J. E., Baglan, R. J., Prassad, S. C., and Blank, W. F. (1981). *Int. J. Radiat. Oncol. Biol. Phys.* **7,** 243–252.

Martins, A. N., Johnston, J. S., Henry, J. M., Stoffel, T. J., and Di Chiro, G. (1977). *J. Neurosurg.* **47,** 336–345.

Match, R. M. (1975). *Arch. Surg.* **110,** 384–386.

Meadows, A. T., Gordon, J., Massari, D. J., Littman, P., Fergusson, J., and Moss, K. (1981). *Lancet* **7** November.

Mikhael, M. A. (1980). *In* "Radiation Damage to the Nervous System" (H. A. Gilbert and A. R. Kagan, eds.), pp. 59–91. Raven, New York.

Montague, E. D., Romsdahl, M. M., Schell, S. R., and Ames, F. C. (1983). *In* "Conservative Management of Breast Cancer. New Surgical and Radiotherapeutic Techniques" (J. R. Harris, S. Hellman, and W. Silen, eds.), pp. 53–59. Lippincott, Philadelphia.

Moss, H. A., Nannis, E. D., and Poplack, D. G. (1981). *Am. J. Med.* **71,** 47–52.

Norman, D., Enzmann, D. R., and Levin, V. A. (1976). *Radiology* **121,** 85–88.

Norton, W. T. (1972). *In* "Basic Neurochemistry" (G. J. Siegel and R. Katzman, eds.), pp. 365–384. Little Brown, Boston.

Ochs, J., Ch'ien, L., Parvey, L., Berg, N., Witaker, J., Coburn, T., Evans, W., Campbell, M., and Bowman, W. P. (1983a). *Proc. ASCO* **2,** 75.

Ochs, J., Parvey, L. S., Whitaker, J. N., Bowman, W. P., Ch'ien, L., Campbell, M., and Coburn, T. (1983b). *J. Clin. Oncol.* **1,** 793–798.

Packer, R., Atkins, T., Littman, P., Rosenstock, J., Sutton, L., Bruce, D., and Schut, L. (1983). *SIOP, Annu. Conf., 16th* (Abstract).

Pallis, C. A., Louis, S., and Morgan, R. L. (1961). *Brain* **84,** 460–479.

Palmer, J. J. (1972). *Brain* **95,** 109–122.

Parker, D., Malpas, J. S., Sandland, R., Sheaff, P. C., Freeman, J. E., and Paxton, A. (1978). *Br. J. Med.* **4,** 554.

Parsons, J. T. (1984). *In* "Management of Head and Neck Cancer. A Multidisciplinary Approach" (R. R. Million and N. J. Cassisi, eds.), pp. 173–207. Lippincott, Philadelphia.

Parsons, J. T., Fitzgerald, C. R., Hood, C. I., Ellingwood, K. E., Bova, F. J., and Million, R. R. (1983). *Int. J. Radiat. Oncol. Biol. Phys.* **9,** 609–622.

Patronas, N. J., Di Chiro, G., Brooks, R. A., DeLaPaz, R. L., Kornblith, P. L., Smith, B. H., Rizzoli, H. V., Kessler, R. M., Manning, R. G., Channing, M., Wolf, A. P., and O'Connor, C. (1982). *Radiology* **144,** 885–889.

Peylan-Ramu, N., Poplack, D. G., Pizzo, P. A., Adornato, B. T., and Di Chiro, G. (1978). *N. Engl. J. Med.* **298,** 815–818.

Pezner, R. D., and Archambeau, J. O. (1981). *Int. J. Radiat. Oncol. Biol. Phys.* **7,** 397–402.

Phillips, T. L., and Buschke, F. (1969). *Am. J. Roentgenol.*

Phillips, T. L., Diener, M. D., Wasserman, T. H., Asbell, S. O., Chang, C. H., Urtasun, R. C., and Moylan, D. J. (1984). *Int. J. Radiat. Oncol. Biol. Phys.* **10,** 145–146.

Price, R. A. (1979). *Am. J. Pediatr. Hematol. Oncol.* **1,** 21–30.

Price, R. A., and Birdwell, D. A. (1978). *Cancer* **42,** 717–728.

Price, R. A., and Jamieson, P. A. (1975). *Cancer* **35,** 306–318.

Raimondi, A. J., and Tomita, T. (1979). *In* "Multidisciplinary Aspects of Brain Tumor Therapy" (P. Paoletti, G. Walker, G. Butti, and Knerich, eds.), pp. 209–222. Elsevier, New York.

Rawlings, G., Arriagada, R., Fontaine, F., Bouhnik, H., Mouriesse, H., and Sarrazin, D. (1983). *Bull. Cancer (Paris)* **70,** 77–83.

Reagan, T. J., Thomas, J. E., and Colby, M. Y., Jr. (1968). *J. Am. Med. Assoc.* **203,** 128–110.

Reinhold, H. S., Kaalen J. G. A. H., and Unger-Gils, K. (1976). *Int. J. Radiat. Oncol. Biol. Phys.* **1,** 651–657.

Reitan, R. M., and Davison, L. (1974). "Clinical Neuropsychology: Current Status and Applications." Winston/Wiley, New York.

Rider, W. D. (1963). *J. Can. Assoc. Radiol.* **14,** 67–69.

Rizzoli, H. V., and Pagnanelli, D. M. (1984). *J. Neurosurg.* **60,** 589–594.

Rubin, P., ed. (1975). "SET R. T. 1: Radiation Oncology. Radiation Biology and Radiation Pathology Syllabus," pp. 132–141. American College of Radiology, Chicago.

Rubinstein, L. J. (1972). "Tumors of the Central Nervous System." Armed Forces Institute of Pathology, Bethesda, Maryland.

Salazar, O. M., Rubin, P., McDonald, J. V., and Feldstein, M. L. (1976). *Int. J. Radiat. Oncol. Biol. Phys.* **1,** 717–727.

Salner, A. L., Botnick, L. E., Herzog, A. G., Goldstein, M. A., Harris, J. R., Levene, M. B., and Hellman, S. (1981). *Cancer Treat. Rep.* **65,** 797–802.

Schultheiss, T. E., Higgins, E. M., and El-Mahdi, A. M. (1984). *Int. J. Radiat. Oncol. Biol. Phys.* **10,** 1109–1115.

Sheline, G. E. (1983). *In* "Oncology of the Nervous System" (M. D. Walker, ed), pp. 223–245. Nijhoff, Boston.

Sheline, G. E., and Tyrell, B. (1984). *In* "Radiation Oncology Annual 1983" (T. L. Phillips and D. A. Pistenmaa, eds.), pp. 1–35. Raven, New York.

Sheline, G. E., Wara, W. M., and Smith, V. (1980). *Int. J. Radiat. Oncol. Biol. Phys.* **6,** 1215–1228.

Shukovsky, L. J., and Fletcher, G. H. (1972). *Radiology* **104,** 629–634.

Smith, B. M., McGinnis, W., Cook, J., and Latourette, H. (1979). *Cancer* **43,** 2239–2242.

Soni, S. S., Marten, G. W., Pitner, S. E., Duenas, D. A., and Powazek, M. (1975). *N. Engl. J. Med.* **293,** 113–118.

Spunberg, J. J., Chang, G. H., Goldman, M., Auricchio, E., and Bell, J. J. (1981). *Int. J. Radiat. Oncol. Biol. Phys.* **7,** 727–736.

Stoll, B. A., and Andrews, J. T. (1966). *Br. Med. J.* **1,** 834–837.

Svensson, H., Westling, P., and Larsson, L.-G. (1975). *Acta Radiol. Ther. Phys. Biol.* **14,** 228–238.

Tamaroff, M., Salwen, R., Allen, J. C., Murphy, M. L., and Nir, Y. (1982). *Proc. ASCO* **1,** 48.

Tamaroff, M., Salwen, R., Miller, D. R., Murphy, M. L., and Nir, Y. (1985). *Proc. ASCO* **4,** 165.

Thomas, J. E., and Colby, M. Y. (1972). *JAMA, J. Am. Med. Assoc.* **222,** 1392–1395.

van der Kogel, A. J. (1979). "Late Effects of Radiation on the Spinal Cord," pp. 118–121. Publication of the Radiobiological Institute of the Organization for Health Research, TNO Rijswijk.

van der Kogel, A. J., and Barendsen, G. W. (1974). *Br. J. Radiol.* **47,** 393–398.

Volk, S. A., Mansour, R. P., Gandara, D. R., and Redmond, J. (1983). *Proc. ASCO* **2,** 185.

Wara, W. M., Phillips, T. L., Sheline, G. E., and Schwade, J. G. (1975). *Cancer* **35,** 1558–1562.

Wara, W. M., Irvine, A. R., Neger, R. E., Howes, E. L., Jr., and Phillips, T. L. (1979). *Int. J. Radiat. Oncol. Biol. Phys.* **5,** 81–83.

Weiss, H. D., Walker, M. D., and Wiernik, P. H. (1974). *N. Engl. J. Med.* **291,** 75–81.

Weiss, H. D., Walker, M. D., and Wiernik, P. H. (1974). *N. Engl. J. Med.* **291,** 127–133.

Westling, P., Svensson, H., and Hele, P. (1972). *Acta Radiol. Ther. Phys. Biol.* **11,** 209–216.

Yoshii, Y., and Phillips, T. L. (1982). *Acta Neurochir.* **64,** 87–102.

Young, D. F., Posner, J. B., Chu, F., and Nisce, L. (1974). *Cancer* **34,** 1069–1076.

Index